Biological Electron Transfer Chains:
Genetics, Composition and Mode of Operation

NATO ASI Series

Advanced Science Institute Series

A Series presenting the results of activities sponsored by the NATO Science Committee, which aims at the dissemination of advanced scientific and technological knowledge, with a view to strengthening links between scientific communities.

The Series is published by an international board of publishers in conjunction with the NATO Scientific Affairs Division

A Life Sciences	Plenum Publishing Corporation
B Physics	London and New York
C Mathematical and Physical Sciences	Kluwer Academic Publishers
D Behavioural and Social Sciences	Dordrecht, Boston and London
E Applied Sciences	
F Computer and Systems Sciences	Springer-Verlag
G Ecological Sciences	Berlin, Heidelberg, New York, London,
H Cell Biology	Paris and Tokyo
I Global Environment Change	

PARTNERSHIP SUB-SERIES

1. Disarmament Technologies	Kluwer Academic Publishers
2. Environment	Springer-Verlag / Kluwer Academic Publishers
3. High Technology	Kluwer Academic Publishers
4. Science and Technology Policy	Kluwer Academic Publishers
5. Computer Networking	Kluwer Academic Publishers

The Partnership Sub-Series incorporates activities undertaken in collaboration with NATO's Cooperation Partners, the countries of the CIS and Central and Eastern Europe, in Priority Areas of concern to those countries.

NATO-PCO-DATA BASE

The electronic index to the NATO ASI Series provides full bibliographical references (with keywords and/or abstracts) to about 50,000 contributions from international scientists published in all sections of the NATO ASI Series. Access to the NATO-PCO-DATA BASE is possible via a CD-ROM "NATO Science and Technology Disk" with user-friendly retrieval software in English, French, and German (©WTV GmbH and DATAWARE Technologies, Inc. 1989). The CD-ROM contains the AGARD Aerospace Database.

The CD-ROM can be ordered through any member of the Board of Publishers or through NATO-PCO, Overijse, Belgium.

Series C: Mathematical and Physical Sciences – Vol. 512

Biological Electron Transfer Chains: Genetics, Composition and Mode of Operation

edited by

G. W. Canters

and

E. Vijgenboom

Leiden Institute of Chemistry,
Leiden University, Leiden, The Netherlands

Springer-Science+Business Media, B.V.

Proceedings of the NATO Advanced Research Workshop on
Biological Electron Transfer Chains: Genetics, Composition and Mode of Operation
Tomar, Portugal
May 3-7, 1997

A C.I.P. Catalogue record for this book is available from the Library of Congress.

ISBN 978-0-7923-5138-2   ISBN 978-94-011-5133-7 (eBook)
DOI 10.1007/978-94-011-5133-7

Printed on acid-free paper

TABLE OF CONTENTS

Preface

From May 3-7, 1997, the NATO Advanced Research Workshop on 'Biological Electron Transfer Chains' was organized in Tomar, Portugal. In the application for support the choice of the topic was justified as follows: "[Until recently efforts] have concentrated on the study of the structure and function of individual redox enzymes and proteins. Enough information is now available to make a start with the study of biological electron transfer (ET) at the next higher level of organization, that of the complete ET chain." The interest in the workshop was high: the majority of participants had registered before the workshop was formally announced, which illustrates the popularity of the topic within the biochemical and biophysical communities. The present volume contains a number of reports based on the lectures presented by the key speakers during the meeting.

The workshop dealt with the following three themes:

a) *Electron transfer*, which is the subject of Chapter 1. The analysis of ET at the molecular level is still fundamental for an understanding of how ET chains operate *in vivo*. After 40 years of research the contours of the subject are becoming clear now.

b) *Bacterial redox chains*. This is the subject of Chapter 2. Its contents show how complicated these chains can be, often involving a number of gene clusters. Our understanding of the regulatory aspects and control mechanisms of these chains is only in its beginning.

c) *Structure and function of redox proteins*. Structural information on proteins remains indispensable for a proper understanding of their function and of the context in which they operate. It is not surprising, therefore, that the two final chapters of these proceedings are devoted to this subject. Chapter 3 deals with oxido-reductases in general, while Chapter 4 treats the cytochrome oxidase family.

It may be interesting to briefly reflect on the contents of the various contributions. A concept that figures prominently in the contributions on ET is that of 'Pathways'. The problem of electronic conduction in the cell has been solved in the course of evolution by the embedding of redox centres at edge-edge distances of 5-15 Å in a non-conducting protein matrix. For mobile ET carriers the ET in an encounter complex takes place over similar distances. The question that has kept many investigators busy is whether in the course of evolution redox proteins and enzymes have evolved such that they offer an electron a well defined and recognizable pathway when it traverses the protein matrix from one redox site to another.

In a semi-empirical manner Beratan and colleagues have attempted to implement the so-called pathway concept to account for the effect of the protein structure on individual ET rates. Their approach has enjoyed a large popularity but the shortcomings of the model are recognised by Beratan himself in his present contribution. They derive from the fact that the pathway model can not account for the phase that is a vital part of the wave description of the electron: when there appears to be more than one viable pathway ('tubes of several pathways') interference may occur, while also reflections at 'dead ends' may be difficult to account for. The way out is to treat the problem quantum-mechanically and determine the electronic coupling between donor and acceptor by calculating the Hamiltonian matrix for the whole protein. It is conceivable that in the structure of this matrix and from the configuration of the non-zero matrix elements a 'path' is recognizable, but the matrix structure may also have a more diffuse character in which case a 'path' may be more difficult to trace.

In this connection it is interesting to note the study by Farver and Pecht. From the effect of point mutations on the internal ET rate in azurin they conclude that there is a recognizable pathway leading from the cystine disulfide at the 'south end' of the protein to the redox active Cu centre at the 'north end'. On the other hand, Verhoeven and coworkers in their elegant studies of ET in a series of covalently linked D-A complexes, find that ET through a stochastically disordered solvent can be more efficient then via a well defined covalent through-bond path. The study of protein complexes with well defined geometries may help us further in dissecting the role of the protein matrix on ET rates. There are only few examples of these, besides the photosynthetic reaction centre, but in the contributions by Adman c.s. and Mathews c.s. two promising cases are presented. Especially the binary and ternary complexes studied by Mathews, although their physiological relevance in part is subject to discussion (see contributions by Duine and Ferguson), offer beautiful opportunities for further study. When electronic interference effects become important, the electronic D-A coupling may become very sensitive to the detailed nuclear configuration of the protein and thus to vibrational effects. Kotelnikov and coworkers stress this point by arguing that the way k_{et} varies with temperature may give a clue as to whether the adiabatic limit possibly obtains.

A useful concept to understand the organization and operation of biological redox chains is that of 'Control'. As Dutton c.s. in their contribution argue, an uphill ET step in a redox chain may function as a point of control of the electron flow. The contributions of Ferguson and Lidstrom also illustrate the complicated network of promoters, signalling cascades, redox sensors and the like that a cell uses to stay in control of its metabolic processes under varying external conditions. The point is emphasized by Westerhoff c.s., who show that metabolic control analysis is just beginning to scratch the surface of the metabolic complexity of the cell.

While the chapter on biological ET focuses on the mechanism of ET, it is at the level of the individual oxido-reductase that we can see the actual biological redox chemistry at work. There is a bewildering diversity and individuality that confronts us here, as is shown by Duine in his review of amine oxido-reductases, and by Ferguson in his discussion of the respiratory pathways of the organism *Paracoccus denitrificans* that has since long been considered a close relative to the primordial endosymbionth from which our mitochondrion derives. The flavocytochromes that are studied by Chapman and coworkers provide beautiful model systems to study the combination of intra-protein ET and all sorts of interesting redox chemistries. Averill c.s. and Kroneck c.s. both concentrate on recent findings with respect to structure and function of nitrite reductases, while Bertini and Xavier and their associates show how the paramagnetism of haem proteins can be used to one's advantage in structure/function research. An interesting property of the cytochrome c_3 studied by Xavier is that it couples electron and proton transfer and may be considered a 'proton thruster' which is operational in energy transduction.

The justification for the final chapter being entirely devoted to the superfamily of the cytochrome c oxidases is firstly, that cytochrome c oxidase (COX) over the years has played a prominent role in the research on biological ET chains, and secondly that the recently published structures of COX from two different sources have provided an extra impulse to the research on this key enzyme. The recently acquired insights on the proton pumping mechanism of COX are reviewed by Gennis, while Brunori c.s. discuss the question whether the internal ET from the haem a to the dinuclear haem a_3-Cu_B centre is the rate limiting step in enzyme turn-over. Ludwig explores the structural and functional relationships between the subunits of cytochrome oxidases from different sources and the evolutionary origin they share with quinol oxidases. The evolutionary aspects are further analysed by Saraste c.s. who argue that COX and the present-day NO reductase share a common ancestor. The contribution by Zumft *et al.* focuses entirely on the later enzyme. They argue convincingly that the catalytically active site of NO reductase is strongly reminiscent of the active site of COX in that the Cu_B site is occupied by a non-haem iron.

The workshop owed its success not only to the presentations by the key speakers, but also to the contributions by the junior speakers and the chairmen, as well as to the participants in the poster sessions. Finally it should be mentioned that the workshop would have been impossible without the generous support of NATO provided through her Special Programme on Supramolecular Chemistry. Important financial assistance was also provided by the Metals in Biology Progamme of the European Science Foundation. We thank both sponsors for their generous help. In addition we gratefully acknowledge support from the Junta Nacional de Investigação Científica e Technológica, Portugal; the Institute ITQB, Lisboa; Bruker, France; the Câmara Municipal de Tomar, Portugal; and the BIOMAC Graduate School, Leiden.

G.W. Canters
E. Vijgenboom

Leiden, February 1, 1998

Chapter 1

Biological Electron Transfer

RESPIRATORY ELECTRON TRANSFER CHAINS

P.L. DUTTON, X. CHEN, C.C. PAGE, S. HUANG, T. OHNISHI, and
C.C. MOSER
*The Johnson Research Foundation and Department of Biochemistry and
Biophysics, University of Pennsylvania, Philadelphia, PA 19104*

1. Abstract

Here we examine the classical electron transfer chains of the mitochondrial respiratory
oxidoreductase complexes. These chains guide electrons between substrates and
energy coupling sites and between the coupling sites themselves. Structural studies on
oxidoreductases are showing that several redox cofactors separated in protein by 4-10
Å are often in the form of a chain. Long chains are evident in respiratory Complex I,
II and in between Complexes III and IV. Calculations show that these chains are able
to transfer electrons over very large distances with high directional specificity. Driven
by small overall free energies, they operate in the microsecond time scale. Close
proximity of the cofactors maintains physiologically productive rates even when
substantial thermodynamically unfavorable steps are encountered in the chain. These
steps offer potential points of regulation.

2. Introduction

The principal parameters naturally selected for engineering electron transfer in the
photosynthetic reaction center have been identified [1] and rules for their engineering
suggested [2]. Although the general applicability of these rules throughout biological
electron transfer systems remains to be established on a case by case basis, the
findings for the photosynthetic reaction center protein (the only system so far
examined that allows any depth of inquiry) are a persuasive demonstration that only
three principal factors are sufficient to provide robust control of reaction rates and
confer directional specificity on the electron transfer. The single most important factor
is the distance between redox cofactors (R), followed by the free energy (ΔG°) and the
response of the cofactors and environment to charge changes (reorganization energy,
λ). How these influence the rate constant (k_{et}) for a single electron transfer is
summarized in a simple empirical rule:

$$\log k_{et} = 15 - (0.6 \text{ Å}^{-1})R - (3.1 \text{ eV}^{-1}) (\Delta G^\circ + \lambda)^2 / \lambda$$

3

G.W. Canters and E. Vijgenboom (eds.),
Biological Electron Transfer Chains: Genetics, Composition and Mode of Operation, 3-8.
© 1998 *Kluwer Academic Publishers.*

Here the electron transfer rate k_{et} is expressed in units of s^{-1}, R in Å and ΔG^0 and λ are expressed in units of eV. The 0.6 $Å^{-1}$ term reflects an exponential dependence of tunneling rate with distance in a typical protein medium, while the 3.1 eV^{-1} term represents a quantum version of a Marcus free energy/rate relationship. This is not the same as the classical room temperature version with a coefficient 4.2 eV^{-1}. For simple electron transfers that connect substrates to oxidoreductase enzymes or to the respiratory and photosynthetic complexes and serve to link the complexes, the large majority of established ΔG^0 values for electron transfers between the cofactors of systems, and in redox enzymes are small, falling between 0 and -0.15 eV. Values for λ are 0.5 to 1.25 eV for electron transfers within protein. Thus typical electron transfers over a redox cofactor edge-to-edge distances of 3-10 Å will be 10^{12} -10^7 s^{-1}, while 10-15 Å transfers should have rates of 10^7 -10^3 s^{-1}. There are very few examples of redox chains in oxidoreductases for which there are high resolution structures, intercofactor ΔG^0 values and time resolved kinetics. One example is the four heme subunit of the bacterial reaction centers of *Rp. viridis* [3] which extends from the edge of the membrane at an angle of 60° some 70 Å out into the aqueous phase. The close positioning of the hemes provides the means for subnanosecond interheme transfer and a demonstrated submillisecond time from the distant end of the heme chain to the light oxidized bacteriochlorophyll dimer [4].

3. Electron Transfer Chain of Complex I

The multiplicity of iron sulfur clusters [5,6] and the flavin associated with Complex I appear to be located exclusively in a long, ~80 Å, promontory extending into the mitochondrial matrix or microbial cytoplasm [7-12], and hence could easily represent the links of an efficient electron transfer chain from NADH to the membrane. The flavin is likely to be reduced by the NADH by atom transfer (H^- i.e., 2-electron reduction) but then, because of a relatively stable single electron reduced state of the flavin [13], it can deliver single electrons into the iron sulfur cluster chain. The site of NADH oxidation in the promontory has not yet been located, but, as is shown in Figure 1, with a connecting redox chain the oxidation point could be at the furthest point without deleterious physiological impact. Thus, if NADH is oxidized at the extreme distant point, an extended chain of the flavin and five iron-sulfur clusters each separated edge-to-edge by 8-10 Å would form a chain ending with cluster N2 at the membrane interface in probable contact with the energy coupling site. The intercofactor edge-to-edge distance suggested are consistent with the distances estimated from paramagnetic spin-spin interactions [5,13]. Thus, as shown in the example at the bottom of Figure 1, even with a modest overall free energy drop along the chain and assuming typical reorganization energies (a value of 1 eV is used throughout this paper), the entire 80 Å distance of the promontory can be readily traversed in microseconds, consistent with experiment [14,15]. Note that this chain includes one significant unfavorable barrier.

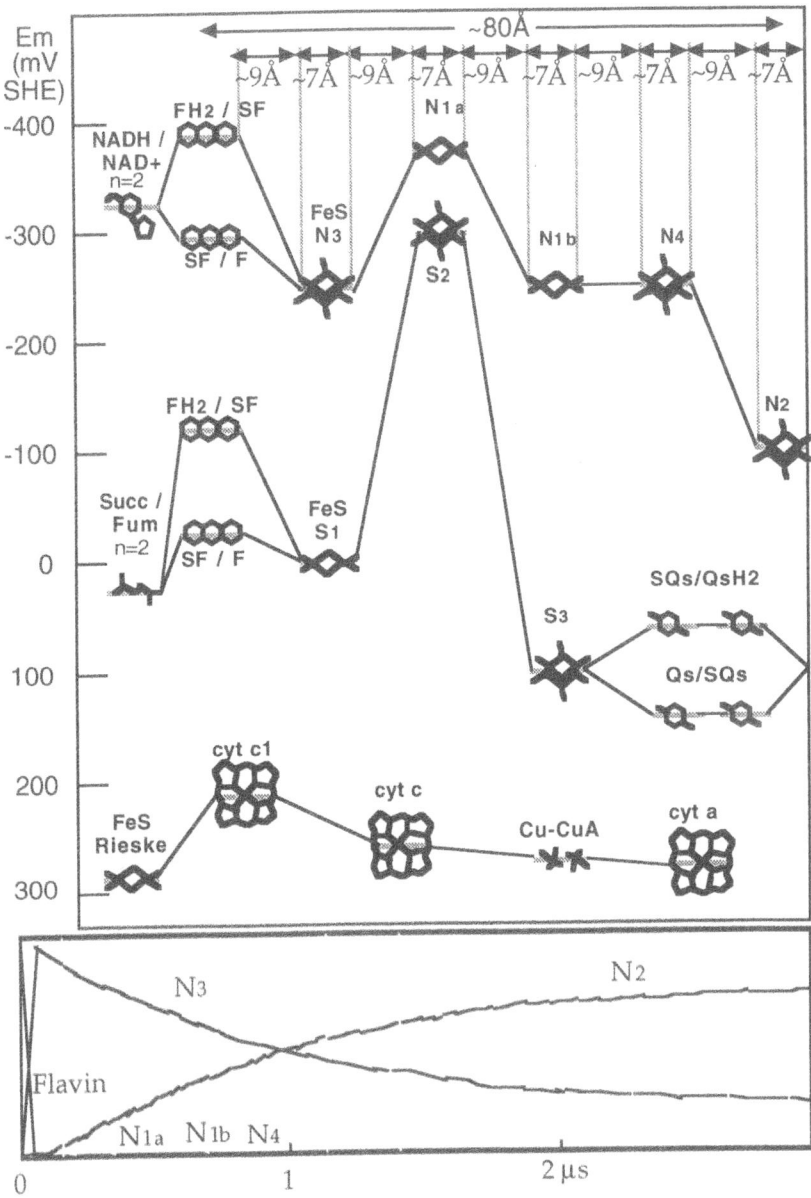

Figure1. Electron transfer chains in mitochondrial respiration. The top panel shows the standard potentials of components that are expected to make up electron guides between energy coupling sites in Complex I, II and III/IV. As an illustration, possible distances between and through redox centers in the proposed chain of Complex I (top) could lead to microsecond electron transfer over 80 Å of a single electron starting on flavin, transiently reducing intermediate clusters, and ending on N2 (bottom). In experiments initiated with NADH, oxidation of NADH may be sufficiently slow that the first species observed to go reduced may be N2. Similar traces can be constructed for the other two chains.

4. Electron Transfer Chain of Complex II

It seems likely that a similar electron transfer chain moves single electrons over a long distance in Complex II from the site of succinate oxidation (or fumarate reduction) in the matrix/cytoplasm to the membrane for the reduction of ubiquinone (Figure 1, center). Analogously, flavin apparently catalyses the oxidation of the strict 2-electron oxidation of succinate and then delivers single electrons into a chain of three iron sulfur clusters (S1, S2 and S3) [16,17] which guide the electrons through the membrane profile to two bound quinones of the Q_s site [18]. Because of their relatively stable semiquinone (SQ) states, these quinones can interface between the single electron FeS centers and the membrane ubiquinone/ubihydroquinone with similar average potential (Q/QH$_2$; E$_{m7}$ 0.09 V [19]). The calculated rate of electron transfer through the Complex II chain, using measured standard free energies and again making modest estimates of reorganization energy and of distances, is microseconds, despite the major thermodynamically unfavorable step (ΔG^0 +0.3 eV) and a small overall ΔG^0 value of -0.07 eV. Again the calculations show that the barrier can easily be overcome by thermal activation provided that the chain components are sufficiently close to promote intrinsically rapid electron tunneling.

5. Electron Transfer Chain Linking Complexes III and IV

Figure 1 shows the components of Complexes III and IV that act as an extended single electron redox chain, carrying electrons from the [2Fe-2S] cluster to cytochrome c_f of Complex III, via cytochrome c (a diffusable element of the chain) to the two coppers of Cu$_A$ and cytochrome a of Complex IV. This chain also operates with a near zero ΔG^0 value. The distances from the resolved structures of Complex III and IV [22-25] are appropriate for rapid electron transfer.

6. Unfavorable Steps in Redox Chains: Potential Sites of Regulation

Thermodynamically unfavorable electron transfer steps apparent in chains of cofactors of the kind discussed for N1a and S2 usually have led authors to seriously question whether the cofactor involved in that step is part of the chain. However, as we show here, thermal activation of electron transfer over substantial thermodynamically unfavorable barriers can be accomplished in the microsecond time scale required for physiological function if the components of the redox chain are in close proximity. A quantitative demonstration is provided by reverse electron transfer from reduced quinone (Q$_A^-$) and the oxidized bacteriochlorophyll dimer (BChl$_2^+$) of photosynthetic reaction centers through the intermediate bacteriopheophytin (BPh). The edge-to-edge distances between the Q$_A$ and BPh, and between BPh and BChl$_2$ are both about 10 Å. Normally this route of charge recombination is barely detectable because of the over 0.5eV uphill step between Q$_A^-$ and BPh. This dominant barrier renders the rate much slower than the 10-100 s^{-1} direct, all downhill recombination rate from Q_A^- to the BChl$_2^+$. Experiments that extracted and replaced the native Q_A with other quinones of different electrochemistry systematically lowered the ΔG^0 of the first uphill step of the BPh route to the charge recombination [20,21,26]. With the Q_A^- to BPh barrier remaining at >0.3 eV submillisecond electron transfer from Q_A^- to BChl$_2^+$ was achieved; moreover, the rate was shown to increase ten fold every 0.06 eV lowering of the barrier.

The four heme subunit of the *Rp. viridis* reaction center offers further physiological insight. The four hemes form a line through the subunit interior and have widely differing redox midpoint potentials. Starting from the one on the end of the 70 Å promontory and moving towards the reaction center BChl$_2$, the E_{m7} values are approximately -0.05, +0.26, +0.02 and +0.36 V [31]. The 0.36 V heme is the immediate donor to BChl$_2^+$, E_{m7} +0.5 V. Thus, for an electron entering at the end, there is an overall downhill trend to the chain with a >0.2eV uphill step between the 0.26 V (second) and 0.02 V (third) heme. Rates from the outer heme through the linear chain to BChl$_2^+$ are calculated to be close to 10^5 s^{-1}, consistent with the measured rates [4,31]. Moreover, it has been suggested that the soluble diffusing cytochrome c_2 (+0.28 V) is oxidized by the outer -0.05 V heme of the four heme chain [31]. Despite presenting a >0.3 eV barrier to that initial step the overall reaction was measured to be 270 s^{-1} [31]. Our own calculation, placing the cytochrome c_2 heme edge a reasonable 8 Å away from the outermost heme edge yields a similar rate. It is very clear from these examples that what may appear to be unlikely thermodynamic barriers in electron transfer chains can be overcome by ambient Boltzman thermal activation when driven by modest overall negative free energy.

We note that such unfavorable steps are common in analogous redox chains. For instance, in the nickel-iron hydrogenase structure, the middle of the chain of three iron-sulfur clusters, an [3Fe-4S] cluster, is some 0.2 V higher than its closely (<6.5 Å) flanking [4Fe-4S] cluster partners [27]. Similarly unfavorable steps are evident in fumarate reductase [16,17] and in the three iron-sulfur cluster of the PSI reaction center [28-29]. These systems, which we will present in detail elsewhere, provide a supporting cast for the notion that the multiplicity of iron sulfur clusters and the flavin associated with Complexes I and II represent the links of an efficient electron transfer chain that can operate at near equilibrium and without significant loss of free energy between substrate and the energy coupling reactions in the membrane. It is possible that the thermodynamically unfavorable steps could modulate electron transfer rates. For instance, if such a link was sensitive to the presence of another protein, metabolite, ions or direct ligand exchange, then reaction rates along the chain could be subject to control.

7. References

1. Moser, C.C., Keske, J.M., Warncke, K., Farid, R.S. and Dutton, P.L. (1992) *Nature* 355, 796-802.
2. Moser, C.C. and Dutton, P.L. (1992) *Biochim. Biophys. Acta* 1101, 171-176.
3. Michel, H., Deisenhofer, J. and Epp, O. (1986) *EMBO Journal* 5, 2445-2451.

4. Shopes, R.J., Holten, D., Levine, L. and Wraight, C.A. (1987) *Photosynthesis Research* 12, 165-180.
5. Ohnishi, T. and Salerno, J.C. (1982) *Iron-Sulfur Proteins*, Vol. 4, Wiley Publishing Co., New York.
6. Beinert, H. and Albracht, S.P. (1982) *Biochim. Biophys. Acta* **683**, 245-277.
7. Weiss, H., Friedrich, T., Hofhaus, G. and Preis, D. (1991) *Eur. J. Biochem.* **197**, 563-576.
8. Walker, J.E. (1992) *Q. Rev. Biophys.* **25**, 253-324.
9. Ohnishi, T. (1993) *J. Bioener. Biomembr.* **25**, 325-330.
10. Yagi, T. (1993) *Biochim. Biophys. Acta* **1141**, 1-17.
11. Fearnley, I.M. and Walker, J.E. (1992) *Biochim. Biophys. Acta* **1140**, 105-134.
12. Leif, H., Sled, V.D., Ohnishi, T., Weiss, H. and Friedrich, T. (1995) *Eur. J. Biochem.* **230**, 538-548.
13. Sled, V.D., Rudnitzky, N.I., Hatefi, Y. and Ohnishi, T. (1994) *Biochemistry* **33**, 10069-10075.
14. Orme-Johnson, N.R., Hansen, R.E. and Beinert, H. (1974) *J. Biol. Chem.* **249**, 1922-1927.
15. Beinert, H. and Palmer, G. (1965) *Advances in Enzymology & Related Areas of Molecular Biology* **27**, 105-98.
16. Ackrell, B.A.C., Johnson, M.K., Gunsalus, R.P. and Ceccini, G. (1992) Chemistry and biochemistry of flavoenzymes, Vol. III, CRC Press, Boca Raton, Florida.
17. Hederstedt, L. and Ohnishi, T. (1992) Molecular Mechanisms in Bioenergetics, Elsevier Science Publishers, Amsterdam.
18. Salerno, J.C. and Ohnishi, T. (1980) *Biochem. J.* **192**, 769-781.
19. Tukamiya, K. and Dutton, P.L. (1979) *Biochim. Biophys. Acta* **546**, 1-16.
20. Gunner, M.R., Robertson, D.E. and Dutton, P.L. (1986) *J. Phys. Chem.* **90**, 3783-3795.
21. Warncke, K. and Dutton, P.L. (1993) *Biochemistry* **32**, 4769-4779.
22. Tsukihara, T. et al. (1995) *Science* **269**, 1069-1074.
23. Tsukihara, T. et al. (1996) *Science* **272**, 1136-1144.
24. Iwata, S., Ostermeier, C., Ludwig, B. and Michel, H. (1995) *Nature* **376**, pp. 660-669.
25. Xia, D. et al. (1997) *Science* **277**, 60-66.
26. Woodbury, N.W.T., et al. (1985) *Biochim. Biophys. Acta* **851**, 6-22.
27. Volbeda, A., Charon, M.-H., Piras, C., Hatchikian, E.C., Frey, M. and Fontecilla-Camps, J.C. (1995) *Nature* **373**, 580-587.
28. Krauss, N. et al (1993) *Nature* **361**, 326-331.
29. Yu, L., Zhao, J., Lu, W., Bryant, D.A. and Golbeck, J.H. (1993) *Biochemistry* **32**, 8251-8258.
30. Golbeck, J.H. and Bryant, D.A. (1991) *Current Topics in Bioenergetics* **16**, 83-177.
31. Knaff, D.B. et al. (1991) *Biochemistry* **30**, 1303-1310.

PROTEIN-MEDIATED ELECTRON TRANSFER:
PATHWAYS, ORBITAL INTERACTIONS, AND CONTACT MAPS
Structure-function relations for protein electron transfer

D.N. BERATAN[1] AND S.S. SKOURTIS[2]

[1] Department of Chemistry, University of Pittsburgh, Pittsburgh, PA 15260 USA

[2] Department of Natural Sciences, University of Cyprus, 1678 Nicosia, Cyprus

1. Introduction

The theoretical framework for describing biological electron transfer (ET) reactions is well established [1,2]. Yet the key ingredients of the theory-reaction free energies, inner and outer sphere reorganization energies, and protein-mediated electronic coupling matrix elements - are not directly measurable and are challenging to compute for macromolecules. The goal of this article is to describe our efforts in the last decade to develop empirical [3-12], semiempirical [13], and *ab initio* methods [14] to compute electron transfer rates in proteins. Our focus is the protein-mediated electron donor-acceptor interaction. These methods were recently combined with powerful tools for analyzing electronic propagation in proteins to develop structure-function relations for bridge mediated electron transfer reactions [15,16]. Contributions in this area from other groups have been considerable as well, and were reviewed recently [2]. Using these tools, a comprehensive view of how primary, secondary, and tertiary structure (as well as protein dynamics) influence ET processes is beginning to emerge.

This article surveys our theoretical methods of analyzing electronic structure influences on electron transfer rates. We review the physical basis of electron tunneling in proteins, a process controlled by the covalent, noncovalent, and hydrogen bonded contacts that intervene between donor and acceptor [1,2]. The simple Pathway model balances these tunneling interactions, and predicts how the folded protein motif influences electron transfer rates. Predictions of ET hot and cold spots [6] and of specific motif effects arising in this simple model were recently confirmed. More detailed molecular orbital models incorporate the effects of multiple tunneling pathways in an explicit manner, but the computed interactions are rather sensitive to the precise three-dimensional structure chosen in the computation. Molecular orbital computations can be analyzed using electron transfer contact maps - the electronic analogue of distance contact maps - to reveal the role played by the protein fold in these more complicated computations. This battery of computational and analytical techniques reveals a richness in the protein electron transfer problem that is not apparent from earlier structureless protein models [17,18].

G.W. Canters and E. Vijgenboom (eds.),
Biological Electron Transfer Chains: Genetics, Composition and Mode of Operation, 9-27.
© 1998 *Kluwer Academic Publishers.*

2. Protein Electron Transfer Rates

The rate of electron transfer (k_{ET}) in the regime of weak donor-acceptor interaction (redox cofactor distances beyond van der Waals contact) is [17]:

$$k_{ET} = \frac{2\pi}{\hbar}|T_{DA}|^2(F.C.) \qquad (1)$$

T_{DA} is the protein-mediated donor-acceptor electronic interaction and (F.C.) is the nuclear Franck-Condon factor. In the high temperature regime, (F.C.) gives rise to the familiar Arrhenius-like activation factor described by Marcus [19. The main focus of this paper is to describe how protein structure controls T_{DA}.

3. Basics of Protein-mediated ET

Protein electron transfer often involves redox cofactors that are disposed at distances well beyond van der Waals contact. In most cases, the cofactors are not sufficiently oxidizing or reducing to transfer electrons to or from the polypeptide itself. However, weak orbital mixing does occur between the cofactors and the protein orbitals. This weak interaction, a consequence of the quantum mechanical nature of matter, is the source of the protein-mediated donor-acceptor interaction. Orbital mixing gives rise to quantum mechanical tunneling of the electron from donor to acceptor, so called because purely classical energy considerations forbid the transfer event. The quantum nature of matter provides for an exponentially small probability of the electron to penetrate the protein and appear on the acceptor.

3.1. THROUGH-BOND AND THROUGH-SPACE TUNNELING

The probability of tunneling depends upon the medium being traversed. Tunneling probabilities through empty space produces approximately exponential decay:

$$|T_{DA}|^2_{vacuum} \propto \exp\left[-2\sqrt{2m_e E_{Bind}}R/\hbar\right] \qquad (2)$$

An electronic binding energy E_{BIND} of 8 eV leads to a decay constant of 2.9 Å^{-1}. Theoretical analysis, as well as experimental probes, of *through-bond* decay (including covalent and hydrogen bonds) leads to exponents that are much smaller, about 1.0-1.5 Å^{-1}. The precise value of the decay constant in bonded systems depends upon donor and acceptor orbital energies and symmetry, bridge orbital symmetry and energetics, and dynamics. *The dramatic difference in exponential decay constants for through-bond vs. through-space propagation set the stage for our more detailed models of protein-mediated electron transfer.*

In the absence of detailed protein structural data, it is of course impossible to build explicit models of electron tunneling that take the two distance scales described above into explicit account. However, "low resolution" models can still be constructed. In models of this kind, the donor-acceptor distance is the critical parameter. The first model of this kind, described by Hopfield, was based upon a 1D structureless tunneling barrier model of the protein [17]. Donor, acceptor, and protein orbital energy estimates defined a tunneling barrier height, and an exponential dependence of rate upon distance. Models of this kind were essential in the 1970's because of the paucity of 3D protein structure data. Hopfield's 1974 model predicted $\beta = 1.4$ Å^{-1} where β is the average exponenetial decay of the ET rate with distance [17]. Dutton's recent analysis [18] of a large body of ET rates in the photosynthetic reaction center found an average decay constant consistent with Hopfield's estimate. Single exponential models capture the general aspects of electron tunneling in proteins, but they necessarily neglect the influence of the three-dimensional protein fold on electronic propagation. It is well known in small molecule chemical ET systems that bridge orbital chemical composition, symmetry, and energetics cause drastic effects on donor-acceptor coupling.

Do we expect the specifics of a protein's primary structure and fold to influence donor-acceptor interactions, or will the protein ET function depend soley upon cofactor distance? This issue can only be addressed by considering detailed protein structures and models that describe the different ranges of electronic propagation through-bond and through-space.

3.2. PATHWAYS: AN EMPIRICAL MODEL FOR PROTEIN-MEDIATED INTERACTIONS

Dramatic progress in protein modification and structure determination motivated the development of atomic-scale theories for protein electron transfer. Gray's Ru-modified proteins provide direct probes of how 3D folded structure influences electron tunneling. In addition, high resolution X-ray structural data on numerous ET proteins provided the essential atomic-scale data required as a starting point for detailed theories of protein electron transfer.

The Pathway model provides the simplest atomic-scale description of electronic propagation in proteins. This model accounts for the very different strengths of through-bond vs. through-space electronic propagation. The exponential decay constant for tunneling through a (1D, square) barrier 8 eV high is 2.9 Å^{-1} (Eq. 2), while the decay constant associated with tunneling through a fully covalent medium is about 1.0-1.5 Å^{-1} [6]. Electron transfer across 5 Å is about 10,000 times faster in the presence of an intervening bonded medium than it would be if the space were empty.

Identifying each pairwise atom-atom interaction in a protein as bonded (C), nonbonded (S), or hydrogen-bonded (H), a penalty factor for propagation between any two regions can be estimated from the strongest connecting pathway, where this pathway strength is proportional to the product of decay factors [2]:

$$|T_{DA}|^2_{protein} \propto \left[Max \left\{ \prod_i^{N_C} \prod_j^{N_S} \prod_k^{N_H} \epsilon_i^C \epsilon_j^S \epsilon_k^H \right\} \right]^2 \tag{3}$$

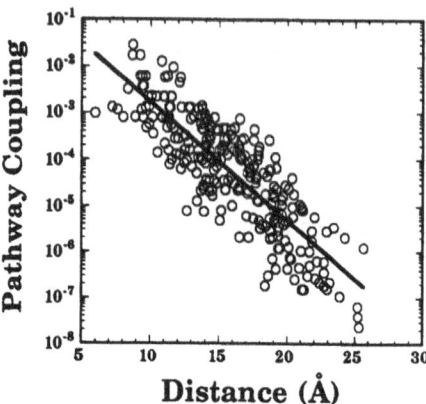

Figure 1. Plot of pathway couplings from the heme of cytochrome c to surface side chains. The protein structure accounts for the scatter of points around an average exponential decay line. It is precisely this scatter that is absent in the simpler structureless barrier models of protein ET

Chemical system experiments, and rough theoretical estimates, suggest:
$\varepsilon^C \approx 0.6$, $\varepsilon^H \approx (\varepsilon^C)^2$ and $\varepsilon^S \approx \varepsilon^C \times \exp[-1.7(r-1.4)]$ [3-8].
The product in this equation represents the combination of penalty factors associated with the *strongest* pathway connecting donor to acceptor (i.e., the largest possible value of the product in Eq. 3. It is a simple matter to assign ε_i values based on a protein's X-ray crystal structure, and to perform a graph search to compute the strongest coupling pathway [11]. Equation 3 emphasizes the strongest pathway. However, pathways exist in bundles of similar strengths [13], so cutting a single path will not, in general, shut down an ET process [11]. Detailed analysis of how pathways within a bundle or *tube* interfere quantum mechanically with one another must be performed with rather more sophisticated methods (*vide infra*) that compute the relative amplitude and phase of electronic propagation along the interfering paths.

Figure 1 shows how the anisotropy of protein structure influences the pathway coupling from the heme in cytochrome c to surface side chains. Note the overall average exponential decay (solid line), with scatter of several orders of magnitude for specific points as a consequence of strong or weak pathway connectivity. The slope of the average line varies from protein to protein, depending on the folded motif. Points falling well above the average line are referred to as hot spots, while those falling well below the line are referred to as cold spots (see Eq. 4).

Pathway predictions reach beyond those accessible with 1D square barrier models. In particular, predictions of hot and cold spots and of secondary motif effects are well documented and tested in Ru-modified proteins [20,21,22]. The predictions are fairly robust; they do not depend on the details of the Pathway parameterization (so long as the range of through-bond decay is much slower than the rate of through-space decay). Analysis of ET rates obtained in Ru-modified cytochromes c clearly support the importance of the Pathway mediation mechanism (see Sect. 3.3).

A critical role is predicted for tunneling pathways dominated by covalent and hydrogen bond steps. The penalty factor for tunneling through a hydrogen bond is predicted to be modest. Hydrogen bonds provide critical shortcuts in the dominant tunneling path. Paths that wind along the peptide backbone may be circuitous and weak mediators of long range tunneling (the beta-barrel proteins are an exception). The prediction of substantial tunneling propagation across hydrogen bonds was confirmed in model compound studies by the Nocera, Therien, and Sessler groups [23], and in proteins by Gray and co-workers [9,10,20,24-28].

The observed overall scale of coupling decay with distance arises from the combination of bonded and non-bonded connectivity in proteins. Coupling through an extended peptide chain (as in a β-strand) is expected to be comparable to that in typical extended chain chemical model compounds according to the Pathway prescription. Decay down an α-helix is somewhat larger for an equivalent donor-acceptor distance because the helix mediated paths are less direct. Distance decay exponents are predicted to be 1.1 Å^{-1} and 1.4 Å^{-1}, respectively, in these structures [6,28]. For ET at 15 Å this difference in average decay constant leads to a 100-fold rate difference. Secondary structure effects of this kind persist even after packing into tertiary motifs [6,7]. This Pathway prediction was recently confirmed by Gray and co-workers [20,21,22,28].

Protein structures are obviously anisotropic, and the Pathway model interprets the consequences of this anisotropy for electron transport processes. While coupling is expected to drop rapidly with donor-acceptor distance, considerable scatter (by several orders of magnitude) about an average exponential line is anticipated. Hot and cold spots for ET are predicted depending upon the directness of coupling pathways between ET cofactors. A dramatic ET cold spot on the surface of cytochrome c was probed in a His 72 modification experiment [27].

Pathways analysis predicts that rates of electron transfer at equal donor-acceptor distances can be dramatically different (by orders of magnitude), depending upon the nature of the intervening medium. To amplify these differences, we defined hot and cold spots quantitatively through the parameter ζ [6]:

$$\zeta = \frac{Max\left\{ \prod_i^{N_C} \prod_j^{N_S} \prod_k^{N_H} \epsilon_i^C \epsilon_j^S \epsilon_k^H \right\}^2}{A \exp[-\beta_{avg} R_{DA}]} \qquad (4)$$

β_{avg} is the average exponential decay constant for the ET rate in a given protein computed from Pathway analysis of couplings between the redox cofactor and a large set of other sites in the protein (e.g., α-carbons, surface accessible side chains, etc.). When $\zeta > 1$ a site is more strongly coupled than average to the redox center given the distance. When $\zeta < 1$ the site is less strongly coupled than average given the distance. As more detailed molecular orbital computations become available, it will be important to compare the predicted hot and cold spots that arise from different methods. Plots of ζ (shown as a color) superimposed on the 3D structure of a protein clearly identify protein hot and cold spots. These plots clearly show how the protein fold establishes these hot and cold spots [6,10].

More importantly, hot and cold spot maps for a range of proteins have shown how secondary and tertiary motifs guide the dominant pathways through a protein structure. Comparison of average pathway mediated decays in a variety of proteins shows clearly that the average decay parameter (β_{avg}) increases with the α-helix content of the protein and decreases with the beta-sheet content as discussed above [12].

Pathway analysis has been used extensively to analyze the electronic interaction between redox cofactors and protein surface groups. In the context of Ru-modified proteins, Pathway analysis has been of use to predict and to analyze relative rates of electron transfer in cytochrome c, myoglobin, cytochrome b_5 and azurin [20,21,22,25]. Because of differences in the cofactor structure among proteins, Pathway analysis best predicts relative rates of electron transfer in systems with similar cofactors.

Pathway calculations have also been used to map protein surface regions that might dominate bimolecular ET in docked complexes. Studies of the cytochrome c / cytochrome c peroxidase (c/ccp) couple [29] and of the cytochrome c_2 / photosynthetic reaction center (c2/prc) couple [30] have been pursued extensively. It is clear that the spatial extent of strongly coupled patches on protein surfaces are of varied sizes. For example, in the c2/prc couple, the strongly coupled prc (acceptor) surface is rather broad, while the c2 (donor) strongly coupled surface is spatially rather limited.

Pathway analysis has also been used to examine how protein structural changes influence electronic couplings throughout the protein. For example, the changes in couplings in cytochrome c upon oxidation/reduction have been surveyed [12].

3.3. EFFECTIVE TUNNELING PATH LENGTH

A fundamental question is whether or not the Pathway analysis is statistically more successful at predicting ET rates than simpler single exponent analysis. To assist our discussion of this issue, we introduce the "pathway effective tunneling length", σ_l [10]. This length is the through-bond length associated with a purely covalent chain (with decay factor 0.6 per bond) that gives rise to a coupling exactly equal to the pathway coupling found in a standard Pathway calculation. (We assume a reference covalent bond length of 1.4 Å.) If the Pathway method worked perfectly, a plot of ln(experimental electron transfer rate) vs. σ_l would be linear. On the other hand, if simple average medium (square barrier) models are adequate, a plot of ln(experimental electron transfer rate) vs. metal-metal separation distance (R_m) would be strictly linear.

The pathway effective tunneling length enters the ET rate as:

$$k_{ET} \propto \left[\prod_i \prod_j \prod_k \epsilon_i^C \epsilon_j^H \epsilon_k^S \right]^2 = (0.6)^{2\sigma_l/1.4} \tag{5a}$$

$$\ln k_{ET} = (2\sigma_l/1.4) \ln 0.6 + const. \tag{5b}$$

This effective length is always larger than the physical distance because: (i) through-space jumps have a much larger effective length than their physical distance, and (ii) paths may never be purely linear. Equation 5b relates σ_l to the pathway coupling. The slopes obtained in simple donor-acceptor distance (R_m) plots give the best fit β_{avg} values for the simple 1D square barrier model. The fixed Pathway parameters *always* predict a slope of 0.73 Å^{-1} for plots of ln k_{ET} versus σ_l, for *every* protein (the slope is exactly $(2/1.4)\ln(0.6)$ and arises from the fixed 0.6 decay per bond parameter and 1.4 Å reference bond length).

In the case of Ru-modified cytochromes c, the Pathway model is far superior, generating correlation coefficients of 0.85 when ln k_{ET} is plotted vs. σ_l as opposed to correlation coefficients of 0.54 when ln k_{ET} is plotted vs simple linear distance [31].

3.4. BIOLOGICAL ET AND TUNNELING PATHWAYS

The pathway model clearly provides a more complete description of protein ET than do the structureless barrier models. However, one must ask whether biology has exploited "pathway effects" in the design of charge transfer systems. One of the most dramatic predictions of the Pathway model is the larger average decay of ET rates through α-helices vs. through β-sheet motifs. This prediction has been subjected to considerable scrutiny by Winkler and Gray for both native and modified systems [21]. Upon observing this effect in Ru-modified proteins they examined two native systems, the photosynthetic reaction center (PRC) and cytochrome c oxidase. It is clear that charge recombination reactions in the PRC are slowed by the helical (indirect) nature of the pathways, increasing the quantum yield for photosynthetic charge separation. Also, the direct wiring of the cytochrome c site to the Cu$_A$ unit in cytochrome oxidase, and of the Cu$_A$ unit to the cytochrome a heme, appear to produce rates very rapid for their driving force, suggesting that nature might be exploiting strong tunneling pathways to achieve rapid electron transfer while preserving the reduction potential difference between cytochrome c and O$_2$ (to drive the proton pumping machinery) [32].

3.5. PATHWAY-LIKE MODELS

The most central aspect of the Pathway model is that it identifies distinct length-scales for the decay of electronic interactions through-bond and through-space. Other empirical models of electron tunneling have incorporated this distinction under a different name. An example is Dutton's binary tunneling model. While the model is neither derived from fundamental physical arguments nor calibrated to experimental systems, it is nearly identical in functional form to the Pathway model. The binary model writes [33]:

$$|T_{DA}|^2 \propto \exp\left\{-(R - 3.6)[f_{atom}\beta_{atom} + (1 - f_{atom})\beta_{vac}]\right\} \qquad (6)$$

where f_{atom} is the fraction of the "medium" composed of atoms and the two β

exponents are characteristic of decay in bonded media vs. decay through empty space. Equation 6 can clearly be written

$$|T_{DA}|^2 \propto \epsilon_C^{2N_C} \epsilon_S^{2N_S} \qquad (7a)$$

$$\epsilon_C^{2N_C} = \exp[-f_{atom}\beta_{atom}(R - 3.6)] \qquad (7b)$$

and

$$\epsilon_S^{2N_S} = \exp[-(1 - f_{atom})\beta_{vac}(R - 3.6)] \qquad (7c)$$

making the binary model equivalent to the Pathway model. The simplest test of any differences between models is simply to plot the ζ parameter for a protein as a color map [6], since this will show exactly where the anomalously strongly and weakly coupled regions of a protein lie according to a given theory.

An open challenge in protein electron transfer is to build molecular orbital - multiple pathway - descriptions of electronic propagation in proteins. Regan et al. [13] and Siddarth and Marcus [34] have used "Pathway guided" methods to analyze the regions of proteins that contribute dominantly to donor-acceptor interactions. In most cases analyzed [2b] only about 10% of the protein needs to be included in a more complete calculation to capture the mediation of the donor-acceptor electronic interaction.

4. Multiple Pathway Methods

The Pathway model is limited in spite of its success. To begin with, Equation 3 involves a proportionality sign. The Pathway model does not directly confront issues of redox cofactor electronic structure, although results of more detailed cofactor analysis have been used in conjunction with pathway computations to estimate coupling prefactors. The most serious limitation of Pathway calculations is the fact that the explicit role of multiple pathways, and their potential constructive and destructive interferences with one another, are not explicitly accounted for. Of course, the numerical values chosen for the decay parameters in the Pathway model (ϵ_i's) are intended to include effects of this kind in an average manner.

Simply stated, the Pathway model captures the difference between through-bond and through-space electronic propagation, and it uses atomic resolution protein structural data as input. Yet, the Pathway model is empirical. Its description of the protein medium does not allow detailed separation of the electronic propagation into interfering pathways, nor does it allow analysis of how atom type, orbital hybridization, or symmetry influence propagation. Figure 2 shows the collection of pathways that contribute to the donor-acceptor couplings in two modified cytochromes c. The pathway approximation is successful when the "strongest"

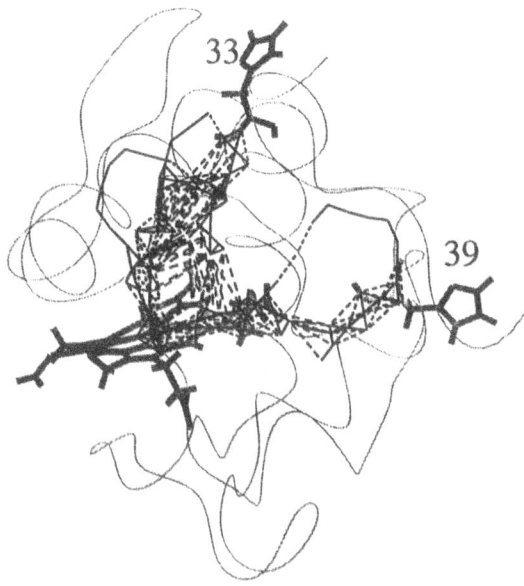

Figure 2. Pathway families that contribute to the donor-acceptor coupling in two Ru-modified cytochromes *c* are shown [13]. Quantum mechanical interferences between electron amplitude propagating through these pathway families, or tubes, can be examined in detail using molecular orbital methods.

pathway is similar in structure to most of the other pathways that make significant contributions to the electronic propagation from donor to acceptor (the so called single tube regime). Regan and Onuchic have analyzed the quantitative relation between protein structural motif and the validity of the Pathway approximation in the context of a simple tight-binding Hamiltonian [2].

Interfering pathways make the task of interpreting the donor-acceptor coupling element in terms of protein structure more complex than a mere search for the shortest pathway. For example, suppose that in a given donor-protein-acceptor system, one finds a shortest (strongest) path of strength $C + 0.5C$ (where C is a constant) a weaker path of strength $-(C+0.1C)$ (note the opposite sign), and several even weaker paths of varying sign and strengths $\sim 0.4|C|$. One might expect the dominant contribution to the donor-acceptor coupling to arise from the strongest paths. However, the opposite signs associated with the two strongest paths leads to destructive interference. If the signs of the two paths were the same, their sum ($2.6C$) would be much larger than the sum of all the weaker pathway contributions, and would dominate the coupling. However, since the two strongest paths have opposite signs, their *sum*, $0.4C$, is smaller than the individual magnitude of each path, and is similar in strength to that of the weaker paths. Therefore, in this example with destructively interfering stong paths, as well as the weaker pathways make a significant contribution to the overall donor-acceptor coupling. Here,

18

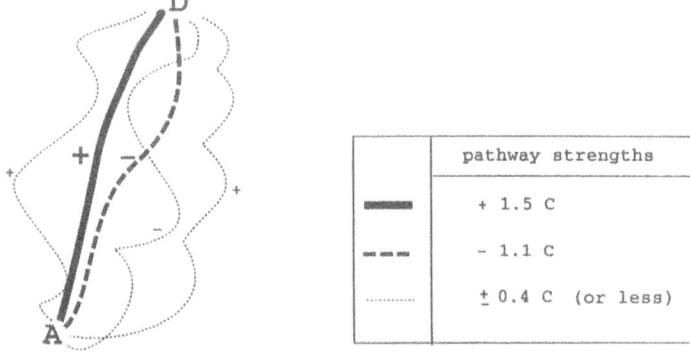

	pathway strengths
▬▬	+ 1.5 C
▬ ▬ ▬	− 1.1 C
··········	± 0.4 C (or less)

Figure 3. Effect of destructive interference between two strong pathways (thick lines) of comparable magnitude and of opposite sign. This destructive interference makes the weaker pathways (longer thin lines) relevant to the overall coupling between donor (D) and acceptor (A).

destructive interference between a few strong (short) paths makes the contributions of many weaker (longer) paths significant. In this situation, a simple strongest pathway search is useful, but not sufficient to compute the donor-acceptor coupling. To analyze the details of the quantum interferences between coupling pathways, it is essential to compute the relative *signs* associated with the pathways to assess the effects of destructive interference between them.

Single and multiple pathway regimes - as described above - can exist in the same protein (between different sites). Skourtis and Beratan [16] recently observed the co-existence of these regimes in small peptide model systems. The tunneling propagation between certain parts of the peptide that was analyzed is dominated by a single pathway mechanism, while the propagation between other parts is controlled by multiple pathways (because of the destructive interferences between the few strongest paths). An important open question is how the internal motion of the peptide influences these interferences. It is likely that dramatic interference effects are weakened by fluctuations of the bridge.

Models of increased quantitative reliability are considerably more complex, but are now becoming available. Such models require a full complement of valence orbitals on each atom. Most valence orbital descriptions of protein ET use an effective electronic Hamiltonian matrix, **H**, for the protein and extract propagation information from it. Computations of this kind were used to examine various aspects of tunneling mediation. For example, we have used used extended-Hückel and extended-Hückel-like Hamiltonians to describe the propagation via both filled and empty molecular orbitals [13]. While simple molecular orbital strategies are appealing in their inclusion of valence atomic orbital mediation, independent electron models treat the energetic and through-space aspects of the problem rather poorly [14]. Extended-Hückel Hamiltonians are qualitatively useful in their description of multiple pathway and interference effects, but it is now well known that they misrepresent important details of the electronic propagation [14,35].

Given a protein electronic Hamiltonian matrix, **H**, built in an atomic orbital basis with overlap matrix **S** [36],

$$k_{ET} \propto \left[(\mathbf{V} - E_t\mathbf{S})_{DP}(\mathbf{H} - E_t\mathbf{S})_{PP}^{-1}(\mathbf{V} - E_{tun}\mathbf{S})_{PA} \right]^2 \qquad (8)$$

V is the matrix of interactions between the donor (D) and acceptor (A) cofactors and the protein (P). This expression lacks the obvious contact with 3D protein structure provided in the Pathway model. However, it does build in the difference between through-bond and through-space propagation through the **V** and **H** elements. Construction and manipulation of these large matrices (consisting of $\sim10^5$ rows and columns) brings forward two new challenges: 1) development of reliable Hamiltonian matrices to describe the long-distance propagation of electronic amplitude in proteins and 2) establishment of new methods to interpret the ET characteristics of a particular protein Hamiltonian matrix.

We recently developed an *ab initio* molecular fragment method that improves the energy and through-space descriptions of electronic propagation in very large systems over that arising in semiempirical methods [14]. The general nature of the predictions in *ab initio* fragment analysis are consistent with those of the Pathway model. For example, the average distance decay parameter found for the idealized α-helix and β-strand are 1.0 Å^{-1} and 1.4 Å^{-1} respectively. However, the new method allows explicit treatment of donor and acceptor structure, dissection of interferences, and analysis of the influence of dynamical fluctuations on the coupling.

5. Quantitative Structure-function Relations for Protein ET: Electron Transfer Contact Maps

A common challenge in structural biology is to represent biomolecular structure in simplified forms in order to emphasize the important structural features. For example, ribbon drawings of proteins help to visualize and to interpret protein secondary and tertiary structural motifs. The task of identifying structural motifs is essentially imposile if all of the side chain atoms are included in the structural representation of the protein. Ribbon drawings are an example of how information reduction can be used to assist structural interpretation. Other examples are Ramachandran plots and distance contact maps [37].

A significant issue of information reduction exists in the field of electron transfer. How can one represent the tunneling mediation properties of a given protein or compare the mediation characteristics of different proteins? Proteins contain thousands of atomic orbitals and the matrix that represents the tunneling propagation from any one atomic orbital to another is very large indeed. It is appealing to seek out meaningful strategies of information reduction for the ET problem.

The hot and cold spot maps described earlier reduce the information associated with a family of simple pathway calculations. However, strategies of this kind do not readily generalize to multiple pathway coupling methods. A more general strategy was recently described by Skourtis and Beratan [16].·Its aim is to analyze

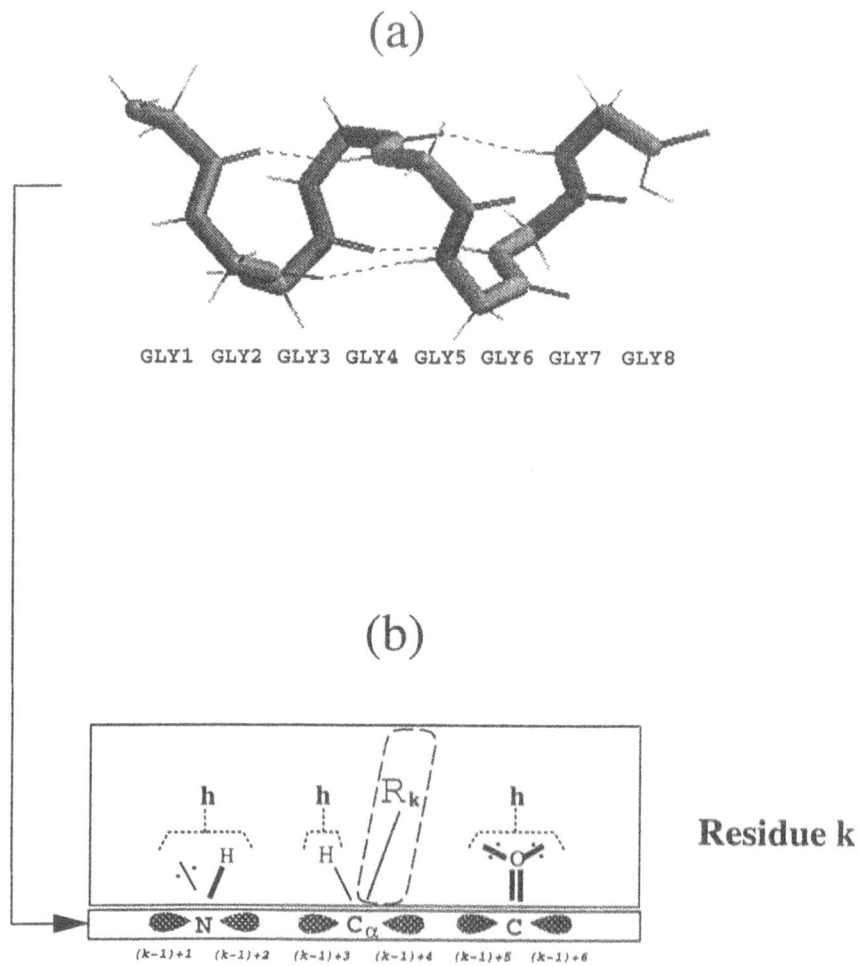

(a)

GLY1 GLY2 GLY3 GLY4 GLY5 GLY6 GLY7 GLY8

(b)

Residue k

Figure 4. Simple example of structures used to generate electron transfer contact maps (ETCMs) for an alpha-helical peptide. (a) α-helical petide $[GLY]_8$. The substructure chosen here for the construction of the ETCMs is the unbranched part of its backbone (shown by the tube). (b) Schematic representation of the atomic orbitals in the unbranched part of the backbone - and of the remaining branched structure - for each residue k of $[GLY]_8$. The branched portion consists of side chain R_k and backbone hanging groups h. The figure also shows the ordering convention (numbering) for residue k of the atomic orbitals in the unbranched backbone (the direction is from the N to the C terminus of the peptide).

(and simultaneously to visualize) how different structural patterns in proteins influence the electron tunneling mediation. The approach utilizes two-dimensional plots called electron transfer contact maps (ETCMs). These plots represent the global tunneling mediation characteristics of a protein structure in a compact and interpretable form.

Suppose that a set of different protein structures are compared to each other. The critical step in this procedure is to choose a substructure in each protein that is common to all proteins of the set in order to make a fair comparison. The most natural choice for this substructure is the backbone of each protein, since the chemical constitution of the backbone is the same for all proteins (the only differences arise from backbone length and possible proline residues). Once the backbone is chosen as the reference substructure, a tunneling propagation matrix is computed for each protein in the set. This matrix $(G_{ij}(E_{tun}))$ describes the exact electron tunneling propagation from any backbone atomic orbital i to any backbone atomic orbital j in the protein [16]. Each G_{ij} computation includes the whole protein even though the initial and final orbitals are backbone orbitals. The final result of this procedure is a set of G_{ij} matrices for the proteins being compared. Although each G_{ij} matrix is different - reflecting differences in primary structure and folded conformation - the i's and j's of each matrix are identical (i.e. backbone atomic orbitals). This means that direct comparisons of the matrices are possible (for example, by subtracting one matrix from the other). The G_{ij} plots (and related quantities) are called electron transfer contact maps (Fig. 4 and 5) in analogy with traditional distance contact maps that plot distances between C_α atoms [38].

ETCM's are reduced descriptions of tunneling mediation in proteins (since they show the tunneling between a fraction of the atomic orbitals in a protein), but they are exact descriptions (since the computation of the tunneling mediation within this set of orbitals incorporates the influence of the entire protein structure). The reduction of information allows the visualization of global tunneling mediation features in the protein, and the direct comparison of tunneling mediations characteristics between different proteins. ETCMs are therefore central for the study of the transferability of electron tunneling propagation characteristics of structural motifs (helix - as in Fig. 4 - and sheet structures, for example). As an example, one could ask the question of how the propagation through a beta-sheet structure in one protein compares to the propagation through a similar β-sheet in another protein. This question can be addressed directly by choosing the common section of the two β-sheets to serve as the refererence substructure for the two proteins, and then computing the ETCMs based on this substructure.

Another use of ETCMs is to analyze the influence of mutations on electron transfer rates. The changes in tunneling propagation induced by a mutation can be probed through importance value matrices, probed through importance value matrices, $R_{ij}(E_{tun}) \equiv (|G_{ij}^{mut}| - |G_{ij}^{wild}|) / |G_{ij}^{wild}|$ [16,40]. Each element R_{ij} of the importance value matrix represents the fractional change in the tunneling propagation between protein atomic orbitals i and j that is induced by the mutation. G_{ij}^{mut} and G_{ij}^{wild} are the exact tunneling propagation between pairs of obitals in the mutated and wild type proteins, respectively. i and j may be backbone atomic orbitals, or atomic orbitals of the residues that are not altered by the mutation. A plot of the importance value matrix indicates the regions (or amino acid pairs) of the protein that show enhanced or reduced electronic coupling as a result of the mutation. Once these regions are identified, a more detailed analysis of the interactions is possible. This analysis utilizes the mutation matrix (see[16]) that interprets all changes in the

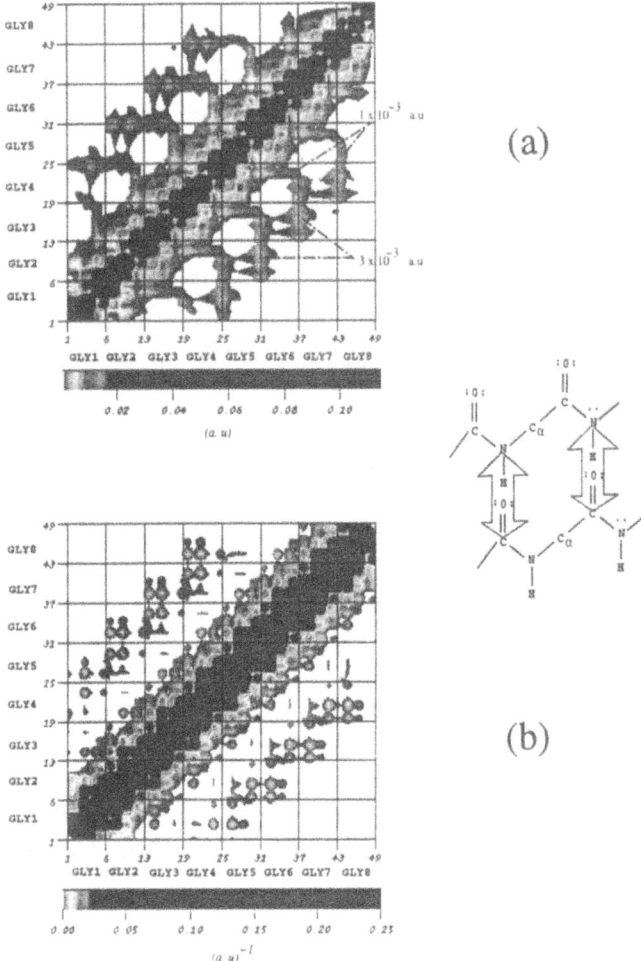

Figure 5. Simple examples of electron transfer contact maps (ETCMs) for the α-helical peptide of Figure 4. (a) ETCM showing electronic coupling between any pair of unbranched-backbone orbitals i, j (only the coupling mediated solely by the branched portion of the peptide is plotted). The x and y axes represent unbranched backbone orbitals ordered as in (b). The grid lines separate adjacent residues. The strongest couplings that connect non-adjacent residues are between all residue pairs k and $k+3(4)$. These are hydrogen bond couplings and they are indicated by the drawing in the middle. (b) Different ETCM showing the full propagation G_{ij} between all pairs of unbranched-backbone orbitals i, j in $[GLY]_8$. This calculation includes the combination of the propagation through the unbranched backbone and the propagation through the branched portion of the peptide. The large peaks between residues k and $k+3(4)$ indicate the persistence of strong hydrogen bond pathways in the presence of alterantive pathways through the backbone. These calculations [16] use a semiempirical CNDO hamiltonian for the peptide and a tunneling energy of -0.2 a.u. (1 a.u = 27 eV).

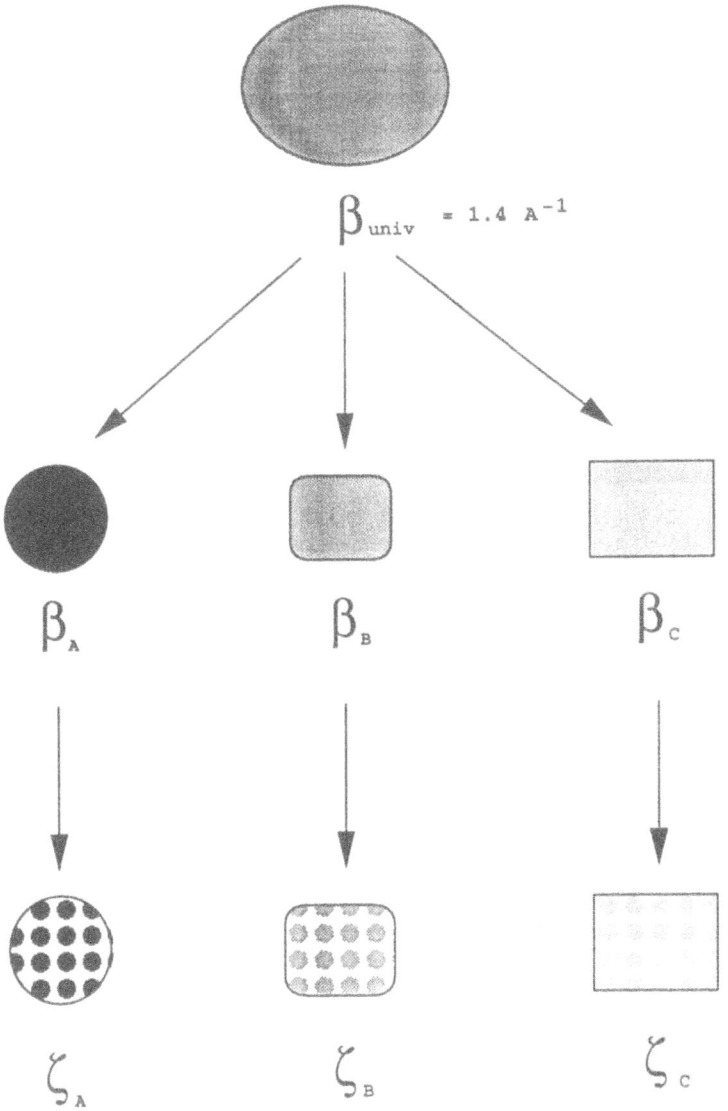

Figure 6. Different levels of description of the tunneling mediation in proteins. The top figure represents a homogenoeus barrier model that is supposed to describe the "average" protein. That is, $T_{DA}(R_{DA}) \propto \exp[-\beta_{avg}R_{DA}]$, where $\beta_{avg} = 1.4$ Å^{-1} is a typical average decay constant for proteins. The middle figures represent homogenous barrier models ($\exp[-\beta R_{DA}]$) that have *different* decay constants ($\beta = \beta_A, \beta_B, \beta_C$) for *different* protein types A, B, C (e.g., azurins, cytochromes, etc.). The bottom figures shows models that can describe fluctuations around the homogenoeus barrier model predictions for each protein type (i.e., the hot and cold spot ζ values of Eq. 4 where $\beta_{avg} = \beta_A, \beta_B, \beta_C$). All three levels can be consistent with one another. The limitation of the top level of theory is that it makes the *a priori* assumption that the bottom two levels are irrelevant. Therefore, it cannot be used to address questions of specificity in structure-function relations.

tunneling propagation between these regions in terms of the new pathways that are created by the mutation. Recent experiments by Farver, Pecht and coworkers [41] showing enhancement of ET rates in azurin upon mutation far removed from the redox cofactors, provide nearly ideal systems for the application of these ETCM methods.

6. Frontiers of Protein ET

Several great challenges remain in protein ET theory. One is to expand the development of reliable protein effective Hamiltonians. *Ab initio* fragment (divide and conquer) methods seem particularly promising in this regard [14]. As methods that explicitly include the influence of multiple pathways become available, the effects of different protein structural motifs on the tunneling propagation can be probed systematically. Interpretive tools such as importance values, electron transfer contact maps, and mutation matrices should be of great value to compare the propagation through several *different* structural motifs. Furthermore, the propagation through any given motif should be studied in different protein structures (containing this motif), in order to understand the *transferability* of this propagation from structure to structure. Structural fluctuations will also need to be confronted more directly in this analysis. That is, quantum interferences between multiple pathways are extremely sensitive to protein geometry. Small fluctuations can cause significant changes in coupling values [42] and appropriate averaging is needed. Quantitative comparisons with experiment will require the computation of electronic interactions in numerous geometries.

Just as our ability to compute electronic interactions is expanding, so is our ability to estimate nuclear Franck-Condon factors. Recent numerical calculations of reorganization energy (based upon numerical solutions of the Poisson equation) show how fluctuations in ET bridge structure may exert dramatic control on the reorganization energy and ET reaction dynamics [43]. Computation of protein reorganization energies with these techniques is warranted.

Electron transfer chains rely upon *multi*-electron events. We have recently developed new theories to describe electronic and nuclear control of multi-electron reactions. The dynamics is rich, and the free energy dependence can be rather different than is seen in single electron processes [44]. Of course, molecular level theories of *coupled* electron/proton transfer in biology remain poorly elaborated.

Finally, the structure-function analysis tools that we developed for protein ET are applicable to other macromolecule ET processes. In particular, our recent attention has turned to the electron mediation characteristics of DNA. Studies from our group suggest the decay of ET rates with distance in DNA to be qualitatively similar that found in proteins (in spite of the unique π-stacked base pairs present in DNA) [45], a view not yet in agreement with a provocative body of recent experiments [46].

The underlying goal of our approach to biological ET is to understand how proteins manipulate donor, acceptor, and bridge structure to control electron transfer rates. It is still unknown whether bridge structure behaves as a very specific electron mediator where structural details are significant, or as a more nearly "average" mediator. This question can only be answered by studying natural and artificial ET systems, and by applying increasingly sophisticated models (containing a minimum of *assumptions* about the propagation) that are then capable of revealing the behavior of a particular system. We may find that biological ET systems operate in regimes that lie between the "highly specific" and the "average". Indeed, it seems likely that the answer will depend upon specifics of the protein structure and dynamics (see Fig. 6).

7. Acknowledgments

We thank I.V. Kurnikov, J.N. Onuchic, S. Priyadarshy, J.J. Regan, S.M. Risser, and L.D. Zusman for their discussion of these issues. D.N.B. is supported by a National Young Investigator Award (National Science Foundation Grant No. CHE-9257093). S.S.S. is supported by the Cyprus government through the research program "From Strong Interactions to Molecular Recognition: Theoretical and Computational Studies" and by the Greek General Secretariat for Research and Technology. The authors may be reached at the email addresses: beratan+@pitt.edu and skourtis@elgreco.ns.ucy.uc.cy.

8. References

1. Bertini I., Gray, H.B., Lippard, S., and Valentine, J.S. (1994) *Principles of Bioinorganic Chemistry*, University Science Books, Mill Valley CA.Lippard, S.J. and Berg, J.M. (1994) *Bioinorganic Chemistry*, University Science Books, Mill Valley, CA.
2. (a) Bendall, D.S. (1996) *Protein Electron Transfer*. BIOS Scientific Publishers, Oxford. (b) Skourtis, S.S. and Beratan, D.N. (1997) Theories of structure function relationships for bridge-mediated electron transfer reactions, *Adv. Chem. Phys.*, in press. (c) Regan, J.J. and Onuchic, J.N. (1997) *Adv. Chem. Phys.*, in press.
3. Beratan, D.N., Onuchic, J.N. and Hopfield, J.J. (1987) Electron tunneling through covalent and non-covalent pathways in proteins, *J. Chem. Phys.* **86**, 4488-4498.
4. Beratan, D.N. and Onuchic J.N. (1989) Electron tunneling pathways in proteins: influences on the transfer rate, *Photosynth. Res.* **22**, 173-186.
5. Onuchic, J.N. and Beratan, D.N. (1990) A predictive theoretical model for electron tunneling pathways in proteins, *J. Chem. Phys.* **92**, 722-733.
6. Beratan, D.N., Betts, J.N., and Onuchic, J.N. (1991) Protein electron transfer rates are predicted to be set by the bridging secondary and tertiary structure, *Science* **252**, 1285-1288.
7. Beratan D.N., Onuchic J.N. and Gray H.B. (1991) Electron tunneling pathways in proteins, in H. Sigel and A. Sigel (eds.) *Metal Ions in Biological Systems*, vol. 27, Marcel Dekker Press, New York, pp. 97-127.
8. Beratan, D.N. and Onuchic, J.N. (1991) Electron transfer - from model compounds to proteins, in J.R. Bolton, N. Mataga, G.L. McLendon (eds.), *Electron transfer in inorganic, organic, and biological systems*, ACS Advances in Chemistry Series vol. 228, ACS Press, Washington, DC, pp 71-90.
9. Beratan, D.N., Onuchic J.N., Winkler, J.R., and Gray H.B. (1992), *Science* **258**, 1740-1741.
10. Onuchic, J.N., Beratan, D.N., Winkler, J.R., and Gray, H.B. (1992) Pathway analysis of protein electron-transfer reactions, *Annu. Rev. Biophys. Biomol. Struct.* **21**, 349-377.
11. Betts, J.N., Beratan, D.N., and Onuchic, J.N. (1992) Mapping electron tunneling pathways: an algorithm that finds the 'minimum length'/ maximum coupling pathway between electron donors and acceptors in proteins, *J. Am. Chem. Soc.* **114**, 4043-4046.

26

12. Beratan, D.N., Betts, J.N., and Onuchic, J.N. (1992) Tunneling pathway and redox state dependent electronic couplings at nearly fixed distance in electron transfer proteins, *J. Phys. Chem.* **96**, 2852-2855.

13. (a) Regan J.J., DiBilio A.J., Langen R., Skov L.K., Winkler J.R., Gray H.B., and Onuchic, J.N. (1995) Electron tunneling in azurin: the coupling across a β-sheet, *Chem. and Biol.* **2**, 489-496. (b) Regan J.J., Risser, S.M., Beratan D.N., and Onuchic, J.N. (1993) Protein electron transport: single versus multiple pathways, *J. Phys. Chem.* **97**, 13083-13088.

14. Kurnikov, I.V. and Beratan, D.N. (1996), *J. Chem. Phys.* **105**, 9561-9573.

15. Skourtis, S.S., Onuchic, J.N. and Beratan, D.N. (1996) A method to analyze multi-pathway effects on protein mediated donor-acceptor coupling interactions, *Inorg. Chim. Acta* **243**, 167-175.

16. Skourtis, S.S. and Beratan, D.N. (1997) Electron transfer contact maps, *J. Phys. Chem.* B **101**, 1215-1234.

17. Hopfield, J.J. (1974) Electron transfer between biological molecules by thermally activated tunneling, *Proc. Natl. Acad. Sci. (USA)* **71**, 3640-3644.

18. Moser, C.C., Keske, J.M., Warncke, K., Farid, R.S. and Dutton, P.L. (1992) Nature of biological electron transfer, *Nature* **355**, 796-802.

19. Marcus, R.A. and Sutin, N. (1987) Some aspects of electron transfer in chemistry and biology, *Biochem. Biophys. Acta* **811**, 265-322.

20. Karpishian, T.B., Grinstaff, M., Komar-Panicucci, S., McLendon, G., Gray, H.B. (1994) Electron transfer in cytochrome *c* depends upon the structure of the intervening medium, *Structure* **2**, 415-422.

21. Gray, H.B. and Winkler, J.R. (1996) Electron transfer in proteins, *Annu. Rev. Biochem.* **65**, 537-561.

22. Langen, R., Colon, J.L., Casimiro, D.R., Karpishin, T.B., Winkler, J.R. and Gray, H.B. (1996) Electron tunneling in proteins: role of the intervening medium, *JBIC* **1**, 221-225.

23. (a) DeRege, P.J.F., Williams, S.A., and Therien, M.J. (1995) Direct evaluation of electronic coupling mediated by hydrogen bonds: implications for biological electron transfer, *Science* **269**, 1409-1413. (b) Turró, C., Chang, C.K., Leroi, G.E., Cukier, R.I. and Nocera D.G. (1992) Photoinduced electron transfer mediated by a hydrogen bonded interface, *J. Am. Chem. Soc.* **114**, 4013-4015. (c) Sessler, J.L., Wang, B., Harriman, A. (1993) Long-range photoinduced electron transfer in an associated but noncovalently linked photosynthetic model system, *J. Am. Chem. Soc.* **115**, 10418-10419.

24. Winkler, J.R., Nocera, D.G., Yocom, K.M., Bordignon, E. and Gray H.B. (1982) Electron transfer kinetics of pentaammine ruthenium(III)-(histidine-33)-ferricytochrome *c* - measurement of the rate of intramolecular electron transfer between redox centers separated by 15 Å in a protein, *J. Am. Chem. Soc.* **104**, 5798-5800.

25. Jacobs, B.A., Mauk, M., Funk, W., MacGillivray, R. Mauk, A.G. and Gray, H.B. (1991) Preparation, characterization, and intramolecular electron transfer in pentaamnieruthenium histidine-26 cytochrome b_5 derivatives - role of the intervening medium in long range donor-acceptor electronic coupling, *J. Am. Chem. Soc.* **113**, 4390-4394.

26. Therien, M.J., Chang, J., Raphael, A.L., Bowler, B.E. and Gray, H.B. (1991) in *Structure and Bonding* vol. 75, 109-129, Springer Verlag, New York.

27. Wuttke, D.S., Bjerrum, M.J., Winkler, J.R. and Gray, H.B. (1992) Electron tunneling paths in cytochrome *c*, *Science* **256**, 1007-1009.

28. Langen, R., Chang, I.J., Germanas, J.P., Richards, J.H., Winkler, J.R. and Gray, H.B. (1995) Electron tunneling in proteins: coupling through a β-strand, *Science* **268**, 1733-1735.

29. Nocek, J.M., Zhou, J.S., DeForest, S., Priyadarshy, S., Beratan, D.N., Onuchic, J.N., Hoffman, B.M. (1996) Theory and practice of electron transfer within protein-protein complexes: application to the multi-domain binding of cytochrome *c* by cytochrome *c* peroxidase, *Chem. Rev.* **96**, 2459-2489.

30. Aquino, A.J.A., Beroza, P., Beratan, D.N. and Onuchic, J.N. (1995) Docking and electron transfer between cytochrome c_2 and the photosynthetic reaction center, Chem. Phys. **197**, 277-288.

31. Balabin, I.A., Onuchic, J.N., and Beratan, D.N. (1997), unpublished data.

32. (a) Winkler, J.R. and Gray, H.B. (1997) Electron tunneling in proteins: role of the intervening medium *JBIC*, in press. (b) Ramirez, B.E., Malmström, B.G., Winkler, J.R. and Gray, H.B. (1995) The currents of life: the terminal electron-transfer complex of respiration, *Proc. Natl. Acad. Sci. (USA)* **92**, 11949-11951.

33. Moser, C.C., Page, C.C., Chen, X. and Dutton, P.L. (1997) Effect of intervening medium on long-range biological electron transfer, *JBIC*, in press.

34. Siddarth, P. and Marcus, R.A. (1993) Correlation between theory and experiment in electron transfer reactions in proteins - electronic couplings in modified cytochrome c and myoglobin, *J. Phys. Chem.* **97**, 13078-13082.

35. Curry, W.B., Grabe, M.D., Kurnikov, I.V., Skourtis, S.S., Beratan, D.N., Regan, J.J., Aquino, A.J.A., Beroza, P. and Onuchic, J.N. (1995) Pathways, pathway tubes, pathway docking, and propagators in electron transfer proteins, *J. Bioenerg. Biomembr.* **27**, 285-293.

36. Priyadarshy, S., Skourtis, S.S., Risser, S.M. and Beratan, D.N. (1996) Bridge-mediated electronic interactions: differences between Hamiltonian and green function partitioning in a non-orthogonal basis, *J. Chem. Phys.* **104**, 9473-9481.

37. T.E. Creighton (1984), *Proteins. Structures and Molecular Properties*, W.H. Freeman Press, New York.

38. The inadequacy of distance contact maps for the description of ET was demonstrated in [39]. ETCMs were used to compare two different peptide structures related by a double mutation. Two calculations were performed. In one calculation the ETCMs were based on the exact computation of the tunneling propagation matrices G_{ij} , using atomic orbital level Hamiltonians that incorporate the details of bonding. In the other, the G_{ij} matrices were approximated by an exponential distance decay model [39] that ignores structural details, i.e. $G_{ij} \propto \exp[-\beta(r_{ij})]$, where $\beta = 1.4$ Å^{-1} and r_{ij} is the distance between backbone orbital i and backbone orbital j. Although the distances between backbone orbitals were not altered by the mutation, the exact ETCMs showed large changes in the propagation between backbone orbitals. These changes were caused by the mutation, and the effects were as large as ~1000 times the original propagation strength. The approximate distance-based ETCMs did not show any changes in the tunneling propagation, since the distances between backbone orbitals remain constant upon mutation. Simple square barrier protein models do not predIct any change in coupling upon mutation.

39. Skourtis, S.S. and Beratan, D. N. (1997) High and low resolution theories of protein electron transfer, *JBIC*, in press.

40. Skourtis, S.S. Regan, J.J., Onuchic, J.N. (1994), Electron transfer in proteins: a novel approach for the description of the donor-acceptor, *J. Phys. Chem.* **98**, 3379-3388.

41. (a) Farver, O. and Pecht, I. (1997) The role of the medium on long range electron transfer, *JBIC*, in press. (b) Farver, O., Skov, L.K., Young, S., Bonander, N., Karlsson, B.G., Vänngård, T. and Pecht, I. (1997), submitted for publication.

42. Wolfgang, J., Risser, S.M., Priyadarshy, S., and Beratan, D.N., (1997) Secondary structure conformations and long range electronic interactions in oligopeptides. *J. Phys. Chem. B.* **101**, 2986-2991.

43. Kurnikov, I.V., Zusman, L.D., Kurnikova, M.G., Farid, R.S., and Beratan, D.N. (1997) Structural fluctuations, spin, reorganization energy, and tunneling energy control of intramolecular electron transfer (1997) *J. Am. Chem. Soc.*, in press.

44. (a) Zusman, L.D. and Beratan, D.N. (1996) Two-electron transfer reactions in polar solvents, *J. Chem. Phys.* **105**, 165-176. (b) Zusman, L.D. and Beratan, D.N. (1997) A three-state model for two-electron transfer reactions, *J. Phys. Chem.*, in press.

45. (a) Beratan, D.N., Priyadarshy, S. and Risser, S.M. (1997) DNA: Insulator or Wire?, *Chem. and Biol.* **4**, 3-8. (b) Priyadarshy, S., Risser, S.M. and Beratan, D.N. (1996) DNA is not a molecular wire: protein-like electron transfer predicted in an extended π-electron system, *J. Phys. Chem.* **100**, 17678-17682.

46. (a) Murphy, C.J., Arkin, M.R., Jenkins, Y., Ghatlia, N.D. Bossmann, S.H., Turro, N.J. and Barton, J.K. (1993) Long-range photoinduced electron transfer through a DNA helix, *Science* **262**, 1025-1029. (b) Brun, A.J. and Harriman, A. (1994). Energy transfer and electron transfer processes involving palladium porphyrins bound to DNA, *J. Am. Chem. Soc.* **116**, 10383-10393. (c) Meade, T.J. and Kayyem, J.F. (1995) Electron transfer through DNA - site-specific modification of duplex DNA with ruthenium donors and acceptors, *Angew. Chem. Int. Ed. Engl.* **34**, 352-354. (d) Olson, E.J.C., Hu, D., Hörmann, A., and Barbara, P.F. (1997) Quantitative modeling of DNA-mediated electron transfer between metallointercalators, *J. Phys. Chem. B* **101**, 299-303.

COUPLING OF ELECTRON TRANSFER AND PROTEIN DYNAMICS

A.I.KOTELNIKOV, V.R.VOGEL, A.V.PASTUCHOV,
V.L.VOSKOBOINIKOV, E.S.MEDVEDEV
*Institute of Chemical Physics, Russian Academy of Sciences,
Chernogolovka, Moscow region, 142432 Russia*

1. Introduction

Electron transfer (ET) reaction with participation of special proteins is the key process of functioning of biological systems. Till recently the high effectiveness and specific features of ET in proteins have been described solely by allowing for a protein's static structure but their molecular dynamics has not been taken into account. However, according to the Marcus theory [1] the dynamic properties of medium constitute one of the most important parameters that govern the effectiveness of electron transfer in condensed media. The dynamic properties of proteins differ appreciably from the analogous parameters of liquids and solids. The local structural adjustment of the active center or the conformational transition of the entire protein globule during the catalytic act is a unique property of enzymes enabling them to optimize and control the chemical process. The aim of this paper is to critically analyze the existing models of long-range ET reactions in proteins. To correct the revealed discrepancies we propose a model, taking into account the molecular dynamics of proteins with a broad distribution of the relaxation times.

2. Application of the Marcus Theory to the Description of Long Range Electron Transfer in Proteins

All original papers and reviews published recently, for example [2-10], analyze the ET in native and chemically modified proteins in the context of the nonadiabatic approach of the Marcus theory. According to the theory under conditions of fast matrix reorganization ($H_{ab}^2 << \{\lambda h \nu / 4\pi\}$) the ET constant K_e is given by the classical Marcus expression [1], derived within the framework of the general theory of radiationless transitions [11].

$$K_e = \sigma \exp[-E_a/RT] \qquad (1)$$

29

G.W. Canters and E. Vijgenboom (eds.),
Biological Electron Transfer Chains: Genetics, Composition and Mode of Operation, 29-49.
© 1998 *Kluwer Academic Publishers.*

where

$$\sigma \sim H_{ab}^2(R) \sim \exp(-\alpha R) \qquad (2)$$

$$E_a = (\Delta G + \lambda)^2 / 4\lambda \qquad (3)$$

In these equations H_{ab}, E_a, ΔG, λ, ν are the tunneling matrix element, the activation energy, the reaction driving force, the nuclear reorganization energy, and the nuclear reorganization characteristic frequency, respectively. In this case K_e has exponential dependence on R and does not depend on ν.

The application of the nonadiabatic approximation depends on the ν and H_{ab} magnitudes that can broadly vary in the case of proteins. As estimated in refs. [12, 13], $H_{ab} = 10^{-1} - 10^{-3} cm^{-1}$ for the ET in myoglobin (mb) and cytochrome c (cyt-c). If the nuclear reorganization is governed by the motion of free water molecules ($\nu = 2 \cdot 10^{12} s^{-1}$) and $\lambda = 1 eV$, the nonadiabatic limit is fulfilled at $H_{ab} << 200\ cm^{-1}$ for all ET reactions in mb and cyt-c [6]. But the relaxation of the myoglobin haem pocket takes place at $\nu \sim 10^8 - 10^4\ s^{-1}$ [14, 15]. If $\nu = 10^4\ s^{-1}$ the nonadiabatic limit is fulfilled at $H_{ab} << 2 \cdot 10^{-2} cm^{-1}$, which is not valid for some ET reactions in mb and cyt-c.

If the condition of nonadiabatic approach is not valid, the adiabatic limit must be used. In accord with the simplest version of the theory [16, 17] if $H_{ab}^2 >> \{\lambda h \nu / 4\pi\}$, K_e is expressed by the equation

$$K_e \sim \nu \exp[-E_a / RT] \qquad (4)$$

In this case K_e does not depend on the matrix element H_{ab} and consequently on the distance R.

The intermediate case between nonadiabatic and adiabatic limits has been considered for a model with a single diffusive mode X [16-18]. The frequency factor ν in Eq. (4) is replaced with

$$\sigma(1 + \sigma \tau / A)^{-1} \qquad (5)$$

where A is a function of H_{ab} and λ, $\tau = 1/\nu$ is the thermally activated relaxation time for the diffusive motion,

$$\tau = \tau_0 \exp(E_{rel} / RT) \qquad (6)$$

with the activation energy E_{rel}.

In more complicated cases, if the times of electron transfer and of matrix reorganization are comparable and there is a wide distribution of the relaxation times, in accordance with adiabatic approach of Rips and Jortner [18]

$$K_e \sim (1/\tau_0)^\beta \, H_{ab}^{2(1-\beta)} \exp[-E_a/RT] \qquad (7)$$

where $0<\beta<1$ is the time distribution parameter of the Davidson-Cole dielectric relaxation spectrum and τ_0 is the characteristic relaxation time, which corresponds to the upper limit of the distribution of τ values. In the last case K_e depends simultaneously on the frequency $\nu_0=1/\tau_0$ and on the distance R.

According to the Marcus theory the ET constant reaches a maximum value K_e^{max}, if $-\Delta G = \lambda$ (see Eq. 3). The values of λ and K_e^{max} can be determined from the $K_e(\Delta G)$-dependence. This opens a possibility for a detailed analysis of ET in proteins and for a choice of one of the three approaches mentioned above. If one considers the ET in the framework of the nonadiabatic approach, then in accordance with Eqs. (1)-(2) log K_e^{max} must be a linear function of the distance R (Fig.1). The existence of some donor-acceptor pairs differently localized in the protein globule (different distances R or different chemical bonds between the centers) makes it possible to determine the coefficient α describing the dependence of H_{ab}^2 on the nature of the protein matrix separating these centers.

In accordance with Eqs. (4), (5) and (7), the absence of $\log K_e^{max}$ dependence on R or a weak dependence on R (in comparison to the nonadiabatic limit) will indicate the adiabatic regime of ET. In the case of the Rips-Jortner limit in accordance with Eq. (7) one can observe the log K_e^{max}-R dependences with different slopes, determined by the parameter β. At $\beta=0$ we have the nonadiabatic limit and at $\beta=1$ a simple adiabatic limit (Fig.1). It is necessary to take into account the possibility of existence of two ET mechanisms in one and the same protein: the nonadiabatic mechanism for a slow electron transfer over a large distance and the adiabatic mechanism for a fast electron transfer over a short distance. In such a case we may observe a nonlinear dependence of $\log K_e^{max}$ on R, as represented in Figure 1.

As has been mentioned above, the typical values of H_{ab} for a long-range ET in proteins are within the range 10^{-1}-10^{-4} cm^{-1}, $\lambda\sim 1$ eV [6,12,13,19]. The characteristic times of protein reorganizations of differet types can vary from 10^{-11} s up to milliseconds [14,15,20-25]. By this reason there is a great ambiguity in an estimation of the nonadiabatic or the adiabatic regime of ET in proteins. In spite of such ambiguity in papers and reviews published recently [2-10,12-13,19,25-35] an analysis of the ET reactions in native and chemically modified proteins has been performed only in the context of the nonadiabatic approach.

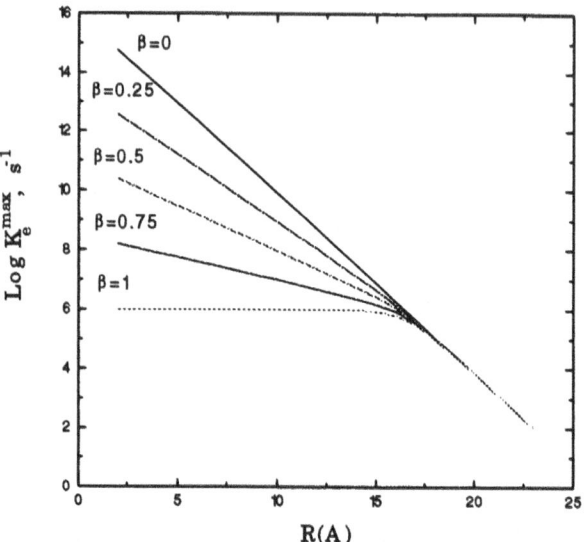

Figure 1. The K_e^{max} dependences on distance R in the framework of the Rips-Jortner approach [18], Eq. (7), for different parameters β. $\tau_0 = 10^6$ s, $K_e^{max}(R=0) = 10^{16}$ s^{-1}.

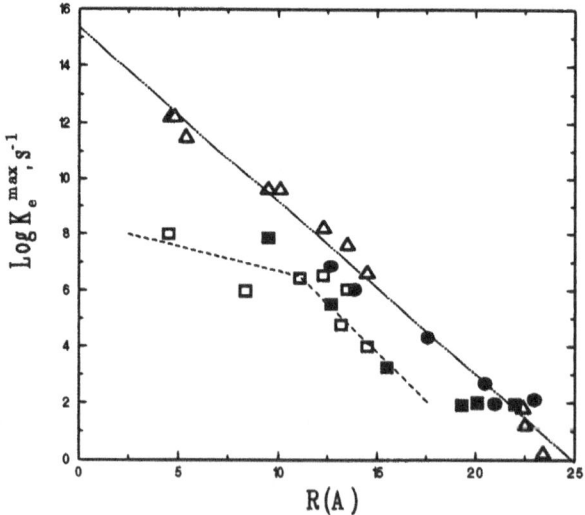

Figure 2. The dependence of K_e^{max} on distance R for ET reactions between donor and acceptor centers in proteins and protein complexes in frames of the one-dimensional square tunneling barrier model: ($\triangle$) reaction centers [7]; ($\blacksquare$) myoglobin [13,25]; ($\square$) cyt-*c* [12,26]; ($\bullet$) azurin [32,33]. The dotted line corresponds to the dependence for cyt-*c* predicted by the Rips-Jortner approach.

The question of whether the nonadiabatic or the adiabatic limits of the ET theory should be used for description of the ET in proteins can be cleared up from the measurements of K_e as a function of distance R between the centers and as a function of frequency ν of the matrix molecular mobility along the reaction coordinate.

3. An Analysis of the Distance Dependence of Electron Transfer in Proteins

Careful experimental studies of the ET kinetics between donor and acceptor centres in some proteins, reaction centres, chemically modified mb, cyt-*c*, hemoglobin, azurin, protein complexes cyt-*c*/cyt *c* peroxidase and cyt-*c*/cyt *b*, have revealed the influence of the structural organization of the protein matrix on the long-range ET efficiency [2-10,12-13,19, 25-35]. The dependence $K_e(R)$ has been analyzed for various proteins using three models.

3.1. ONE-DIMENSIONAL SQUARE TUNNELING BARRIER MODEL

According to this model the long-range ET in proteins is considered as tunneling of an electron through a single square barrier created by the protein matrix along the straight line connecting the edges of the donor and the acceptor centers [2,4,6-9,19,33]. The protein is treated as a structureless matrix, the particular protein structure not being taken into account. Undoubtedly, this is a rough approximation to the polypeptide matrix. Only for RC the majority of experimental data are well described (according to Eqs. (1) - (2)) in semi logarithmic coordinates $\log K_e^{max}$ - R by a single linear dependence with the coefficient $\alpha=1.4$ Å (Fig. 2). A rather good coincidence with the above dependence is observed also for the ET reactions in Mb and azurin. At the same time a weak dependence of $\log K_e^{max}$ on R for the ET reactions in cyt-*c* at distances of 8-12 Å may be an indication of the adiabatic limit. The increment of $\log K_e^{max}$ for distances of 12-15 Å is the same as for other proteins. At these distances the regime of ET in cyt-*c* may be also nonadiabatic as for reaction centres, mb and azurin. In this case the characteristic frequency ν for the adiabatic ET reaction in cyt-*c* at room temperature may be taken to be 10^6 s^{-1}. This value is quite reasonable if one takes into account the structural reorganization times reported in [24].

On the other hand, the substantional spread in the K_e values, their deviations from the linear dependence may be due to a particular structural organisation of protein macromolecules, i.e. electron transfer along the assigned pathways through a chain of chemical bonds.

3.2. TUNNELING PATHWAY MODEL

Based on the strong dependence of H_{ab}^2 on the distance between the donor and acceptor centers, Beratan and Onuchic have developed a more sophisticated model of the ET in proteins along the most effective pathway [26-27, Beratan & Skourtis this volume]. They considered this way as a combination of steps of three types: covalent σ-bonds, hydrogen bonds and non-bonded contacts. A step of each type is described by a special coefficient α. The distance σ_l of the ET is represented as the sum of the σ-bonds lengths and of the effective lengths of the other bonds recalculated to the equivalent length of σ-bonds of the same conductivity.

This approach is relatively simple and allows one to take into account the structural details of the protein under consideration. In the subsequent paper [28] this model has been improved by considering multiple parallel pathways. But the experimental curves for mb and cyt-c in the coordinates $\log K_e^{max}$ - σ_l are described by a considerably different dependence. For example the correlation for mb is worse than in the previous model. This means that the empirical parameters of the nonadiabatic approach used in this model are not universal for all proteins. The deviations from a linear dependence may also be considered in terms of the adiabatic approach under the assumption of characteristic frequencies near 10^6 s^{-1}.

3.3. PATHWAY-SEARCH MODEL COMBINED WITH ARTIFICIAL INTELLIGENCE-SUPEREXCHANGE METHOD

A more general approach to the ET in proteins was introduced by Siddarth and Marcus [36]. They calculated the K_e^{max}(calc) using an artificial intelligence-superexchange method. The correlations between $\log K_e^{max}$(exp), experimental results, and $\log K_e^{max}$(calc) were examined for mb and cyt-c. The value of $\log K_e^{max}$(calc) can be considered as a parameter reflecting the effective distance R_{ef} between the donor and the acceptor centers with due regard for the influence of protein matrix on the exchange interactions. In comparison to the Beratan-Onuchic model, the linear correlations are much better, but the slopes for mb and cyt-c exhibit more than two-fold difference (Fig.3). Different slopes of mb- and cyt-c-dependences in Siddarth-Marcus' model cannot be explained in the framework of the nonadiabatic theory by difference of particular protein structures because the static structure of proteins has been taken into account in calculation of $\log K_e^{max}$(calc).

The observed differences between the $\log K_e^{max}$ -dependences for Mb and cyt-c can easily be explained in terms of Rips-Jortner adiabatic limit. When substituting Eq. (2) in Eq. (7) the rate for ET reactions in viscous medium with a broad distribution of the relaxation times becomes:

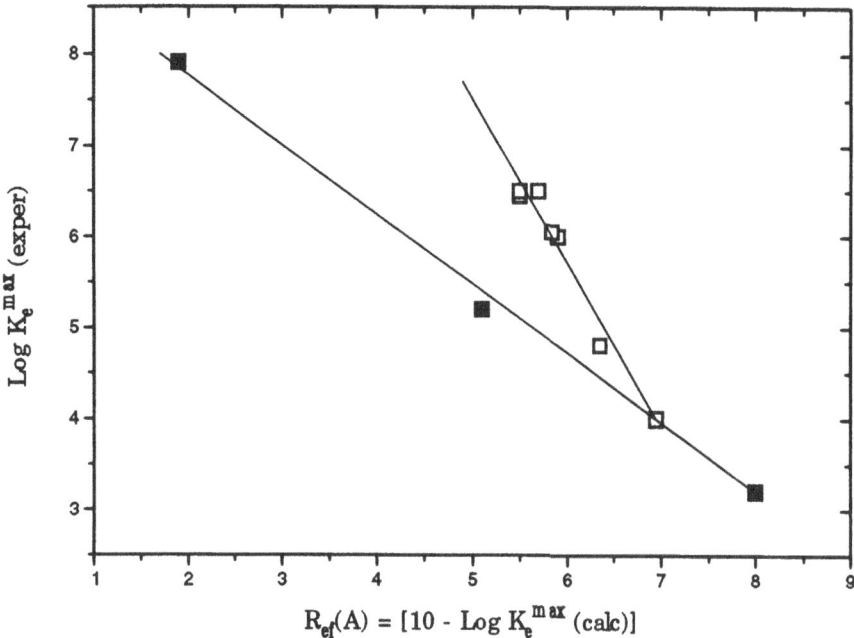

Figure 3. The correlations between $\log K_e^{max}$ (exp) and $\log K_e^{max}$ (calc) for reactions of the ET between the donor and acceptor centres in mb and cyt-*c* in accordance with the Siddarth-Marcus model: (■) mb; (□) cyt-*c* (adapted from [36]).

$$K_e \sim H_{ab}^{2(1-\beta)}(R) \sim \exp[-\alpha(1-\beta)R] \qquad (8)$$

As seen from Eq. (5), in addition to the parameter α the parameter β determines the slope of curves $\log K_e(R)$. The difference between the dependences of $\log K_e^{max}$ on the effective distance R for mb and cyt-*c* in all three models can be explained by different dynamic behaviour of these two proteins in the microsecond and millisecond time domain and in the case of Rips-Jortner model it can be described by different values of parameters β and τ_0.

4. The Influence of Intraprotein Dynamics on ET

4.1. SINGULARITIES OF INTRAMOLECULAR PROTEIN DYNAMICS

Can proteins exhibit relaxation processes on the same time scale as the ET processes? Molecular motions in proteins occur in a very broad frequency range: fast motion of small surface groups $(10^{-11}-10^{-9}$ s) and slower relaxational intramolecular transitions, e.g., in the sites of the haem groups of mb and cyt-c $(10^{-9}-10^{-3}$ s) [14-15,20-24]. A broad distribution of relaxation times in proteins at room temperature in a water solution can be explained by using the block-hinge model of the internal molecular mobility of proteins [22]. The protein globule can be represented as a set of relatively large blocks, connected to each other by flexible hinges and weak bonds. This system is imbedded into a viscous amorphous medium, composed of amino acid residues and surface molecules. The reorganization of protein can be reduced to a complex cooperative change of interactions of such blocks in viscous medium. The characteristic times of the Brownian motion of protein macromolecules in a viscous solution and membranes at room temperatures are known to be $10^{-5}-10^{-4}$ s [37,38]. Evidently, motion of this kind is characteristic only of the natively organized macromolecules. The frequency distribution may appreciably depend on the protein nature and consequently determine the protein ability to the ET reactions. Motion of such kind cannot exist in denaturated proteins and synthetic polymers.

Relatively large times of structural reorganization of proteins in redox reactions can be also a consequence of selective protonation of particular protein groups. With the proton concentration in a solution amounting to $[H^+]=10^{-7}$ M and the diffusive constant of the interaction equal to 10^{10} $M^{-1}s^{-1}$ the characteristic time of the reaction will be 10^{-3} s. Possibly, the same reason explains the fact that the most of enzymes operate at rates of 10^2-10^5 cycles per second.

In native proteins, which are highly organized molecular systems, the reorganization along some functionally important coordinates can take place at frequencies differing considerably from their statistically averaged values at equilibrium. Such relaxation seems to exist only during the enzyme action and is absent in a protein equilibrium state. Due to this fact such motions can be of a very low statistical weight in the total thermal motion of a polypeptide chain. To observe such processes one has to employ the pulse techniques enabling the registration of transient processes in an active center in a microsecond time span.

From this point of view the method of triplet labels - the registration of the phosphorescence spectra relaxation shift - is a most suitable one for investigating the phenomenon of protein molecular mobility during ET in the time range of $10^{-7}-10^{-3}$ s [39-43].

4.2. ADIABATIC REGIME OF ELECTRON TRANSFER REACTIONS AT LOW TEMPERATURES

The adiabatic approach can be applied with much greater success for an analysis of ET in proteins under conditions of lower molecular mobility: in large protein complexes, in membranes, at low temperatures, and at low moistures. The well-known experiment of DeVault and Chance [44], as well as experiments on the temperature dependences of ET in reaction centres, chemicaly modified haem-proteins and model systems [25,31,35, 40-43,45-46] may be examples. In all cases ET was either absent or small and temperature independent at low temperatures (T<120 K), but ET rates increased at T>120-220 K when the matrix molecular mobility was raised. In many cases nonexponentiality of the electron transfer kinetics can be observed. These facts combined may bear witness to the adiabatic mechanism of the observable reactions.

At the same time a detailed analysis of the temperature dependences of the ET reactions in reaction centres and reactions involving haem-proteins mb, hb, cyt-*c* and cyt *c* peroxidase/cyt-*c* complex exhibits a number of important peculiarities which necessitate an additional analysis.

4.2.1. *Electron Transfer Reactions in Haem-Proteins at Low Temperatures*

The temperature dependence of the kinetics of electron phototransfer between Zn-substituted porphyrin and a Ru(His48) complex have been reported [25]. It was found that no transfer ($K_e < 10^2$ s^{-1}) occurred within the temperature range of 77-170 K, then it became more intensive at T>180 K, and reached $K_e = 7 \cdot 10^4$ s^{-1} at room temperature. Approximately in this temperature range one can observe a change in the rate of the recombination ET from the reduced haem of hemoglobin to the photooxidized Zn-porphyrin of the neighbouring subunit [31]. At the same time in cyt-*c* the reaction analogous to that in mb is observable already at T=125 K [46] whereas the ET from the triplet-excited Zn-porphyrin of cyt c peroxidase to cyt-*c* haem at similar rates manifests itself only beginning with 220 K [35]. So, the regions where the ET reactions significantly depend on temperature may differ by more than 100 degrees for haem-proteins with very similar structural and redox parameters.

An analogous shift of the temperature dependences of the ET reactions between the reduced haem c_{559} of tetrahaem cytochrome and the oxidized primary donor P$^+$ in the reaction centres was observed in [45] under various oxidation states of the other cytochrome haems c_{552} and c_{556}.

The different temperature dependences of the ET between structurally identical centers in the aforementioned proteins may be attributed to the different intramolecular dynamics of protein globules and to the adiabatic behaviour of the ET reactions limited by a particular dynamics. Quantitative correlations between the ET kinetics and the molecular dynamics of protein fragments in the region of donor and acceptor groups under similar conditions may be an experimental proof of this suggestion. However, no such studies have been performed so far.

38

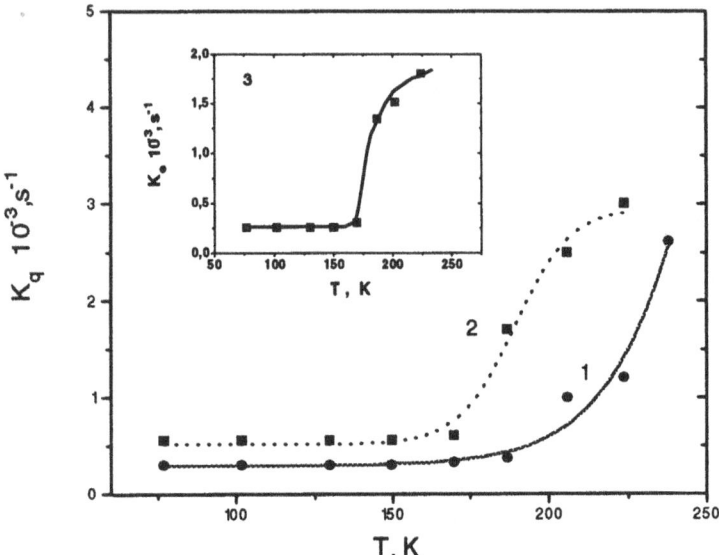

Figure 4. The temperature dependence of the phosphorescence quenching rate constant K_q for apo-Mb (1), Mb (2), and of K_e for the ET reaction from triplet excited EITC attached to the N-terminus of Mb to the haem group in 50% ethyleneglycol (3).

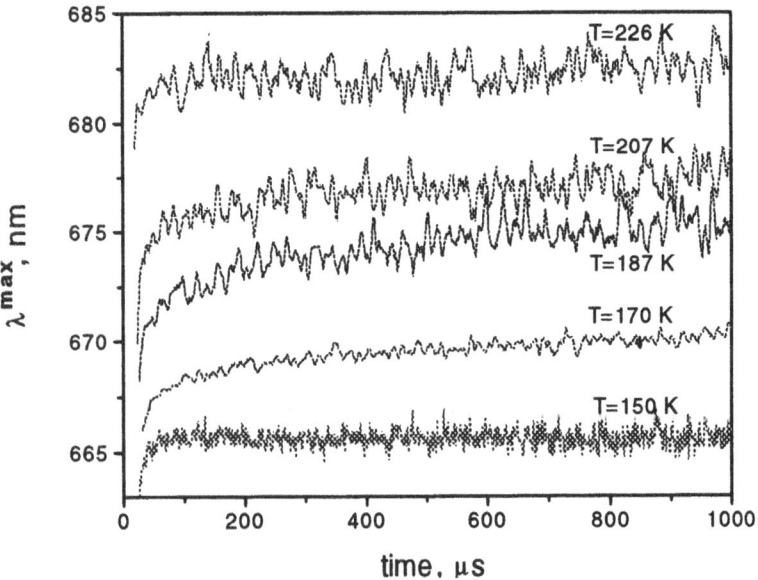

Figure 5. The effect of temperature on kinetics of Stokes shifts of phosphorescence spectra of EITC attached to N-termini of apo-Mb in 50% ethyleneglycol.

The relationship between the ET effectiveness in protein and the molecular dynamics of the medium has been studied by us for the system close to that described in [25]. The ET between the triplet excited eosin isothiocianate (EITC) covalently attached to the terminal amino group of mb and the haem group was registered from quenching of the EITC phosphorescence. The kinetics of the EITC phosphorescence quenching in the absence of ET was determined by registering the quenching of EITC attached to apoprotein (Fig. 4).

The molecular dynamics of the water-protein matrix surrounding the triplet-excited EITC was measured in a microsecond time range by registering the kinetics of Stokes shifts of phosphorescence spectra. The experimental results are given in Figure 5. Similar to what has been reported by Cowan *et al.* [25] we have found that K_e for the ET between the triplet excited chromophore and the haem group is equal to $0.25 \cdot 10^3$ s^{-1} in the temperature range of 77-170 K and increases within 170-224 K from $K_e = 0,25 \cdot 10^3$ s^{-1} to $K_e = 1,8 \cdot 10^3$ s^{-1} (Fig. 4). One has to note here a significant nonexponentiality of the ET kinetics.

Also, no relaxational adjustment of the medium surrounding an excited chromophore which is detectable from a maximum position of an instantaneous phosphorescence spectrum of the chromophore participating in the reaction has been found in the temperature range of 77-170 K. But a subsequent small increase of temperature within 170-206 K gives rise to a drastic intensification of the relaxation processes (Fig. 5). The characteristic relaxation times of the matrix at 178 and 187 K are close to the ET times. A significant nonexponentiality of the relaxational process is also found. A direct coincidence of the temperature points of the initiation of the ET processes and the relaxation processes in the donor site, the similarity of their kinetic characteristics provide strong evidence to suggest that the mechanism of the ET in mb is adiabatic at temperatures from 170 to 230 K.

An analogous explanation of the temperature dependences of the ET reactions within 180-280 K is possible for an intraglobular electron transfer in hemoglobin or in the protein complex cyt-c/cyt c peroxidase. One may suggest that the intensity of relaxation processes in the interior of large protein complexes is much lower compared to the mobility of surface protein fragments. Due to this fact the temperature transitions in hemoglobin and in cyt-c/cyt c peroxidase complex are shifted toward higher temperatures.

However, the temperature variations of the ET rate at T=125 K for cyt-c and even more so the observation of the ET reaction between the haem of the cytochrome and the primary acceptor of reaction centres at temperatures of 50-100 K (Fig. 6) are unlikely to be related to a particular orientational reorganization of the matrix. To analyse these curves one has to employ some other mechanisms.

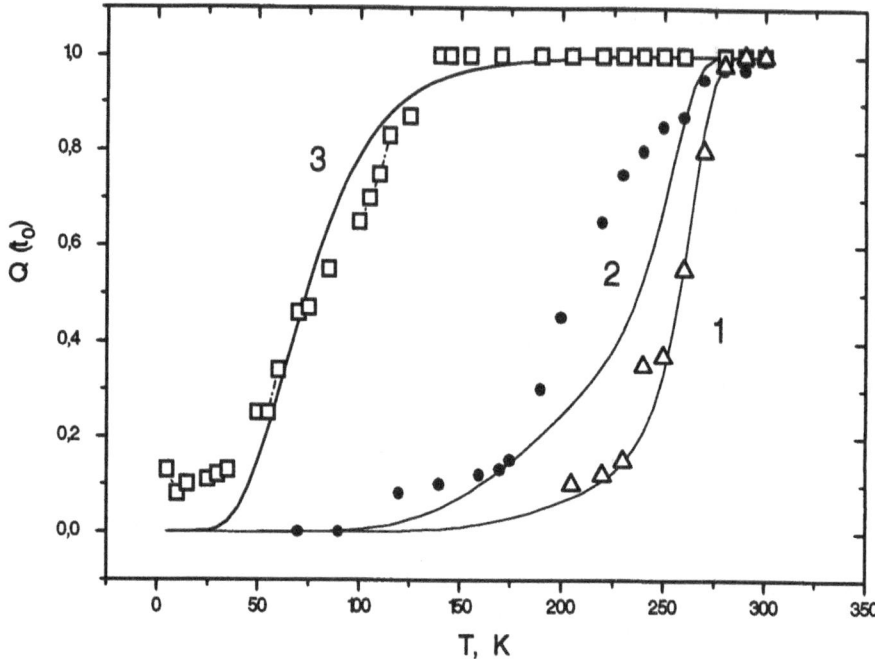

Figure 6. The temperature dependences of the ET efficiency $Q(t_0)$ of the P^+-P regeneration due to c_{559}-P^+ ET at different redox conditions when only the c_{559} heme is reduced (Δ), c_{559} and c_{556} are reduced ($\bullet$), or c_{559}, c_{552} and c_{556} are reduced ($\square$) (taken from [45]). The solid lines correspond to calculations by Eqs. (16)-(20) with parameters from Table 1, Set 1.

4.2.2. *Analysis of Temperature Dependences of ET Reactions Between Cytochrome and Primary Acceptor of a Reaction Centre*

In [34] a shift of the temperature curves of the ET efficiency between c_{559}-haem of tetrahaem cytochrome and primary donor P^+ in reaction centres of *Rhodopseudomonas viridis* under a change in the oxidation state of haems c_{556} and c_{552} (Fig. 6) was attributed to a change in the reaction free energy due to electrostatic interaction of the electron being transferred with additional electrons which appear on haemes c_{556} and c_{552} upon their reduction. Inhibition of the transfer reactions with decreasing temperature is related to freezing out of the molecular dynamics of proteins.

In order to describe the course of the temperature curves, given in Figure 6, in a broad temperature range of 50-300 K, it appears necessary to invoke the ideas of the possibility of energy dissipation in the system not only by diffusional but also vibrational degrees of freedom.

Obviously, along with the diffusive motion, in proteins there are vibrational (nondiffusive) motions including intramolecular vibrations, internal rotations, etc. Such motions have very short relaxation times and never freeze out completely even at low temperatures. These motions contribute to the reorganization energy provided there is a difference in bond lengths or angles between reactant and product of the ET reaction. When these motions are ignored, as above, the rapid decrease of the relaxation frequency with T as described by Eqs. (4)-(6) is equivalent to the sharp increase in the activation energy of the reaction in Eq. (1). As a result, the transition state cannot be reached at low temperature because of too high activation energy at a fixed initial value of the diffusive coordinate X. On the other hand, when a vibrational motion q is taken into account, the transition state is located at an intersection of twodimensional reactant's and product's potential energy surfaces. Therefore, when the temperature decreases beyond the freezing point of the X motion, the transition state moves from the absolute minimum at the intersection line, E_a, to a point of conditional minimum at a fixed initial value of X, E_a'. The low-temperature transition state may still have a moderate activation energy E_a', depending upon the redox potential, and hence may be reached at low temperature.

To be more specific, let us consider the harmonic reactant and product potential energy surfaces

$$
\begin{aligned}
V^r(q,X) &= \tfrac{1}{2}aq^2 + \tfrac{1}{2}X^2, \\
V^p(q,X) &= \tfrac{1}{2}a(q-q_0)^2 + \tfrac{1}{2}(X-X_0)^2 + \Delta G,
\end{aligned}
\tag{9}
$$

where q_0 and X_0 are the equilibrium values of q and X for the product. The intersection line is given by equation

$$
X_0 X + a q_0 q = \lambda + \Delta G, \tag{10}
$$

where the total reorganization energy

$$
\lambda = \lambda_q + \lambda_X \tag{11}
$$

is a sum of partial reorganization energies for modes q and X,

$$
\lambda_q = \tfrac{1}{2}aq_0^2, \quad \lambda_X = \tfrac{1}{2}X_0^2
$$

The minimum on the intersection line (10) is reached at $q = q_*$ and $X = X_*$ where

$$q_* = q_0 \frac{\lambda + \Delta G}{2\lambda}, \quad X_* = X_0 \frac{\lambda + \Delta G}{2\lambda}. \tag{12}$$

The activation energy for this transition state, $E_a = V^r(q_*, X_*) - V^r(0,0)$, calculated by Eqs. (9) and (12) is given by Eq. (3).

On the other hand, when the X motion is frozen at the initial value of $X=0$, the intersection occurs at a single configuration $q = q'_*$ and $X = X'_* = 0$ where

$$q'_* = q_0 \frac{\lambda + \Delta G}{2\lambda_q} \tag{13}$$

The corresponding activation energy is

$$E'_a = V^r(q'_*, X'_*) - V^r(0,0) = \frac{(\lambda + \Delta G)^2}{4\lambda_q} \tag{14}$$

Obviously, $E'_a > E_a$ since $\lambda_q < \lambda$. Nevertheless, the transition state may still be accessible at low temperature provided the reorganization energy for the q motion is not very small.

The twodimensional model specified by Eq. (9) was rigorously treated by Sumi and Marcus [47]. Although the theory is quite complicated, the net physical result is very transparent. In general, the decay of the initially prepared state is multiexponential. When it can be approximated with a single exponential, the ensuing rate has the thermally activated form of Eq. (1) were both the activation energy and the frequency factor depend upon the relaxation time τ for the motion along the diffusive X coordinate. In the low-temperature limit they tend to their values corresponding to a higher transition state which is reached along the q coordinate, keeping the frozen X coordinate constant.

The above-mentioned theories were developed for short-range ET reactions where donor and acceptor are in close contact to each other. In this case the matrix element H_{ab} is associated with a direct overlap of electronic orbitals of donor and acceptor, and therefore it is essentially a constant independent of nuclear coordinates (the so-called Condon approximation [11]).

On the other hand, in biological objects ET reactions proceed with a significant rate over long distances of tens of Ångströms. In a long-range reaction the intervening biological medium serves as a bridge between donor and acceptor. A nonvanishing H_{ab} for such reactions arises due to multiple successive overlaps in the chain:

donor - bridge group 1 - bridge group 2 - ... - acceptor.

Evidently, all overlaps strongly depend on nuclear coordinates of the bridge since they occur at tails of electronic wavefunctions. Hence, H_{ab} is a function of the bridge coordinates. In other words, the tunneling electron interacts with the bridge vibrations on its way from donor to acceptor. Recent numerical simulations of molecular dynamics in Ru-modified azurin have shown [48] that this interaction is strong enough and must significantly affect the reaction rate. The ET theory accounting for this interaction has been developed in [48] and [49]. An important physical result of these studies is that the tunneling electron is capable of emitting/absorbing vibrational quanta to/from the bridge. These inelastic tunneling processes modify both the reorganization energy and the driving force of the reaction in such a way that the reaction becomes activationless at long enough distances.

Both the vibrational motion of the ligands and the energy exchange of the tunneling electron with bridge vibrations act in a common direction, making the reorganization energy nonvanishing in the low-temperature limit. This effect can be described phenomenologically by introducing temperature dependent quantities related to the transition state of the ET reaction under consideration. Let us introduce a function $f(T)$ which serves as a switch between the high-temperature (nonadiabatic) and low-temperature (adiabatic) limits,

$$f(T) = 2\left[1 + \exp\left(\frac{E_{rel}}{RT} - \frac{E_{rel}}{RT_1}\right)\right]^{-1} \qquad (15)$$

where $T_1 = 295K$ is room temperature. The switching is governed by the activation energy E_{rel} for diffusional relaxation, see Eq. (6). For proteins τ decreases from 140 ns at room temperature to 1 ms at 180 K, which corresponds approximately to $E_{rel}/R = 4500\,K$.

Further we assume that the ET rate is given by an expression equivalent to Eqs. (1) and (2) but with the reorganization energy, the driving force, and the frequency factor which depend on temperature via the function $f(T)$,

$$K_e = \sigma(T)\exp\left[-E_a(T)/RT\right] \qquad (16)$$

$$\sigma(T) = \sigma\left[p_\sigma + (1 - p_\sigma)f(T)\right] \qquad (17)$$

In Eq. (17) we assume that the frequency factor at room temperature takes a nonadiabatic value σ whereas in the adiabatic limit it is σp_σ where $p_\sigma < 1$. Similarly,

$$E_a(T) = \frac{[\Delta G(T) + \lambda]^2}{4\lambda(T)} \tag{18}$$

$$\Delta G(T) = \Delta G\left[p_g + (1 - p_g)f(T)\right] \tag{19}$$

$$\lambda(T) = \lambda\left[p_\lambda + (1 - p_\lambda)f(T)\right] \tag{20}$$

The first term in Eq. (20), λp_λ, represents a sum of constant contributions from vibrational modes q and inelastic tunneling processes (denoted above as λ_q) whereas the second is the temperature dependent reorganization energy of the X mode (denoted above as λ_X). Note, however, that the numerator of Eq. (18) contains the constant reorganization energy corresponding to the room-temperature limit of Eq. (20). This is because we have to switch from the high-temperature activation energy, Eq. (3), to the low-temperature activation energy, Eq. (14). The quantity measured in experiment is the number of centers which have undergone the reaction till the moment t_0,

$$Q(t_0) = 1 - \exp(-K_e t_0) \tag{21}$$

where $t_0 = 10\mu s$.

The results of the calculations are presented in Figure 6. The observable $Q(t_0)$ was calculated for the reaction centers at different redox conditions (curves 1, 2 and 3). The temperature dependent driving force for curve 1, $\Delta G_1(T)$, was calculated by Eq. (19) with $\Delta G = -0.14eV$. For other curves, the following relations were used:

$$\Delta G_2(T) = \Delta G_1(T) - 0.04eV,$$
$$\Delta G_3(T) = \Delta G_1(T) - 0.15eV.$$

Other parameters are given by Set 1 in Table 1. A few additional sets of parameters are also included in the table for which the quality of the fit is similar to that shown in the figure. The calculated room-temperature values of the rates for the three curves $(K_{e1}, K_{e2}, \text{ and } K_{e3})$ are included as well.

The agreement with the experimental data looks quite satisfactory. The values of the parameters seem to be reasonable. Thus, we conclude that phenomenologically treating the effects of dynamical ligand modes and inelastic ET processes allows us to explain the temperature dependence of the ET reaction in a wide region from room temperature down to 50 K.

TABLE 1. Sets of parameters in Eqs. (16)-(20)

Set #	λ, eV	p_λ	$v, \mu s^{-1}$	p_R	p_v	$K_{e1}, \mu s^{-1}$	$K_{e2}, \mu s^{-1}$	$K_{e3}, \mu s^{-1}$
1	0.29	0.7	2.5	0.1	0.6	1.2	1.7	2.5
2	0.32	0.6	8	0.1	0.6	3.0	4.4	7.8
3	0.45	0.6	17	0.6	0.6	2.1	3.5	9.7

5. Conclusions

Currently the effect of molecular dynamics on the electron transfer in proteins has been studied for individual cases [34,40-43,48-53]. From the above it is concluded that protein dynamics has its own singularities and may be as important in the ET reactions as the spatial protein structure.

The dielectric parameters of a protein around the active centre can be affected appreciably both by the vibrational interactions at the active site and the relaxation of the protein globule as a whole.

An approximate equality of the characteristic times of the ET reactions and the relaxational reorganization times of the protein $(10^{-11}-10^{-3}$ $s^{-1})$ will always pose the question of whether the transfer regime is nonadiabatic or adiabatic. The transition from the former to the latter can decrease the ET rate by several orders. As seen from Figure 1 this will be particularly noticeable for the ET reactions over relatively short distances of 5-7 Å. In this case the ET rate may go down from $10^{12}-10^{11}$ s^{-1} to 10^6-10^3 s^{-1} due to the absence of optimal protein relaxational reorganization.

This may be of importance for example for ordering cyt-c reactions with its redox partners. Once a free cyt-c in solution possesses the molecular organisation ensuring a relatively slow adiabatic ET to a nonspecific redox partner, then its interaction with a specific redox partner followed by formation of a strong protein complex can transfer the reaction to the nonadiabatic regime, thus increasing its effectiveness significantly.

The account of the extent of adiabaticity of the reaction is important for the correct determination of H_{ab} values which characterize the electron capability for tunnelling through a protein globule. The H_{ab} values reported in [6,12-13,19] and cited in this paper were determined under the assumption that the transfer is nonadiabatic. These estimations may be reduced significantly if in the real situation the reactions proceed in the adiabatic regime. In the end this will result in the incorrect estimation of the conducting properties of the protein matrix, of the role of chemical bonds and the spatial organisation of the protein globule in the electron transfer over long distances.

The experimental results analyzed in this work are suggestive of the adiabatic character of some ET reactions in proteins at room and lower temperatures. The difference between the kinetics of the intramolecular ET reactions in cyt-c, mb and other proteins, (the most clearly defined in the Siddarth-Marcus model), is the kinetical evidence for the functional importance of the microsecond molecular dynamics of proteins. These dynamical properties are not inherent of the proteins as usual polymers, they are the outcome of the block-hinge organization of proteins and can change considerably according to the particular protein structures. Thus the dynamical organization of protein globules is as important as the chemical structure of active centres at the ET reactions in proteins and at other enzyme reactions.

6. Acknowledgements

These investigations have been supported by the Russian Foundation for Basic Research.

7. References

1. Marcus, R.A., Sutin, N. (1985) Electron transfer in chemistry and biology, *Biochim.Biophys.Acta* **811**, 265-322.
2. Mayo, S.L., Ellis, W.R., Crutchley, R.J., Gray H.B. (1986) Long-range electron transfer in heme proteins, *Science* **233**, 948-952.
3. Gray, H.B., Malmström, B.G. (1989) Long-range electron transfer in multisite metalloproteins, *Biochemistry* **28**, 7499-7505.
4. Kotelnikov, A.I. (1991) Triplet labels in investigation of biological systems, *Doctorate review*. Moscow State University, Moscow.
5. Hoffman, B.M., Natan, M.J., Nocek, J.M., Wallin, S.A. (1991) Long-range electron transfer within metal-substituted protein complexes, *Structure and bonding* **75**, 85-108, Springer-Verlag, Berlin, Heidelberg.
6. Winkler, J.R., Gray, H.B. (1992) Electron transfer in ruthenium-modified proteins, *Chem.Rev.* **92**, 369-379.
7. Moser, C.C., Keske ,J.M., Warncke, K., Farid, R.S., Dutton, P.L. (1992) Nature of biological electron transfer, *Nature* **355**, 796-802.
8. McLendon,G., Hake,R. (1992) Interprotein electron transfer, *Chem.Rev.* **92**, 481-490.
9. Kotelnikov, A.I. (1993) Analysis of the experimental data on the electron conductivity of globular proteins, *Biophysics* **38**, 217-220.
10. Ramirez, B.E., Malmström, B.G., Winkler, J.R., Gray, H.B. (1995) The currents of life: the terminal electron transfer complex of respiration, *Proc.Natl.Acad.Sci.USA* **92**, 11949-11951.
11. Medvedev, E.S., Osherov, V.I. (1995) *Radiationless Transitions in Polyatomic Molecules*, Sect.6.5, Springer, Berlin.
12. Casimiro, D.R., Richards, J.H., Winkler, J.R., Gray, H.B. (1993) Electron transfer in ruthenium-modified cytochromes c. σ−Tunelling pathways through aromatic residues, *J.Phys.Chem.* **97**, 13073-13077.
13. Casimiro, D.R., Beratan, D.N., Onuchic, J.N., Winkler, J.R., Gray H.B. (1994) Donor-acceptor electronic coupling in ruthenium-modified heme proteins, *Mechanistic Bioinorganic Chemistry*. (Thorp, H.H. and Pecorado,V. eds) Advances in Chemistry, Series. American Chemical Society, Washington. DC.
14. Bashkin, J.S., Mc Lendon, G., Mukamel, S., Marohn, J. (1990) Influence of medium dynamics on solvation and charge separation reactions: comparison of a simple alcohol and protein "solvent", *J.Phys.Chem.* **94**, 4757-4761.
15. Beece, D., Eisenstein, L., Frauenfelder, H., Good, D., Marden, M.C., Reinisch, L., Reynolds, A.H., Sorensen, L.B., Yue, K.T. (1980) Solvent viscosity and protein dynamics, *Biochemistry* **19**, 5147-5157.
16. Zusman, L.D. (1980) Outer-sphere electron transfer in polar solvents, *Chem.Phys.* **49**, 295-304.
17. Alexandrov, I.V.(1980) Physical Aspects of Charge Transfer Theory, *Chem. Phys.* **51**, 449-457.
18. Rips, I., Jortner, J. (1987) The effect of solvent relaxation dynamics on outer-sphere electron transfer, *Chem.Phys.Lett.* **133**, 411-414.
19. Bjerrum, M.J., Casimiro, D.R., Chang, I-Jy, Di Billio, A.J., Gray, H.B., Hill, M.G., Langen, R., Mines, G.A., Skov, L.K., Winkler J.R., Wuttke, D.S. (1995) Electron transfer in Ru-modified proteins, *J.Bioenergetics and Biomembranes* **27**, 295-302.
20. Blumenfeld, L.A., Davidov, R.M.(1979) Chemical reactivity of metalloproteins in conformationally out-of-equilibrium states, *Biochem.Biophys.Acta.* **549**, 255-280.
21. Pierce, D.W., Boxer, S.G. (1992) Dielectric relaxation in protein matrix, *J.Phys.Chem.* **96**, 5560-5566.
22. Likhtenstein, G.I., Kotelnikov, A.I.,(1983) The study of fluctuational intermolecular lability of proteins by physical labeling, *Molekuljarnaja biologija* **17**, 505-517.
23. Kitagawa, T., Sakan, Y., Nagai, M., Ogura, T., Fraunfelder, F.A., Mattera, R., Ikeda-Saito, M. (1993) Time-resolved resonance Raman studies of recombination intermediates of CO-photodissociated myoglobin, hemoglobin and their E7 mutants, *J.Inorg.Biochem.* **51**, 217.
24. Pascher, T., Chesick, J.P., Winkler, J.R., Gray, H.B. (1996) Protein folding triggered by electron transfer, *Science* **271**, 1558-1560.

48

25. Cowan, J.A., Upmacis, R.K., Beratan, D.N., Onuchic, J.N., Gray ,H.B. (1988) Long-range electron transfer in myoglobin, *Ann.N.Y.Acad.Sci.* **550**, 68-84.
26. Beratan, D.N., Onuchic, J.N., Betts, J.N., Bowler, B.E., Gray, H.B. (1990) Electron tunneling pathways in rutenated proteins, *J.Am.Chem.Soc.* **112**, 7915-7921.
27. Beratan, D.N., Onuchic, J.N., Winkler, J.R., Gray, H.B. (1992) Electron-tunneling pathways in proteins, *Science* **258**, 1740-1741.
28. Casimiro, D.R., Wong, L.L., Colon, J.L., Zewert, T.E., Richards, J.H., Chang, I.J., Winkler, J.R., Gray, H.B. (1993) Electron transfer in ruthenium/zinc porphyrin derivatives of recombinant human myoglobins. Analysis of tunelling pathways in myoglobin and cytochrome c, *J.Am.Chem.Soc.* **115**, 1485-1489.
29. Liang ,N., Kang, Ch.H., Ho, P.S., Margoliash, E., Hoffman, B.M. (1986) Long-range electron transfer from iron(II)-cytochrome c to, (Zn-cytochrome c peroxidase) (+) within the 1:1 complex, *J.Am.Chem.Soc.* **108**, 4665-4666.
30. Natan, M.J., Hoffman, B.M. (1989) Long-range [Fe^{2+}(heme)-(M(porfirin))$^+$] electron transfer within [M, Fe] (M=Mg, Zn)hemoglobin gybrids, *J.Am.Chem.Soc.* **111**, 6468-6470.
31. Kuila, D., Baxter, W.W., Natan, M.J., Hoffman, B.M.(1991) Temperature-independent electron transfer in mixed-metal hemoglobin hybrides, *J.Phys.Chem.* **95**, 1-3.
32. Farver, O., Pecht, I. (1989) Long-range intramolecular electron transfer in azurins, *Proc.Natl.Acad.Sci.USA.* **86**, 6968-6972.
33. Langen, R., Colon,J.L., Casimiro, D.R., Karpishin, T.B., Wikler J.R., Gray, H.B. (1996) Electron tunelling in proteins: role of the intervening medium, *J.Biol.Inorg.Chem.* **1**, 221-225.
34. Frolov, E.N., Goldanskii, V.I., Birk, A., Parak, F. (1996) The influence of electrostatic interactions and intramolecular dynamics on electron transfer from the cytochrome subunit to the cation-radical of the bacteriochlorophyll dimer in reaction centers from Rps.viridis, *Eur.Biophys.J.* **24**, 433-438.
35. Nocek, J.M., Liang, N., Wallin, S.A., Mauk, A.G., Hoffman, B.M. (1990) Low-temperature conformational transition within the [Zn-cytochrome c peroxidase, cytochrome c] electron transfer complex, *J.Am.Chem.Soc.* **112**, 1623-1625.
36. Siddarth, P., Marcus, R.A. (1993) Correlation between theory and experiment in electron transfer reactions in proteins: electronic couplings in modified cytochrome c and myoglobin derivatives, *J.Phys.Chem.* **97**, 13078-13082.
37. Cherry, R.J., Schneider, G. (1976) A spectroscopic technique for measuring rotational diffusion of macromolecules. 2: Determination of rotational correlation times of proteins in solution. *Biochemistry*, **15**, 3657-3661.
38. Moore, C., Boxer, D., Garlaund, P. (1979) Phosphorescence depolarisation and the measurement of the rotational motion of proteins in membranes, *FEBS Letters*, **108**, 161-166.
39. Likhtenstein, G.I., Kulikov, A.V., Kotelnikov, A.I., Levchenko, L.A. (1986) Methods of physical labels - a combiend approach to the study of microstructure and dynamics in biological systems, *J.Biochem.Biophys.Methods* **12**, 1-28.
40. Kotelnikov, A.I., Vogel, V.R., Kochetkov, V.V., Likhtenstein, G.I., Noks, P.P., Grishanova, N.P., Kononenko, A.A., Rubin, A.B. (1983) *Molekuljarnaja biologija* **17**, 846-854.
41. Noks, P.P., Bystrjak, I.M., Kotelnikov, A.I., Shaitan, K.V., Kononenko, A.A., Zacharova, N.I., Likhtenstein, G.I., Rubin, A.B (1989) The Influence of glycerol and sugars on electron phototransfer in a system of quinone acceptors of reaction centers of purple bacteria, *Ann.Acad.Sci.USSR, Biol.* **5**, 651-659.
42. Likhtenshtein, G.I., Bystrjak, S.M., Kotelnikov, A.I. (1990) Role of medium molecular dynamics in electron transfer reactions in viscous mediums, *Chemical Physics (Russian)* **9**, 697-706.
43. Rubtsova, E.T.,Vogel, V.R., Khudjakov, D.V., Kotelnikov, A.I., Likhtenstein ,G.I., (1993) Influence of molecular dynamics of protein matrix on the photoinduced electron transfer kinetics, *Biophysics*, **38**, 211-216.
44. DeVault, D., Chance, B.(1966) Studies of photosynthesis using a pulsed laser. 1. Temperature dependence of cytochrome oxidation rate in chromatium: Evidence for tunneling. *Biophys.J.* **6**, 825-847.

45. Ortega, J.M., Mathis, P. (1993) Electron transfer from tetraheme cytochrome to the special pair in isolated reaction centres of *Rhodopseudomonas viridis*, *Biochemistry* **32**, 1141-1151.
46. Zang, L.H., Maki, A.H. (1990) Photoinduced electron transfer in the Zn-substituted cytochrome *c* Ru(NH$_3$)$_5$(His-33) derivative studied by phosphorescence and optically detected magnetic resonance spectroscopy, *J.Am.Chem.Soc.* **112**, 4346-4351.
47. Sumi, H., Marcus, R.A. (1986) Dynamical effects in electron transfer reactions, *J.Chem.Phys.* **84**, 4894-4914.
48. Daizadeh, I., Medvedev, E. S., Stuchebrukhov, A. A. (1997) Effect of protein dynamics on biological electron transfer, *Proc. Natl. Acad. Sci. USA*. **94**, 3703-3708.
49. Medvedev, E. S., Stuchebrukhov, A. A.(1997) Inelastic tunneling in long-distance biological electron transfer reactions, *J. Chem. Phys.* **107**, 3821-3831.
50. Vogel, V.R., Rubtsova, E.T., Likhenshtein, G.I., Hideg, K. (1994) Factors affecting photoinduced electron transfer in a donor-acceptor pair (D-A) incorporated into bovine serum albumin, *J.Photochem.Photobiol. A: Chem.* **83**, 229-236.
51. Kotelnikov, A.I., Vogel, V.R., Pastuchov, A.V. (1995) Analysis of electron transfer in proteins in the framework of adiabatic approach of outer-sphere electron transfer theory, *J.Inorg.Biochem.* **59**, 268.
52. Likhtenshtein, G.I. (1996) Role of orbital overlap and local dynamics in long-distance electron transfer in photosynthetic reaction centers and model systems, *J.Photochem.Photobiol. A: Chemistry* **96**, 79-92.
53. Kotelnikov, A.I., Vogel, V.R. (1996) Analysis of the experimental data on the kinetics of electron transfer in metal-containing proteins within the context of an adiabatic approximation, *Biophysics* **41**, 597-605.

RECENT SURPRISES IN THE STUDY OF PHOTOINDUCED ELECTRON TRANSFER: COVALENT VERSUS NON-COVALENT PATHWAYS

J.W. VERHOEVEN*, M. KOEBERG, M.R. ROEST,

Laboratory of Organic Chemistry, The University of Amsterdam, Nieuwe Achtergracht 129, 1018 WS Amsterdam, The Netherlands
M.N. PADDON-ROW*, J.M. LAWSON,

School of Chemistry, University of New South Wales, Sydney, NSW 2052, Australia

1. Abstract

Stepwise photoinduced charge separation in a rigid trichromophoric system with a U-shaped conformation is described. In the thus produced totally charge-separated state the plus and minus charges reside on the termini of the U-shaped molecule with a separation of about 16 Å measured 'through-space' against about 27 Å as measured via the shortest 'through-bond' pathway. Charge recombination to the ground-state is found to be accompanied by charge transfer fluorescence. Both from the overall rate of this recombination and from the radiative rate constant of the accompanying charge transfer fluorescence it appears that the interaction between the termini is much stronger than can be accounted for by a pure through-bond mechanism.

2. Introduction

Over the last two decades intramolecular electron-transfer has been studied extensively in a variety of D(onor)-bridge-A(cceptor) systems in which the bridge constitutes a saturated hydrocarbon moiety with a rigidly extended configuration [1-10].

Variation of the length, structure and configuration of the bridge has amply demonstrated the possibility of fast long-range electron transfer across such bridges

51

G.W. Canters and E. Vijgenboom (eds.),
Biological Electron Transfer Chains: Genetics, Composition and Mode of Operation, 51-61.
© 1998 *Kluwer Academic Publishers.*

and at the same time confirmed [9,10] the importance of through-bond interaction as a mechanism providing the electronic coupling between D and A.

The latter notion has apparently induced a swing in opinion regarding the mechanism of electron-transfer in such complex systems as redox proteins. Earlier approaches to describe electron transfer between distant redox-centres in such systems often relied upon the supposed mediating properties of e.g. aromatic amino-acid side chains. More recently, however, electron transfer pathway descriptions have been proposed [11-13] that rely heavily upon through-bond mechanisms following a covalently bonded pathway along the peptide backbone, while pathways involving 'jumps' across hydrogen bridges and especially those involving the 'open space' between non-bonded atoms are considered to be strongly disfavoured. Although such pathway analyses are appealing in providing a means to predict and visualise the eventual anisotropy of electron transport in a protein matrix, it has recently been argued especially by Dutton et al. [14] that the evolutionary significance of such anisotropy is minor if any.

In this communication we wish to present some recent data on model systems which are designed to allow competition between through-bond interaction (TBI) across a relatively long covalent pathway and through-solvent/space interaction (TSI) across a shorter pathway with the latter, however, still greatly exceeding direct Van der Waals contact between the chromophores.

3. Results and Discussion

Scheme 1 shows the structure of DMN[8]DCV which is one of a series of D-[bridge]-A systems studied by us extensively before.

It contains a 1,4-dimethoxynaphthalene (DMN) donor, a 1,1-dicyano vinyl (DCV) acceptor and a bridge with an effective pathlength of eight sigma bonds.

Earlier studies involving e.g. changes in bridge configuration have provided convincing evidence for dominant through-bond interaction (TBI) in DMN[8]DCV and its higher and lower homologues [9,10]. This TBI allows for the occurrence of rapid (~ 30 ps) and quantitative intramolecular charge separation in DMN[8]DCV following excitation of the DMN chromophore.

It is important to note that this charge separation time remains virtually constant upon changing the temperature [15] or the solvent [16] which can be understood from the fact that its energetics are close to the Marcus optimal conditions which makes it an (almost) barrierless process. In contrast charge recombination to the ground-state involves a very large energy gap and therefore displays a typical 'inverted region' behaviour speeding up dramatically in more polar solvents [17] that stabilise the charge separated state and therefore reduce the energy gap to the ground-state.

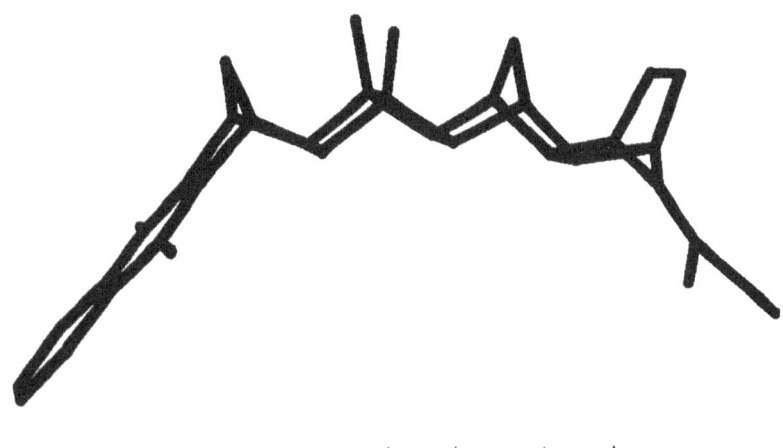

DMN[8]DCV

[2]DMN[8]DCV

Scheme 1. Structures of DMN[8]DCV and [2]DMN[8]DCV as well as a 3D-model ("tube display") of the former.

It has thus been found [17] that at room temperature the lifetime of the charge separated state of DMN[8]DCV decreases from 58 ns in t-decaline to 32 ns in benzene and 2.5 ns in 1,4-dioxane. In addition to its strong solvent dependence another important property of the recombination process is its partially radiative nature. Thus for DMN[n]DCV systems with n = 4-10 a weak and broad long wavelength fluorescence has been observed [18] identified as charge-transfer fluorescence (hv_{ct}) emerging from a radiative component of the charge-recombination process ($D^+A^- \rightarrow DA + hv_{ct}$).

Also displayed in Scheme 1 is the structure of [2]DMN[8]DCV which contains an additional alkyl substituent on the DMN in order to constitute a better model for the system presented in Scheme 2 (see below).

In fact - as reported earlier [19,20,21] - the behaviour of [2]DMN[8]DCV is very similar to that of DMN[8]DCV including the occurrence of CT-fluorescence [21]. As shown in Fig. 1 this CT-fluorescence occurs at ~ 450 nm in methylcyclohexane as solvent. Measurements of the fluorescence lifetime (τ = 70 ns) allowed a convenient determination of the lifetime of the charge transfer state in [2]DMN[8]DCV, a result confirmed by independent transient absorption and time resolved microwave conductivity measurements.

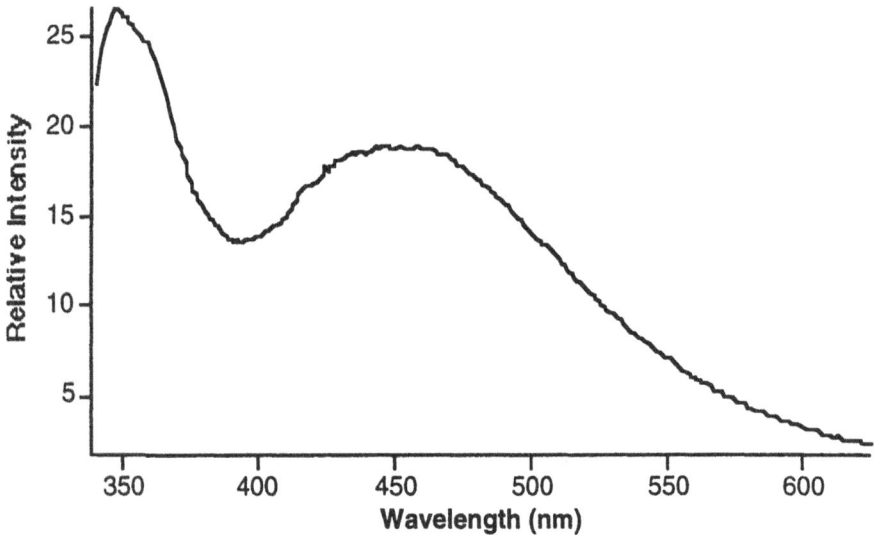

Figure 1. CT fluorescence of [2]DMN[8]DCV in methylcyclohexane at room temperature (λ_{max} = 450 nm, τ = 70 ns). Note that at shorter wavelength some residual fluorescence of the DMN chromophore occurs.

In Scheme 2 the structure is presented of a trichromophoric system that extends those of Scheme 1 with an additional electron donor of the dimethylaniline type (DMA) connected to DMN by a 6-bond bridge which is oriented *syn* with respect to the 8-bond bridge connecting DMN and DCV. As also shown in Scheme 2 the *syn*-orientation of the two bridges causes the molecule to have a striking U-shape [22]. As a result the terminal DMA and DCV chromophores are now separated by a much shorter TS-pathway (~ 16 Å atom to atom) than the shortest TB-pathway, the latter adding up to 14 bridge-bonds plus 4 aromatic bonds of DMN with an overall pathlength of about 27 Å.

It should be noted that this differentiation is significantly more pronounced than in some systems recently investigated with the same purpose by Zimmt [23]

syn-DMA[6]DMN[8]DCV

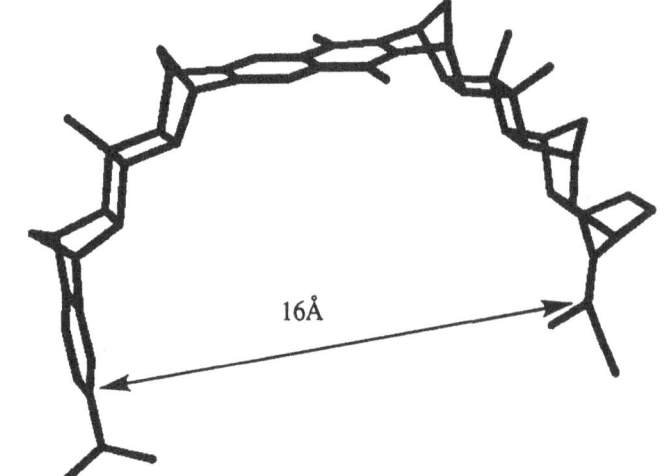

Scheme 2. Structure and 3D-model ("tube display") of *syn*-DMA[6]DMN[8]DCV

As we reported earlier[21,24] photoexcitation of *syn*-DMA[6]DMN[8]DCV at room temperature leads to virtual complete population of a fully charge separated state. Formation of this state occurs largely or even exclusively (see below) in a two-step sequence starting with the well known (see above) electron transfer from excited DMN to DCV (k_{cs1}) followed by hole migration (k_{cs2}) from DMN⁺ to the lower oxidation potential DMA site (see Fig. 2). Although the driving-force for the latter step is rather small especially in alkane-solvents, its rate constant (k_{cs2}) is still high enough at room temperature to make the intermediate CS-1 state undetectable by nanosecond transient absorption spectroscopy. At lower temperature, however, k_{cs2} slows down sufficiently to detect in addition to the typical absorption of the DMA radical cation around 470 nm [19,20] a clear feature around 408 nm known

[19,20] to be typical for the DMN radical cation (see Fig. 3), thus supporting the intermediacy of CS-1 in the formation of CS-2.

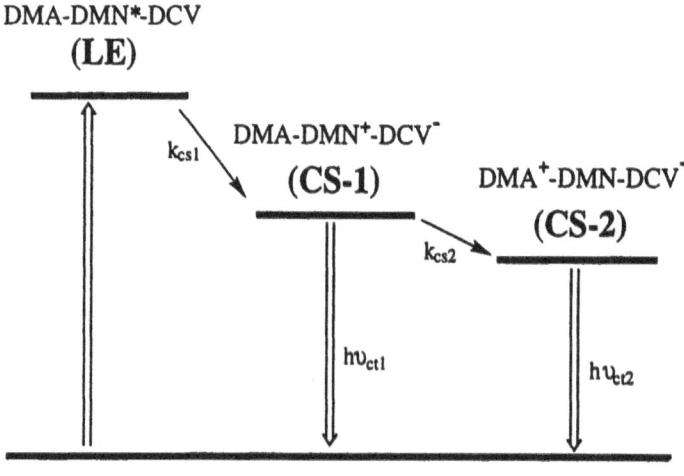

DMA-DMN*-DCV
(LE)

k_{cs1}

DMA-DMN⁺-DCV⁻
(CS-1)

DMA⁺-DMN-DCV·
(CS-2)

k_{cs2}

hν$_{ct1}$

hν$_{ct2}$

DMA-DMN-DCV

Figure 2. Reaction scheme for photophysical processes in syn-DMA[6]DMA[8]DCV

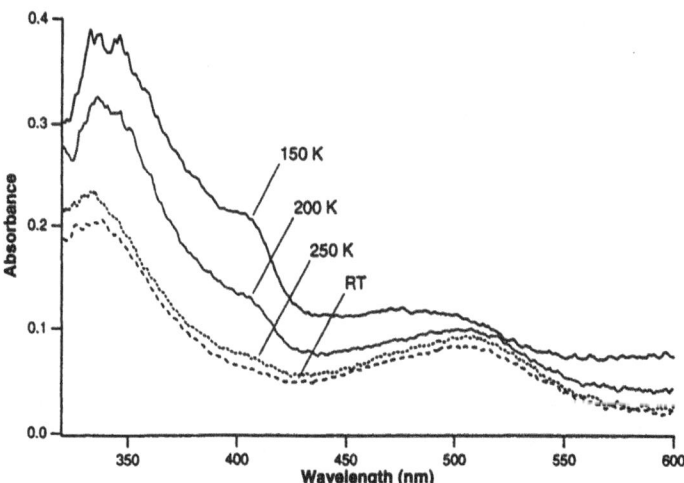

Figure 3. Nanosecond transient absorption spectra obtained in methylcyclohexane at various temperatures for syn-DMA[6]DMN[8]DCV (excitation at 308 nm, observation window 5ns immediately following the laser pulse)

The observations described above were fully expected during the design of the syn-DMA[6]DMN[8]DCV system, but we were rather baffled to find that the lifetime of the CS-2 state is dramatically shortened as compared to that of the CS-1 state in [2]DMN[8]DCV. The latter amounts to 70 ns in MCH (see Fig. 1) while the TA experiments (see Fig. 3) suggested a CS-2 lifetime for syn-

DMA$^+$[6]DMN[8]DCV$^-$ of only a few ns in the same solvent at room temperature [25]. It should be noted that these data exclude the possibility that charge recombination from CS-2 involves back electron transfer via CS-1.

Thus it appears that in the CS-2 state significant electronic coupling between the DMA$^+$ and DCV$^-$ units exists allowing (sub)nanosecond charge recombination, notwithstanding the unduly long TB-pathway separating these sites!

The existence of such direct interaction is confirmed by some quite unexpected results of a careful study of the fluorescence spectra of *syn*-DMA[6]DMN[8]DCV. Initially our expectation was that at room temperature only some residual local fluorescence in the <350 nm region would appear and that at lower temperature, where TA spectroscopy (see Fig. 3) showed build-up of CS-1, charge transfer fluorescence emerging from it might be detectable around 450 nm (see Fig. 1).

As shown in Fig. 4 the latter expectation is substantiated by observation of a broad fluorescence maximum at ~ 455 nm in a spectrum measured at 200 K. As also expected this band virtually disappears at room temperature but remarkably it is then substituted by another broad emission at even longer wavelength (~ 550 nm).

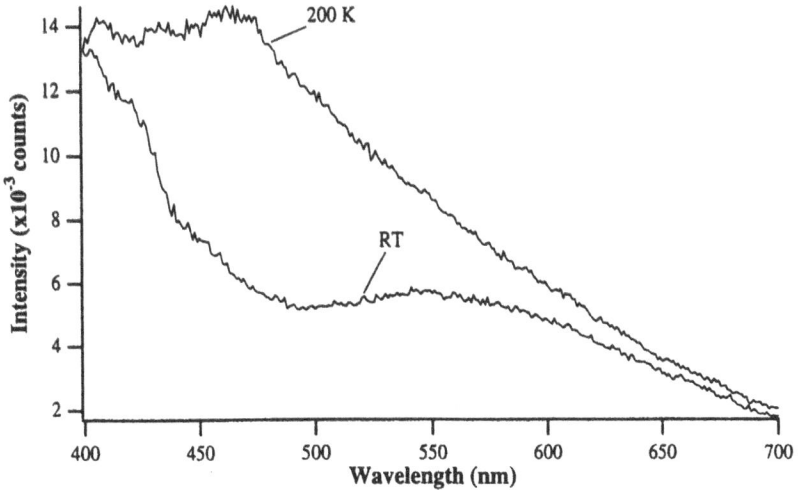

Figure 4. Fluorescence spectra of *syn*-DMA[6]DMN[8]DCV in methylcyclohexane at 293 K and at 200 K (λ_{exc} = 310 nm).

Preliminary time resolved fluorescence studies indicate that the decay time of the 550 nm emission band coincides with the lifetime of the CS-2 species detected by TA spectroscopy and furthermore show that it is kinetically linked to the emission at 455 nm attributed to the CS-1 species. Representative decay curves obtained at three wavelengths and two temperatures are shown in Fig. 5.

The wavelengths (400, 470 and 600 nm) were chosen in regions where the overall fluorescence is expected to be dominated by respectively local fluorescence of the DMN (as well as of the DMA) chromophore, by emission of the CS-1 state and by emission corresponding to the 550 nm maximum.

Upon inspection it is clear that the main component of the 400 nm fluorescence decays very rapidly (~ 30-40 ps) which is consistent with the expected high rate of quenching of local emission (LE) by formation of the CS-1 state both at room temperature and at 200 K. The fluorescence at 470 nm rises equally fast as the decay at 400 nm as expected in view of its CS-1 bond nature.

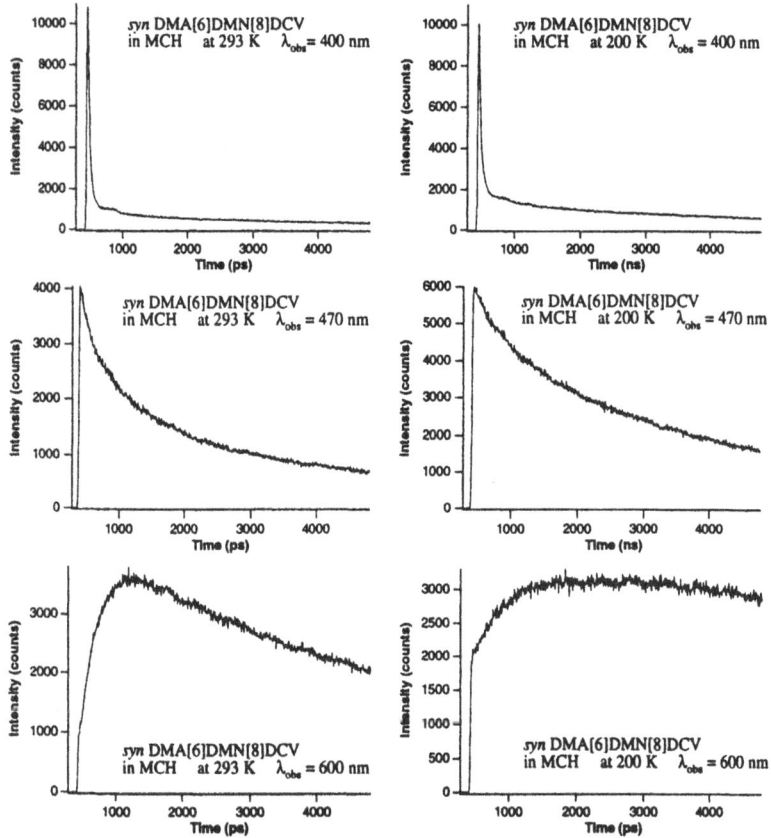

Figure 5. Fluorescence decay curves measured by time correlated single photon counting at three wavelengths and two temperatures for *syn*-DMA[6]DMN[8]DCV in methylcyclohexane (λ_{exc} = 315 nm).

More important, however, is the observation that the 470 nm emission decays biexponentially with time constants that reappear at 600 nm as a growth and a decay component respectively. This strongly suggests that the emission in the ≥ 450 nm region is composed of two kinetically linked species with that emitting at shorter wavelength (ca. 455 nm) being the precursor of that emitting at longer wavelength (ca. 550 nm).

While as stated above the former species is clearly CS-1, we now have to accept that the latter is CS-2, which quite unexpectedly also appears to decay via a partially radiative process.

Global analysis of decay curves obtained at 10 nm intervals across the 400-630 nm region could indeed be fitted quite well to a model containing three linked emissive states (LE → CS-1 → CS-2, see Fig. 2) with the decay times compiled in Table 1.

TABLE 1. Fluorescence decay times (ns) obtained by global fitting over the 400-630 nm region for *syn*-DMA[6]DMN[8]DCV in methylcyclohexane at 293K and at 200K.

	LE	CS-1	CS-2
293 K	0.030	0.3	5
200 K	0.040	1.7	10

It is interesting to note that in contrast to the weak temperature dependence of the τ_{LE} both τ_{CS-1} and τ_{CS-2} increase substantially at lower temperature. This implies that both the interconversion from CS-1 to CS-2, which is the main process determining τ_{CS-1} and the charge-recombination in CS-2, slow down at lower temperature.

The most important result in the present context, however, is that strong evidence has now been obtained for direct and radiative charge recombination between the DMA and DCV chromophores. This implies the existence of a sizable electronic coupling between these distant sites.

Although an accurate determination of the radiative rate constant has yet to be made, comparison of the CS-1 fluorescence intensity and lifetime for [2]DMN[8]DCV (see Fig. 1) with that of the CS-2 fluorescence observed for *syn*-DMA[6]DMN[8]DCV at room temperature (see Fig. 4 and Table 1) suggests that the ratio of the CS-1 and CS-2 radiative rate constants is <10.

It has been observed earlier [18] that for the type of bridges involved here the radiative rate constant of through-bond mediated CT fluorescence decays exponentially with the number (n) of bonds with a damping factor ~ 0.8 per bond, which is closely related to the damping factor for the rate of electron transfer. Even neglecting the aromatic bonds of the DMN unit, the number of bonds in the TB pathway between DMA and DCV outnumbers that between DMN and DCV by six, which should decrease the radiative rate constant by ≈ 400 times.

We are thus forced to conclude that the main electronic coupling between DMA and DCV in the CS-2 state of *sýn*-DMA[6]DMN[8]DCV cannot be through-bond in nature.

The most plausible mechanism seems then to be a through-solvent interaction involving in particular those solvent molecules that enter the U-shaped cleft of the molecule (see Scheme 2). While in a related study [23] this cleft was so small as to allow a single solvent molecule to span it, in our case clearly a number of solvent molecules is required. We cannot at present say whether the electric field constituted across that cleft in the CS-2 state plays a role in polarizing these solvent molecules to enhance through solvent coupling. Nonetheless the present results provide a remarkable example of competition between TB and TS-pathways of electron transfer, which appears to be clearly decided in favor of the shorter TS pathway (~ 16 Å) over the longer TB pathway (~ 27 Å). This should be taken as a warning against assuming an excessive preference for TB-pathways if shorter TS-pathways are available.

4. Acknowledgements

The present research was supported by the Netherlands Foundation of Chemical Research (SON) with financial aid from the Netherlands Organization for Scientific Research (NWO). We furthermore acknowledge support from the Australian Research Council (ARC) and the award of an ARC postgraduate scholarship to JML.

5. References

1. Pasman, P., Rob, F., and Verhoeven, J.W. (1982), *J. Am. Chem. Soc.* **104**, 5127.
2. Pasman, P., Koper, N.W., and Verhoeven, J.W. (1982), *Recl. Trav. Chim. Pays-Bas* **101**, 363.
3. Calcaterra, L.T., Closs, G.L., and Miller, J.R. (1983), *J. Am. Chem. Soc.* **105**, 670.
4. Hush, N.S., Paddon-Row, M.N., Cotsaris, E., Oevering, H., Verhoeven, J.W., and Heppener, M. (1985), *Chem. Phys. Lett.* **117**, 8.

5. Oevering, H., Paddon-Row, M.N., Heppener, M., Oliver, A.M., Cotsaris, E., Verhoeven, J.W., and Hush, N.S. (1987), *J. Am. Chem. Soc.* **109**, 3258.
6. Closs, G.L., and Miller, J.R. (1988), *Science* **240**, 440.
7. Knapp, S., Dhar, G.M., Albaneze, J., Gentemann, S., Potenza, J.A., Holten, D., and Schugar, H.J. (1991), *J. Am. Chem. Soc.* **113**, 4010.
8. Stein, C.A., Lewis, N.A., and Seitz, G. (1982), *J. Am. Chem. Soc.* **104**, 2596.
9. Oliver, A.M., Craig, D.C., Paddon-Row, M.N., Kroon, J., and Verhoeven, J.W. (1989), *Chem. Phys. Lett.* **150**, 366.
10 Paddon-Row, M.N. (1994), *Acc. Chem. Res.* **27**, 18.
11. Betts, J.N., Beratan, D.N., and Onuchic, J.N. (1992), *J. Am. Chem. Soc.* **114**, 4043.
12. Kurnikov, I.V., and Beratan, D.N., (1996), *J. Chem. Phys.* **105**, 9561.
13. Siddarth, P. and Marcus, R.A. (1993), *J. Phys. Chem.* **97**, 2400.
14. Moser, C.C., Keske, J.M., Warncke, K., Farid, R.S., and Dutton, P.L. (1992), *Nature* **355**, 796, see also Baum, R.M. (1993), *Chem. & Eng. News* **Febr. 22**, 20.
15. Kroon, J., Oevering, H., Verhoeven, J.W., Warman, J.M., Oliver, A.M., and Paddon-Row, M.N. (1993), *J. Phys. Chem.* **97** 5065.
16. Kroon, J., Verhoeven, J.W., Paddon-Row, M.N., and Oliver, A.M. (1991), *Angew. Chem. Int. Ed. Engl.* **30**, 1358.
17. Warman, J,M., Smit, K.J., de Haas, M.P., Jonker, S.A., Paddon-Row, M.N., Oliver, A.M., Kroon, J., Oevering, H., and Verhoeven, J.W. (1991), *J. Phys. Chem.* **95**, 1979.
18. Oevering, H., Verhoeven, J.W., Paddon-Row, M.N., and Warman, J.M.(1989), *Tetrahedron* **45**, 4751.
19. Roest, M.R., Lawson, J.M., Paddon-Row, M.N., and Verhoeven, J.W., (1994), *Chem. Phys. Lett.* **230** , 536.
20. Roest, M.R., Verhoeven, J.W., Schuddeboom, W., Warman, J.M., Lawson, J.M., and Paddon-Row, M.N. (1996), *J. Am. Chem. Soc.* **118**, 1762.
21. Roest, M.R., PhD Dissertation, University of Amsterdam, 1996.
22. Lawson, J.M., Craig, D.C., and Paddon-Row, M.N. (1995), *Tetrahedron* **51**, 3841.
23. Zimmt, M.B. (1997), *Chimia* **51**, 82.
24. Lawson, J.M., Paddon-Row, M.N., Schuddeboom, W., Warman, J.M., Clayton, A.H., and Ghiggino, K.P. (1993), *J. Phys. Chem.* **97**, 13099.
25. A further shortening to about 150 ps has been observed for *syn* -DMA[4]DMN[8]DCV in di-n-butylether [21] and in benzene [24].

MECHANISMS AND CONTROL OF ELECTRON TRANSFER PROCESSES IN PROTEINS

O. FARVER° AND I. PECHT*

° Institute of Analytical Chemistry, Royal Danish School of Pharmacy, DK-2100 Copenhagen, Denmark, *Department of Immunology, The Weizmann Institute of Science, Rehovot 76100, Israel

1. Abstract

Specific rates and activation parameters of electron transfer processes were determined for two types of copper-containing proteins which serve as model systems for resolving the role of the polypeptide matrix in these processes. In azurins which are well characterized blue single copper proteins predominantly consisting of a β-sheet polypeptide matrix, the above parameters were determined for the intramolecular long range electron transfer between the pulse radiolytically generated disulphide radical-anions and the copper(II) centre, as a function of driving force and nature of the separating medium in 19 different wild type and single site mutated proteins. The internal ET from the type-1 Cu(I) to the trinuclear Cu(II) centre of ascorbate oxidase is part of this enzyme's catalytic cycle. We investigated the temperature and pH dependence of this process. Results obtained from studies of both types of proteins correlate with a model based on electron transfer proceeding through well defined pathways using a through-bond tunneling mechanism.

2. Introduction

A central question addressed in studies of long range electron transfer (LRET) in proteins is how structure of the medium and distance separating the electron donor and acceptor determine the reaction rates. Understanding the experimental results in terms of the chemical and structural properties of the polypeptide matrix is still a major challenge. Specific questions being whether the shortest direct electron transfer (ET) path is the one employed or do structural and electronic properties of the medium provide more facile, though longer routes [1-5].

We have investigated intramolecular LRET in azurins which are well characterized β-sheet structured blue, single copper proteins [6-13]. They contain two potential redox centres; One, the copper ion, is coordinated directly to amino acid residues and a second the disulphide group is present at the opposite end of the molecule, at a direct distance of 2.65 nm [13-17]. Significantly, the function of the disulphide bridge is most probably structural rather than of redox activity, which is confined to the copper site only. The

63

G.W. Canters and E. Vijgenboom (eds.),
Biological Electron Transfer Chains: Genetics, Composition and Mode of Operation, 63-74.
© 1998 Kluwer Academic Publishers.

medium separating these two redox centres thus provides a model system in which specific structural changes can be made and their impact, on the ET reactivity examined. As the direct distance between the electron donor and acceptor sites is most probably maintained constant in all azurins studied, results of structure-reactivity analysis provide a very useful basis for rationalizing the dependence of the internal ET rate constants on specific structural changes introduced in the separating medium.

The second system studied is the intramolecular ET between the electron uptake site of the blue copper enzyme, ascorbate oxidase (AO) and the centre where the oxidising substrate, O2, binds and is reduced to H2O. This is the prototype catalytic cycle of those enzymes that reduce dioxygen to water [18,19]. Therefore it is a system where an evolutionarily selected LRET process takes place. In spite of the short peptide stretch connecting the above two centers in AO, the rate of ET is relatively slow probably due to a rather low driving force. In addition, however, nuclear rearrangements caused by the redox changes and substrate binding have been resolved and suggest that gating may also be involved in this LRET process.

3. Electron-Transfer Within Azurins

We have shown earlier that ET between the two redox centres of azurin can be induced following pulse radiolytic reduction of the disulphide group to the RSSR⁻ radical-anion which is then followed by the reaction:

$$Az[Cu(II)RSSR^-] \longrightarrow Az[Cu(I)RSSR]$$

It should be stressed that both processes, i.e. reoxidation of RSSR⁻ and reduction of Cu(II) occur synchronously without any resolvable intermediates, hence the formation of any amino acid radical intermediates is excluded and supports the notion that the ET proceeds by electronic coupling. Rate constants of this intramolecular ET were determined as a function of temperature and pH for a large number of different wild type (WT) and single site mutated azurins [6-12]. The effect of the specific structural changes that were introduced into azurin on the LRET rates and activation parameters were analyzed and are summarized in Table 1.

The semiclassical Marcus theory for non-adiabatic processes predicts that intramolecular ET rates are governed by the standard free energy of reaction (ΔG^0), the nuclear reorganization energy (λ), and the electronic coupling (H_{DA}) between electron donor (D) and acceptor (A) at the transition state [20]:

$$k = \frac{2\pi}{h} \frac{H_{DA}^2}{(4\pi\lambda RT)^{1/2}} e^{-(\Delta G^0 + \lambda)^2/4\lambda RT} \tag{1}$$

TABLE 1 Kinetic and thermodynamic data for the intramolecular reduction of Cu(II) by RSSR⁻ in azurins

Azurin	k_{298} s^{-1}	E' mV	$-\Delta G^0$ kJ mol^{-1}	$\Delta H^{\neq}$ kJ mol^{-1}	$\Delta S^{\neq}$ J K^{-1} mol^{-1}
Wild Type					
Ps. aer.[a]	44 ± 7	304	68.9	47.5 ± 2.2	-56.5 ± 3.5
Ps. fluor.[b]	22 ± 3	347	73.0	36.3 ± 1.2	-97.7 ± 5.0
Alc. spp.[a]	28 ± 1.5	260	64.6	16.7 ± 1.5	-171 ± 18
Alc. faec.[b]	11 ± 2	266	65.2	54.5 ± 1.4	-43.9 ± 9.5
Mutant					
D23A[e]	15 ± 3	311	69.6	47.8 ± 1.4	-61.4 ± 6.3
F110S[f]	38 ± 10	314	69.9	55.5 ± 5.0	-28.7 ± 4.5
F114A[c]	72 ± 14	358	74.1	52.1 ± 1.3	-36.1 ± 8.2
H35Q[d]	53 ± 11	268	65.4	37.3 ± 1.3	-86.5 ± 5.8
I7S[f]	42 ± 8	301	68.6	56.6 ± 4.1	-21.5 ± 4.2
M44K[d]	134 ± 12	370	75.3	47.2 ± 0.7	-46.4 ± 4.4
M64E[f]	55 ± 8	278	66.4	46.3 ± 6.2	-56.2 ± 7.2
M121L[c]	38 ± 7	412	79.3	45.2 ± 1.3	-61.5 ± 7.2
V31W[g]	285 ± 18	301	68.6	47.2 ± 2.4	-39.7 ± 2.5
W48A[g]	35 ± 7	301	68.6	46.3 ± 5.9	-58.3 ± 6.0
W48F[g]	80 ± 5	304	68.9	43.7 ± 6.7	-61.9 ± 9.7
W48S[g]	50 ± 5	314	69.9	49.8 ± 4.9	-44.0 ± 3.5
W48Y[g]	85 ± 5	323	70.7	52.6 ± 6.9	-30.2 ± 3.6
W48L[c]	40 ± 4	323	70.7	48.3 ± 0.9	-51.5 ± 5.7
W48M[c]	33 ± 5	312	69.7	48.4 ± 1.3	-50.9 ± 7.4

a) Ref. [6]; b) Ref. [7]; c) Ref. [9]; d) Ref. [8]; e) Ref. [11]; f) Ref. [10]; g) Ref. [12];

The electronic coupling energy, H_{DA}, is expected to decay exponentially with the distance separating D and A as:

$$H_{DA} = H_{DA}^0 e^{-0.5\beta(r-r_0)} \qquad (2)$$

This distance $(r-r_0)$ may be considerable in proteins (≥ 1.0 nm), leading to a rather low electronic coupling. Still, intramolecular ET processes have been observed over distances of 2.0 nm or more as is the case for the azurins [21].

For most azurins studied, the activation enthalpy $\Delta H^{\neq}$ given by the following relation was found to be constant within experimental error (cf. Table 1) [20]:

$$\Delta H^{*} = \frac{\lambda}{4} + \frac{\Delta H^{0}}{2}(1 + \frac{\Delta G^{0}}{\lambda}) - \frac{(\Delta G^{0})^{2}}{4\lambda} \tag{3}$$

This finding supports our previous assumption that as long as mutations are not introduced at or near the redox centres, the reorganization energy of this LRET does not change significantly.

The activation entropy includes a contribution from the distance dependence of the electronic coupling [20] (cf. equation 2):

$$\Delta S^{\neq} = \Delta S^{*} - R\beta(r - r_{0}) \tag{4}$$

and ΔS^{*} is related to the standard entropy of reaction, ΔS^{0}:

$$\Delta S^{*} = \frac{\Delta S^{0}}{2}(1 + \frac{\Delta G^{0}}{\lambda}) \tag{5}$$

Using the experimentally determined rate constants and activation parameters for 19 different wild type and mutated azurins (Table 1) and assuming that the reorganization energy does not vary significantly amongst them, we calculate $\lambda = 99.4 \pm 5$ kJ mol^{-1} and $\beta(r-r_0) = 24.6 \pm 1.2$.

Moser et al. [2,3] have analyzed data derived from a large body of intramolecular electron transfer reactions and found that the free energy optimized rate constants for these processes correlate well with the edge-to-edge distance between donor and acceptor, with a decay factor, $\beta = 14$ nm^{-1}. The distance $(r-r_0)$ separating S_γ of Cys3 in the disulphide bridge and the copper ligating S_γ of Cys112, which represents the shortest edge-to-edge distance between electron donor and acceptor, is calculated from the refined (0.19 nm) *Pseudomonas aeruginosa* azurin structure to be 2.51 nm [13]. This yields a $\beta = 9.8 \pm 0.8$ nm^{-1}. The difference between the above two β-values is too large to be accounted for by experimental errors. Moreover, the calculated maximum (i.e. for $\lambda = -\Delta G^0$) LRET rate constant for azurin is 120 s^{-1} while using the correlation line of Farid *et al.* [3] the rate constant should be smaller by at least two orders of magnitude. The high-resolution structures available for azurins (and ascorbate oxidase) and the fact that no cofactors or other extraneous components separate the redox centers, make these copper proteins most appropriate model systems. In contrast, a considerable number of proteins analyzed by Dutton *et al.* [2,3] are characterized by having cofactors rather than the polypeptide separating the ET partners. In fact, the protein provides in

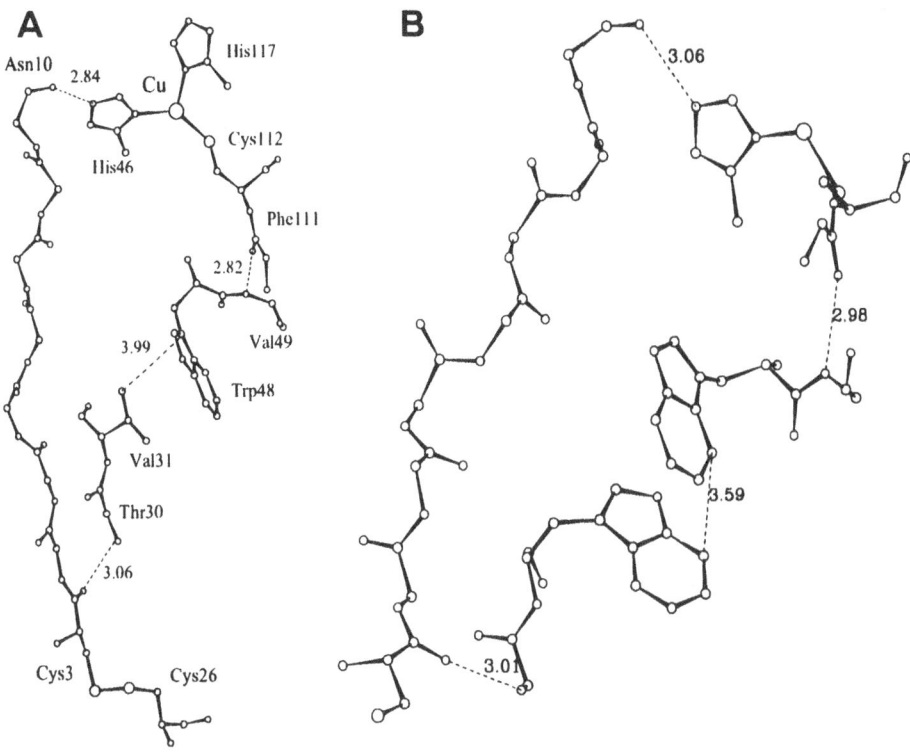

Figure 1. Calculated pathways for ET from the sulphur of Cys3 to the copper centre in the wild type (A) and the V31W mutant (B) of *Pseudomonas aeruginosa* azurin. Some interconnecting distances (3 H-bonds and one van der Waals contact) are given (in Å). In the V31W mutant the closest distance between the two tryptophans (3.5 Å) occurs between W48 $C_{\zeta3}$ and W31 $C_{\varepsilon3}$. The coordinates were taken from the Brookhaven Protein Databank. The V31W pathway (B) was calculated on the basis of NMR data and energy minimization [13].

many of these cases only the surrounding matrix. This obviously has its impact, yet not in mediating the electron transfer process. Significantly, the above calculated distance decay constant, $\beta = 9.8$ nm^{-1}, is in excellent agreement with that predicted for the coupling decay along a β-sheet strand [22], as experimentally found by Gray and co-workers [23] for ruthenium modified cytochrome *c* and azurin. Taken together, these results clearly illustrate the role that the protein medium plays in intramolecular LRET.

Beratan and Onuchic introduced a theoretical approach for examining the medium dependence of protein LRET [24, 25] identifying potential routes based on available three-dimensional protein structures. We have employed this model to calculate potential electronic coupling pathways in *P. aeruginosa* azurin and its mutants [13-17]. For mutants for which crystallographic structures are not available, structures based on 2D-NMR studies and energy minimization calculations were employed. Significantly, pathway calculations suggest that the same two dominant electron transfer routes, shown in Figure 1, operate in all azurins examined [6-12]: One path is through the peptide chain to the copper-ligating imidazole of His46 while another, more direct one leads through the polypeptide chain via the buried indole ring of Trp48 to the copper coordinating S_γ of Cys112. The electronic coupling factors were found to be $\pi\varepsilon = 2.5 \times 10^{-7}$ and 4.7×10^{-8}, respectively.

The effective tunneling distances defined by Langen et al. [26] can be calculated for azurin and include hydrogen bonds and through-space contacts. It is $\sigma_1 = 4.07$ nm for the "His46-pathway" (27 covalent bonds and 1 H-bond) while it becomes 4.62 nm for the "Trp48 pathway" (19 covalent bonds, 2 H-bonds and 1 van der Waals contact). Note that all distances reported here for azurins are defined as those between S_γ of Cys3 and the atom coordinating to the copper ion. Introducing the above distance along the His 46-pathway into the $\beta(r-r_0)$ we obtain $\beta = 6.0 \pm 0.4$ nm-1. This is in good agreement with both theoretical calculations for electron tunneling through a saturated aliphatic $(CH2)_n$ chain ($\beta = 6.5$ nm^{-1}) [27] and with results of Gray and co-workers [22,26] studying Ru-modified azurins (7.3 nm^{-1}, based on the σ_1 covalent path). Caution should however be taken in distance comparisons since in the latter systems different donors and/or acceptors are often involved.

A more detailed examination of the two pathways is instructive. A comparison between covalent tunneling lengths of the above two pathways (or, equivalently, their two electronic coupling factors) in the different azurins suggests that the "Trp48 pathway" is less effective. However, electronic interactions between the Cu(II) ion and its ligands were not taken into account in this pathway analysis. Theoretical studies have recently proposed that a high degree of anisotropic covalency exists in the blue single-copper protein, plastocyanin which would promote ET through the thiolate ligand [28,29]. By similar arguments, from the Ψ_{HOMO} ligand coefficients in azurin obtained by Larsson et al. [30] it can be estimated that ET through the thiolate would be enhanced ~150 fold more than that via one of the His ligands. This means that the "Trp48" and "His46" pathways would have about equal contributions to the process (i.e. within an order of magnitude). It is also noteworthy that the pathway analysis treats the van der Waals contact between Val31 and Trp48 only as a single point interaction between $C_{\gamma 1}$ of valine and $C_{\delta 2}$ of tryptophan. The refined structure coordinates of azurin [13] show however that at least six of the indole ring atoms are in close contact (≤ 0.43 nm) with the valine side chain. Finally the "Trp pathway" involves an aromatic residue raising the possible involvement of a conjugated π-orbital system in the electronic coupling. So far, the pathway calculation, however, only takes σ-bonds into account.

The possible involvement of aromatic residues in ET was examined in single-site mutated azurins where Trp48 has been substituted by other amino acids, with both aromatic and non-aromatic residues [9, 12]. The results are shown in Table 1 together with the standard free energies of reaction (ΔG^0), and demonstrate that substitution of Trp48 with other amino acids has only a negligible effect on the kinetic parameters when corrected for changes in driving force. More recently, however, an additional mutant was prepared where Val31 was substituted by Trp. This "double-Trp" azurin (V31W) [12] has the two indole residues placed in neighbouring positions (cf. Figure 1). 2D-NMR measurements of this mutant demonstrated that the region located between Trp 48 and the copper center of this protein maintains the same structure as its

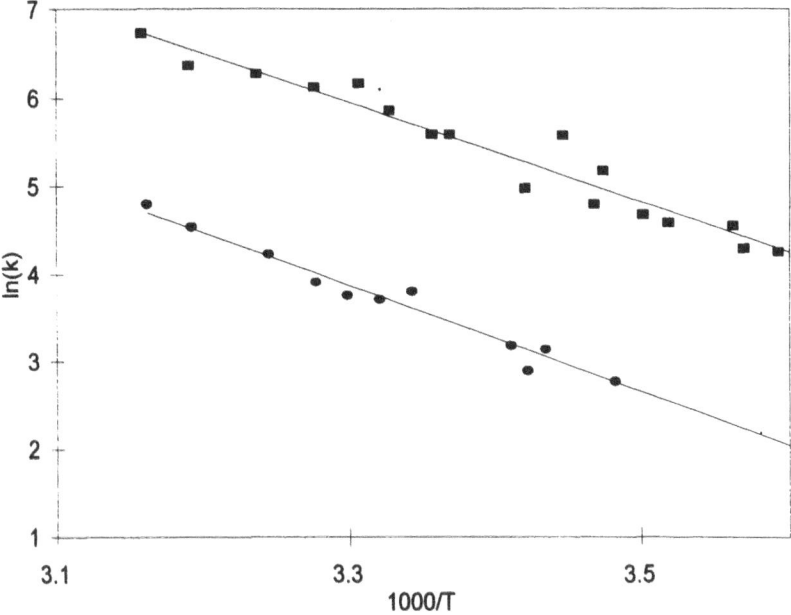

Figure 2. Eyring plots of the rate constants determined for the intramolecular ET from RSSR⁻ to the Cu(II) in wild type (●) and the V31W mutant (■) of *P. aeruginosa* azurin. Measurements were done in aqueous solution (10 mM Na phosphate buffer, 100 mM Na formate, saturated with N_2O, pH 7.0).

wild-type protein equivalent. Moreover, energy minimization calculations have also been performed and show a close (van der Waals contact) of the two indole rings consistent with observed NOEs between protons of the two indoles (cf. Figure 1B) [12]. Like in all other azurins studied so far, the LRET between the disulfide and the Cu(II) ion is also taking place in the V31W azurin mutant and was found to proceed with a rate constant of 285 s⁻¹ at 298 K and pH 7.0, i.e. several fold faster than any other azurin studied so far [6-12]. Figure 2 illustrates the temperature dependence of this reaction. Thus, in addition to suggesting a role for aromatic residues in the ET process, it is also in line with the dominant ET route being the "Trp48" pathway, since the alternative pathway through His46 would not be affected by this mutation.

The activation parameters for ET in V31W azurin provide some further new insights: The increase in rate in V31W azurin is due to a more advantageous entropy of activation (Table 1) which is larger by 16.8 J K⁻¹mol⁻¹ compared with wild type (WT) azurin. Since ΔS^0 is most probably the same for intramolecular ET in WT and V31W azurins, the increase in activation entropy would according to Equation 4 cause a <u>decrease</u> in $\beta(r-r_0)$ from the previously calculated value of 24.6 for WT to 22.6 in V31W mutated azurin. As the separation distance is the same in both azurins, a smaller electronic decay factor, β, in the mutant is implied and is also reflected in the electronic coupling energy, H_{DA}, between the electron donor and acceptor which we have calculated to be 2.1×10^{-7} eV on the basis of the experimentally determined rate constant.

This is a 2.6 fold improvement relative to the value calculated for WT azurin (H_{DA} = 0.8×10^{-7} eV). In contrast, a calculation of the covalent tunneling length gave σ_1 = 5.0 nm for the "double Trp pathway" in V31W azurin as compared with σ_1 = 4.62 nm for the "Trp48 pathway" in WT azurin (*vide supra*). Obviously, a different tunneling length cannot explain the observed increase in rate in this mutant.

The above results suggest that the presence of the two tryptophans and their proximity may enhance the interaction between donor (D) and acceptor (A) since the ring systems are in van der Waals contact which may provide a considerable electronic overlap and give rise to a resonance-type tunneling through their indole rings. Aromatic residues placed in appropriate positions may enhance ET through proteins by a more effective coupling through their extended π^*-orbitals since the energy gap between that of the tunneling electron and the aromatic π-system is significantly smaller than that between the electron tunneling energy and σ-orbitals. The lack of any effect of the single tryptophan residue placed midway between D and A in a predominantly σ-ET pathway suggest that it is not advantageous by itself. The reason being that $\sigma \rightarrow \pi \rightarrow \sigma$ ET will be energetically unfavourable. However, several aromatic residues placed in consecutive, proximal positions or aromatic molecules in direct contact with either D or A may act as an extended relay system, possibly enhancing the electronic coupling.

4. LRET Among Active Sites Of Blue Copper Enzymes

One of the central issues debated in the field of LRET within proteins is whether evolutionary pressure has lead to specific preferential pathways. This obviously has to be examined in systems where intramolecular electron transfer is part of functional redox cycles. Cytochrome *c* oxidase and ascorbate oxidase belong to a class of enzymes which catalyze multiple sequential ET steps that are part of a cycle leading to dioxygen reduction to water. This makes these oxidases a relevant system for investigating this problem, distinct from the wide range of other redox enzymes catalysing atom transfer, e.g. the dehydrogenases.

High resolution (0.19 nm) three-dimensional structures are now available for ascorbate oxidase in both its oxidised and reduced states [31,32]. Like all other blue oxidases, half the catalytic cycle is constituted of four sequential single electron transfer steps from a donor to the type-1 (T1) Cu(II) site while dioxygen coordinates and is reduced at a trinuclear, type-2/type-3 (T2/T3) copper centre [33]. Thus, intramolecular ET is a central element of this enzyme's function. Under anaerobic conditions, the rate constant for intramolecular electron transfer from T1Cu(I) to T2/T3Cu(II) in AO was found to be only ~200 s^{-1} at 25^{0}C [18] although the two redox centers are connected by a direct chemical bond pathway (cf. Figure 3) with σ_1 = 1.26 nm between Cys507(S_γ) and His506/508(N_e). The possible cause for the observed slow ET rate is at least two-fold: (A) Under anaerobic conditions the reactions driving force is close to zero, and (B) the reorganisation energy was found to be considerable, 142 ± 10 kJ mol^{-1} [18]. This suggests that the enzyme catalyzed process of dioxygen reduction is gated by either substrate (O_2) binding and/or redox induced conformational changes at the trinuclear site. Indeed, both O_2 binding to the trinuclear centre [19] and oxidising the fully reduced

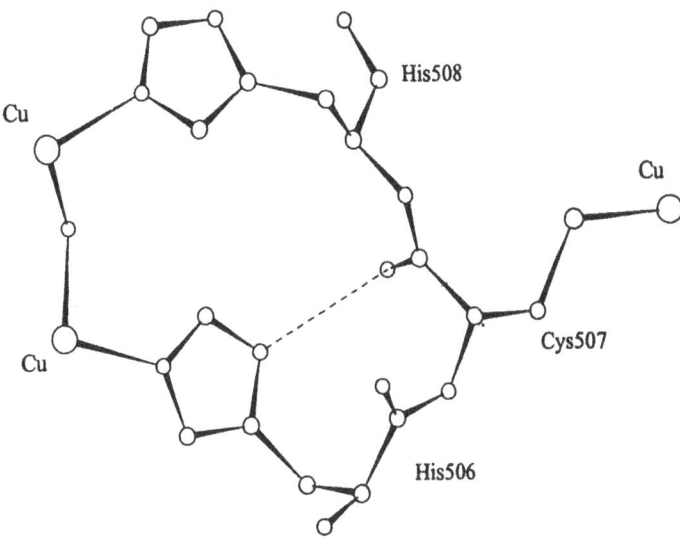

Figure 3. Calculated pathways for ET from T1Cu(I) (right, coordinated to Cys507) to the T3Cu(II) centre (left, coordinated to His506/508) in ascorbate oxidase [18]. The coordinates were taken from the Brookhaven Protein Databank [34].

enzyme [34] were found to markedly enhance rates of intramolecular ET.

While the structural changes that were resolved between fully oxidised and fully reduced AO may make some of the internal ET steps adiabatic, this may not be the case for the initial ET step of cytochrome *c* oxidase reduction. For the latter case a rate constant of 2×10^4 s^{-1} has been observed [35] whereas the analogous step in AO is 100 fold slower. The shortest distance separating CuA and haem *a* in cytochrome *c* oxidase is 2.0 nm through a hydrogen bonding system including His504 and Arg438/439 [36] which is a considerable distance for an intraprotein ET process. This further accentuates the contrast with the fact that in AO, the T1-T3 distance separating Cys507(S$_\gamma$) and His506/508(N$_\varepsilon$) is only 1.3 nm.

An unusual and to our knowledge unprecedented feature of the intramolecular LRET in AO is its distinct multiple phases: When monitored either at the 610 nm band of the T1 Cu(II) site or the 330 nm band of the T3 centre the same two to three reaction phases are observed [18,19]. We have excluded the possibility that these are reflecting reactions of AO molecules at different reduction states. An alternative cause considered for the existence of distinct AO species differing in their reactivity is the following: reactivity of laccase was claimed to be affected by the binding of an OH$^-$ ion to its T2 Cu(II) site. The high resolution structure of AO established the presence of an oxygen containing species (OH$^-$ or water) bound to the T2 Cu(II). We have therefore considered the possibility that the multiple reaction phases could be due to distinct protonation states of one or more of the reactive centres involved in the LRET and investigated their pH dependence. As Figure 4 shows practically no effect on either of the two dominant

Ascorbate Oxidase
pH effect on ET

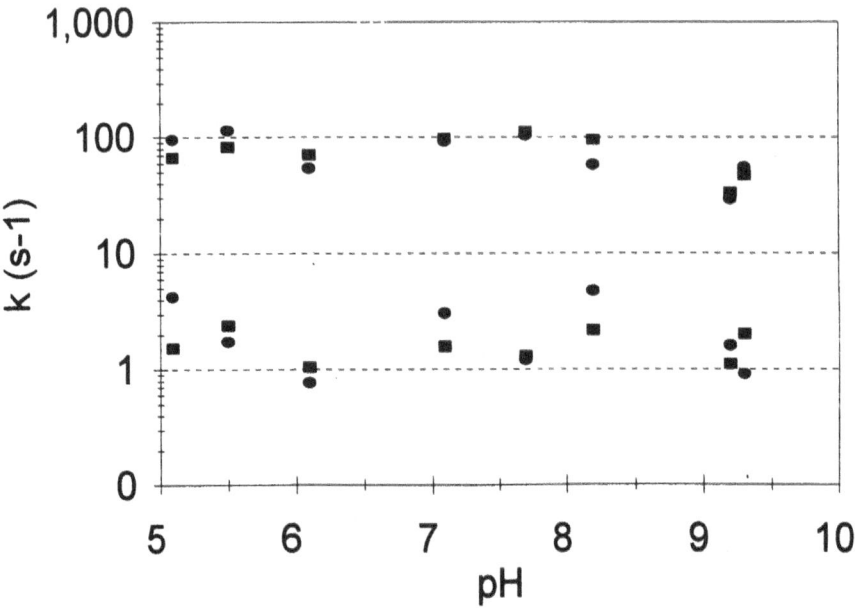

Figure 4. First order rate constants observed for the intramolecular electron transfer between the T1 Cu(II) and the trinuclear centre of AO monitored at 610 nm (■) or 330 nm (●) bands at 18°C as a function of pH. Experiments were carried out anaerobically in 10 mM sodium phosphate and 100 mM Na formate saturated with N_2O producing the CO_2^- radicals by pulse radiolysis. AO concentration was 10 μM.

LRET reaction phases could be observed in the pH range of 5.0 to 9.2. This excludes the possibility of a pH dependent equilibrium between reactive species in AO as the reason for the observed reaction pattern. Still, these results do not exclude other causes for the existence of distinct AO species. These could be the distinct forms for which Reinhammar et al. [37] have recently provided spectroscopic evidence.

In conclusion; azurins turned out to be very useful for examining the parameters which control LRET in a β-sheet polypeptide matrix, in a domain that was not a product of functional selection. Ascorbate oxidase provides an example for an ET process that is an intrinsic element of a catalytic cycle. Three-dimensional tructures of both proteins are available at high resolution. They predominantly consist of a β-sheet polypeptide matrix, to which the redox centres are directly connected without any intervening cofactors. This, and the fact that in both cases single electron transfer processes take place, enable a rigorous definition of separation distances and a detailed quantitative analysis. The increasing body of data for intramolecular LRET in blue copper proteins convincingly supports mechanisms where defined pathways are operative.

5. Acknowledgement

This research has been supported by the Danish Natural Science Research Council and the German-Israeli Foundation (I 320-211-05) which is gratefully acknowledged. We wish to thank Dr. Lars K. Skov for performing the energy minimization calculations and for producing the figures for this paper.

6. References

1. Beratan D.N., Onuchic J.N., Winkler J.R., Gray H.B. (1992) *Science* **258**, 1740-1741
2. Moser C.C., Keske J.M., Warncke K., Farid R.S., Dutton P.L. (1992) *Nature* **355**, 796-802
3. Farid R.S., Moser C.C., Dutton P.L. (1993) *Curr. Opin. Struct. Biol.* **3**, 225-233
4. Siddarth P., Marcus R.A. (1993) *J. Phys. Chem.* **97**, 13078-13082
5. Gray H.B., Winkler J.R. (1996) *Ann. Rev. Biochem.* **65**, 537-561
6. Farver O., Pecht I. (1989) *Proc. Natl. Acad. Sci. USA* **86**, 6968-6972
7. Farver O., Pecht I. (1992) *J. Am. Chem. Soc.* **114**, 5764-5767
8. Farver O., Skov L.K., van de Kamp M., Canters G.W., Pecht I. (1992) *Eur. J. Biochem.* **210**, 399-403
9. Farver O., Skov L.K., Pascher T., Karlsson B.G., Nordling M., Lundberg L.G., Vänngård T., Pecht, I. (1993) *Biochemistry* **32**, 7317-7322
10. Farver O., Skov L.K., Gilardi G., van Pouderoyen G., Canters G.W., Wherland S., Pecht I. (1996) *Chem. Phys.* **204**, 271-277
11. Farver O., Bonander N., Skov L.K., Pecht I. (1996) *Inorg. Chim. Acta* **243**, 127-133
12. Farver O., Skov L.K., Young S., Bonander N., Karlsson B.G., Vänngård T., Pecht I. (1997) Submitted for publication
13. Nar H., Messerschmidt A., Huber R., van de Kamp M., Canters, G.W. (1991) *J. Mol. Biol.* **221**, 765-772
14. Baker E.N. (1988) *J. Mol. Biol.* **203**, 1071-1095
15. Nar H., Messerschmidt A., Huber R., van de Kamp M., Canters G.W. (1991) *J. Mol. Biol.* **218**, 427-447
16. Romero A., Hoitink C.W.G., Nar H., Huber R., Messerschmidt A., Canters G.W. (1993) *J. Mol. Biol.* **229**, 1007-1021
17. Hammann C., Messerschmidt A., Huber R., Nar H., Gilardi G., Canters, G.W. (1996) *J. Mol. Biol.* **255**, 362-366
18. Farver O., Pecht I. (1992) *Proc. Natl. Acad. Sci. USA* **89**, 8283-8287
19. Farver O., Wherland S., Pecht I. (1994) *J. Biol. Chem.* **269**, 22933-22936
20. Marcus R.A., Sutin N. (1985) *Biochim. Biophys. Acta* **811**, 265-322
21. Clarke M.J. et al. (eds) (1991) *Long Range Electron Transfer in Biology; Struct. Bond.* 75
22. Winkler J.R., Gray H.B. (1997) *JBIC* **2**, 399-404
23. Langen R., Colón J.L., Casimiro D.R., Karpisin T.B., Winkler J.R., Gray H.B. (1996) *JBIC* **1**, 221-225
24. Regan J.J., Risser S.M., Beratan D.N., Onuchic J.N. (1993) J. *Phys. Chem.* **97**, 13083-13088
25. Skourtis S.S., Regan J.J., Onuchic J.N. (1994) *J. Phys.Chem.* **98**, 3379-3388
26. Langen R., Chang I-J., Germanas J.P., Richards J.H., Winkler J.R., Gray H.B. (1995) *Science* **268**, 1733-1735
27. Broo A., Larsson S. (1990) *Chem. Phys.* **148**, 103-115
28. Christensen H.E.M., Conrad L.S., Mikkelsen K.V., Nielsen M.K., Ulstrup, J. (1990) *Inorg. Chem.* **29**, 2808-2816
29. Lowery M.D., Guckert J.A., Gebhard M.S., Solomon E.I. (1993) *J. Am. Chem. Soc.* **115**, 3012-3013
30. Larsson S., Broo A., Sjölin L. (1995) *J. Phys. Chem.* **99**, 4860-4865
31. Messerschmidt A., Ladenstein R., Huber R., Bolognesi M., Avigliano L., Marchesini A., Petruzzelli R., Rossi A., Finazzi-Agró A. (1992) *J. Mol. Biol.* **224**, 179-205
32. Messerschmidt A., Luecke H., Huber R. (1993) *J. Mol. Biol.* **230**, 997-1014
33. Farver O., Pecht I. (1997) Electron Transfer Reactions, in A. Messerschmidt (ed.) Multi-Copper Oxidases, World Scientific Publications, in Press
34. Hazzard J.T., Marchesini A., Curir P., Tollin G. (1994) *Biochim. Biophys. Acta* **1208**, 166-170

35. Nilsson T. (1992) *Proc. Natl. Acad. Sci. USA* **89**, 6497-6501
36. Tsukihara T., Aoyama H., Yamashita E., Tomizaki T., Yamaguchi H., Shinzawaitoh K., Nakashima R., Yaono R., Yoshikawa S. (1996) *Science* 272, 1136-1144
37. Reinhammar, B., Aasa, R., Vänngård, T., Maritano, S. and Marchesini, A. (1997) *Biochim. Biophys. Acta* **1337**, 191-197.

Chapter 2
Redox Chains: Composition and Control

THE *PARACOCCUS DENITRIFICANS* ELECTRON TRANSPORT SYSTEM: ASPECTS OF ORGANISATION, STRUCTURES AND BIOGENESIS

S.J. FERGUSON
Department of Biochemistry, University of Oxford,
South Parks Road, Oxford, OX1 3QU, U.K.

1. Introduction

The electron transport system of *Paracoccus denitrificans* (then known as *Micrococcus denitrificans*) first attracted significant interest in the 1960's when it was shown that at least some of its components required for aerobic respiration are similar to those found in mitochondria. The identification of cytochrome c_{550} as a protein with significant similarity to mitochondrial cytochrome c led to an early X-ray diffraction structure determination. However, attention was strongly focused on the *P. denitrificans* electron transfer chain following the publication in 1975 of a review article by John and Whatley [32] that developed the thesis that the aerobic electron transport chain of the organism is more similar to that of the mitochondrion than that found in any other bacterium. The corollary of this thesis was that *P. denitrificans* is related to an aerobic symbiont from which the present-day mitochondrion is derived [32]. In subsequent years a great deal has been learned about the *P. denitrificans* electron transport system, much of it confirming at the molecular level that many of its components are indeed closely related to, but often possessing a simpler subunit structure than, those of the mammalian mitochondrion. Furthermore, in the years since 1975 a great deal of insight into other aspects of the electron transport chain has been obtained. This brief review aims to collect together current aspects of this electron transport system. The treatment cannot be comprehensive and does not attempt to relate work with *P. denitrificans* to that on other bacteria. However, observations on the organism *Thiosphaera pantotropha* are included. It is presently argued that the latter should be renamed as a *P. denitrificans* strain, although there are arguments against doing this [28]. Nevertheless, *T. pantotropha* is clearly very closely related to *P. denitrificans* and so it is appropriate to refer to aspects of its electron transfer proteins here.

1.1 THE OVERALL ORGANISATION OF THE ELECTRON TRANSPORT CHAIN

As was first argued some years ago [19], it is very probable that the overall *P. denitrificans* chain should be regarded as organised as a set of large multi-polypeptide complexes which are connected by two mobile carriers, either a pool of ubiquinone or a pool of periplasmic c-type cytochromes and other low molecular weight redox proteins.

G.W. Canters and E. Vijgenboom (eds.),
Biological Electron Transfer Chains: Genetics, Composition and Mode of Operation, 77-88.
© 1998 *Kluwer Academic Publishers.*

This organisation has not been fully confirmed but is adopted for a working model, as shown in Figure 1. This figure is certainly incomplete. Recently it has been shown that *T. pantotropha* (called elsewhere *P. denitrificans* GB17) catalyses ammonia monooxygenase and hydroxylamine oxidase activities and the enzymes have been purified [38-39]. How these enzymes might fit into the scheme shown in Figure 1 is presented elsewhere [9]. Figure 1 also does not show a pathway for thiosulphate oxidation which, at least in some strains *eg P. denitrificans* GB17/*T. pantotropha*, involves a periplasmic molybdoprotein system that donates electrons to the *c*-type cytochrome level of the electron transport chain [67].

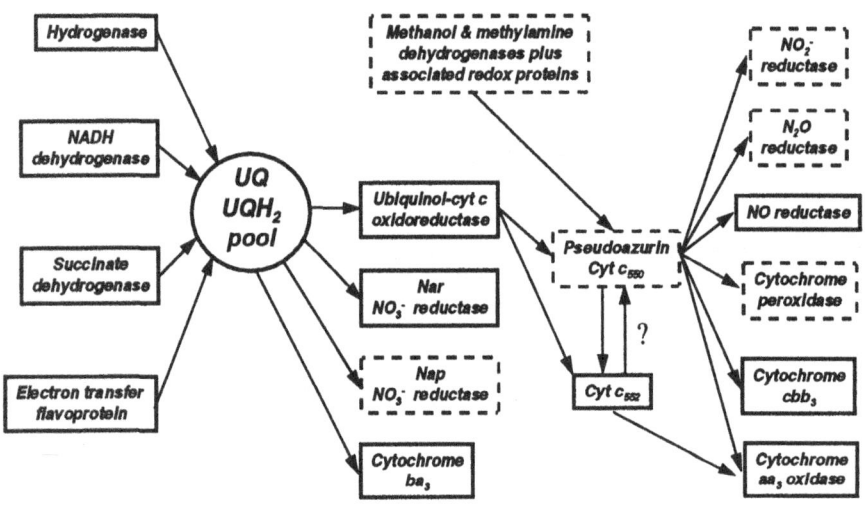

Figure 1. Schematic representation of *Paracoccus dentrificans* electron transport systems. Components in solid boxes are intregral membrane proteins, those in dotted boxes are periplasmic.

2. Pathways for Reduction of Ubiquinone

There appear to be at least four routes for feeding electrons into the electron transport chain at the level of ubiquinone, NADH dehydrogenase, succinate dehydrogenase, electron transfer flavoprotein and hydrogenase. Of these NADH dehydrogenase has attracted perhaps the most interest because of the indications that it is functionally similar to the highly complex enzyme in mitochondria. Following original indications that the *P. denitrificans* NADH dehydrogenase shared similar FeS centres with the mitochondrial enzyme, as well as rotenone and piericidin sensitivity, it was shown that the enzyme had related polypeptides as well as FMN by immunochemical methods [23-25]. However, attempts over many years to purify a highly active and rotenone sensitive enzyme have failed. A full set of genes, 14 in number, for the enzyme have been identified and sequenced [57]. Many have been expressed individually in *E. coli* [57,68].

The succinate dehydrogenase of *P. denitrificans* has been purified [47] and shown to have four subunits each of which has been sequenced [18]. The enzyme contains covalently bound flavin, iron sulphur centres and cytochrome *b*, showing overall considerable similarity to its mitochondrial counterpart.

P. denitrificans possesses a two subunit electron transfer flavoprotein which can act as an electron acceptor from glutaryl-CoA dehydrogenase as well as porcine acyl-CoA dehydrogenases [5]. This flavoprotein, which shows significant sequence similarity to its mammalian counterpart [5], presumably transfers electrons to a so far uncharacterised electron transfer flavoprotein-ubiquinone oxidoreductase which the organism needs for growth on fatty acids.

3. Pathways for Ubiquinol Oxidation

3.1 CYTOCHROME bc_1 COMPLEX

In common with many Gram negative bacteria and mitochondria, *P. denitrificans* possesses the cytochrome bc_1 complex (otherwise known as complex III or ubiquinol-cyt *c* oxidoreductase; see Fig. 1). Electron transfer through this complex can be diagnosed by inhibition with antimycin or myxothiazol. The subunit structure is relatively simple compared with the mammalian mitochondrial counterpart. As shown in Figure 1, the cytochrome bc_1 complex serves to connect, and to couple to proton translocation, oxidation of ubiquinol to the reduction of oxygen by two oxidases, cytochromes cbb_3 and aa_3, hydrogen peroxide, nitrite, nitric oxide and nitrous oxide. It is not expected that the cytochrome bc_1 complex will be the direct donor to these enzymes, but rather that other electron transfer proteins will play this role. As will be explained later, it is presently not clear which or how many such proteins can act as electron acceptors from the cytochrome bc_1 complex. *In vitro* studies appear only to have established its ability to pass electrons to horse heart cytochrome *c* [54].

3.2 CYTOCHROME ba_3 OXIDASE

It has long been recognised that *P. denitrificans* possesses a pathway for electron flow to oxygen that is insensitive to antimycin and therefore independent of the cytochrome bc_1 complex. On the basis of little experimental evidence this pathway was frequently attributed to a cytochrome *o*. However, the molecular species involved has now been identified as a four subunit cytochrome ba_3 oxidase (suggestions that it is a bb_3 oxidase presently seem unlikely) [69,70]. Nevertheless, it turns out that the cytochrome ba_3 is related to the cytochrome *bo* quinol oxidase of *E. coli* and is a proton pump [52].

3.3. Nar AND Nap NITRATE REDUCTASES

For many years it was thought that *P. denitrificans* possessed one type of respiratory nitrate reductase, a membrane enzyme that is directly reduced by ubiquinol and often known as Nar. The active site faces the cytoplasm and the movement of electrons through two *b*-type haems in the integral membrane subunit is argued to be instrumental in the generation of a protonmotive force. A model for the organisation of this haem-bearing subunit has been presented [8] and its merits discussed [10,59]. A long-standing puzzle concerning this enzyme is how nitrate gains access from the cell exterior to the active site and how the product nitrite is exported to the periplasm, the site of its respiratory reductase [46].

During the 1980's it was realised that some species of bacteria possess a second type of respiratory nitrate reductase, known as Nap, that is located in the periplasm. This has now been identified in a 'traditional' strain of *P. denitrificans* [53] but has been extensively characterised at the molecular level from *T. pantotropha* [7]. The enzyme purifies as a two subunit species with the NapA protein carrying both the catalytic site, where molybdopterin is found, and an FeS centre. The second subunit known as NapB contains two *c*-type haem centres. A tetrahaem *c*-type cytochrome known as NapC is believed to be involved in passing electrons from the ubiquinol pool to the NapB subunit [7,9]. The periplasmic location of Nap nitrate reductase strongly suggests that the electron flow to it from ubiquinol is not linked to generation of a protonmotive force. What is the function of this periplasmic nitrate reductase? Since the expression of the enzyme is higher when cells are grown on a relatively reduced carbon source such as butyrate, one role would be to provide a non-proton translocating step to permit the loss of excess reducing equivalents from the cell [50]. It is notable that the periplasmic nitrate reductase is expressed under aerobic conditions whereas the membrane-bound enzyme is expressed only under anaerobic conditions.

A puzzling feature of the Nap system is the presence in the NapA subunit of the FeS centre with redox potential -160 mV [9]. This is too low a value to fall logically on the electron transfer pathway from a *c*-type haem centre of NapB, with redox potential +80 mV, to the active site of the enzyme. However, a thermodynamically uphill electron transfer from NapB to the FeS centre, followed by electron transfer to the active site of NapA may well provide a much faster rate of electron transfer than would be obtained by the electron moving directly over a much longer distance from the haem to the active site. The periplasmic nitrate reductase is related to a formate dehydrogenase for which the crystal structure shows its FeS centre to be within a reasonable distance for rapid electron transfer from the Mo atom of the active site [11]. An FeS centre with much the same redox potential may function in both formate dehydrogenase, where its redox potential may be in line with thermodynamic direction of electron transport out of the enzyme, and Nap because its presence in the latter will give sufficient catalysis of the rate of electron transport from NapB to the active site to satisfy the rates of chemical steps associated with catalysis of nitrate reduction. Thus whilst evolution might have provided an FeS centre of higher redox potential in Nap, and thus accelerated electron transport through the enzyme to an even greater extent, other kinetic restraints within the enzyme have provided no pressure for this to happen.

4. Pathways for the Oxidation of the Cytochrome bc_1 Complex.

Electrons passing through the cytochrome bc_1 complex are destined for either one of two oxidases, a peroxidase or three of the reductase enzymes of denitrification (Fig. 1). As was mentioned above and will be discussed further below, the exact acceptors of electrons from the cytochrome bc_1 complex have not been fully established. In some circumstances a membrane bound cytochrome c_{552} (coded for by the $cycM$ gene) appears to be the electron acceptor. Thus antibodies against this protein inhibit electron transport from NADH to oxygen in presumably leaky membrane vesicles and this cytochrome can copurify in a super-complex with cytochrome bc_1 and cytochrome aa_3 [58]. Whether it acts as an electron donor to both cytochrome aa_3 and cytochrome cbb_3, as well as perhaps to nitric oxide reductase and other enzymes is presently unclear. These two oxidases are both believed to be proton pumps, although there is some uncertainty about this for cytochrome cbb_3 [15,17].

The cytochrome c peroxidase is a periplasmic dihaem c-type cytochrome of molecular weight approximately 42,000 [27].

Nitrite and nitrous oxide reductases are both periplasmic proteins. The former is of the cytochrome cd_1 type whilst the latter contains two types of copper of centre, one of which, Cu_A, it has in common with cytochrome aa_3 as the accepting centre from electron donor proteins. Nitric oxide reductase is related to certain members of the cytochrome oxidase family, especially the cytochrome cbb_3 and it contains b and c type haem centres [12,21,29]. Despite the similarity to the oxidases no copper is found in the enzyme, but it appears to contain Fe which is associated with neither haem nor iron sulphur centres [29]. It seems highly likely that this iron is part of the active site, possibly forming a binuclear centre with the b-type haem.

5. Oxidation of Methanol and Methylamine

Methanol dehydrogenase is a periplasmic enzyme that contains PQQ as the redox active moiety at the active site. From both *in vitro* experiments [33] and disruption of its gene [62], which is located adjacent to other genes for methanol oxidation, cytochrome c_{551i} (sometimes known also as cytochrome c_L) appears to be the immediate electron acceptor from methanol dehydrogenase. From this cytochrome electrons would normally be expected to pass on to the cytochrome aa_3 or cbb_3 oxidases. There is some evidence that cytochrome c_{550} may play a role in this onward electron transfer to the oxidases, but it cannot be an obligatory carrier for reasons discussed below.

The first step in the oxidation of methylamine is catalysed by the periplasmic methylamine dehydrogenase which has the unusual TTQ cofactor at its active site. TTQ is made formed by cross-linking two tryptophan side chains and oxidising the ring of one of them to a quinone form. From methylamine dehydrogenase the electrons are passed to a type 1 copper protein, amicyanin. This role of the protein is supported by both the failure of a mutant carrying a knock out in the amicyanin gene to grow on methylamine [61] and the formation of a binary crystal of amicyanin and methylamine dehydrogenase [13]. From amicyanin the electrons are proposed to pass to cytochrome c_{551i}. This is based on kinetic studies, the induction of this cytochrome during growth on methylamine, and the formation of a ternary complex of methylamine dehydrogenase, amicyanin and cytochrome c_{551i} [14]. However, this pathway would mean that cytochrome c_{551i} is able to act both on the pathways of methanol and methylamine oxidation, and it cannot be an obligatory carrier in the latter case because disrupting the gene does not, in contrast to the consequence for methanol oxidation, significantly compromise methylamine oxidation *in vivo* [63]. It is also perhaps relevant that in the ternary crystal the distance from the copper of the amicyanin to the haem Fe of cytochrome c_{551i} is approximately 25 Å, which is generally regarded as being a little long to permit biologically relevant rates of electron transfer. Whatever the role of cytochrome c_{551i}, the question of how electrons eventually reach a cytochrome oxidase also arises. In common with methanol oxidation, cytochrome c_{550} may play a non-obligatory role in methylamine oxidation and it has been proposed that amicyanin can reduce either cytochrome c_{550}, cytochrome c_{552} or the cytochrome bc_1 complex [16] (and see also below). It is further proposed that whereas cytochrome c_{550} can reduce either cytochrome aa_3 or cytochrome cbb_3, cytochrome c_{552} is an electron donor only to the former [16].

6. Structures

A number of high resolution of crystal structures have been obtained for redox proteins from *P. denitrificans* and *T. pantotropha*. These are cytochrome cd_1 (nitrite reductase) [22], methylamine dehydrogenase [13], both of which contain a β-propeller structure, cytochrome c_{550}, pseudoazurin [66], amicyanin [13], cytochrome c_{551i} [14], and the piece de resistance, cytochrome aa_3 oxidase [31], one of the still small number of integral membrane proteins to have had its structure solved.

7. Enigmatic Periplasmic Electron Transfer Proteins

There are at least three periplasmic electron transfer proteins for which no function can definitely be ascribed at present. The first of these is cytochrome c_{550}. Originally it was expected that this would serve to connect the cytochrome bc_1 complex with cytochrome aa_3 oxidase as well as, in all probability, some of the periplasmic dehydrogenases and reductases. The role in carrying electrons from cytochrome bc_1 to cytochrome aa_3 was simply expected on the basis that cytochrome c_{550} has significant similarity to mitochondrial cytochrome c and that therefore a comparable role to that firmly established for the latter might be expected. However, several lines of evidence [54] including, as mentioned above, the formation of a supercomplex of cytochrome bc_1 and cytochrome aa_3 plus the membrane-anchored c-type cytochrome known as cytochrome c_{552} (*cycM*) have suggested that the latter rather than cytochrome c_{550} is the intermediary between cytochrome bc_1 and cytochrome aa_3. The expectation that cytochrome c_{550} is the electron carrier [34] between the cytochrome bc_1 complex and the various reductases (Fig. 1) of denitrification that accept electrons at this level of the respiratory chain was disturbed by the finding that knocking out the gene for cytochrome c_{550} had no major effect on denitrification [60]. There was a similar lack of effect on utilisation of methylamine or methanol [60]. Oxidation of the latter two compounds in the periplasm was also expected to involve cytochrome c_{550}. However, the observation of a close to null phenotype as a consequence of disrupting the gene for cytochrome c_{550} does not automatically mean that this redox protein plays no role in denitrification or the oxidation of C1 compounds. It may be that its absence can be compensated by another protein. Indeed, the electron transport to certain reductases of the denitrification pathway appeared sensitive to the presence of a copper-chelator, presumably because the chelator removes the copper from pseudoazurin, a cupredoxin for which a firm role has never been proposed although *in vitro* it can donate to nitrite and nitrous oxide reductases. Based on this finding Moir and Ferguson [36] suggested that pseudoazurin can substitute for cytochrome c_{550}. This proposal needs to be confirmed or contradicted by construction of a double mutant deleted in both pseudoazurin and cytochrome c_{550}. Even the successful construction of such a mutant will not necessarily be definitive because a third redox protein may be able to participate in these electron transfer reactions. An interesting aspect to the proposal that cytochrome c_{550} and pseudoazurin are interchangeable is that of how two such structurally different proteins can interact with multiple partners [66]. The concept of pseudospecifity has been advocated [66]. Recently evidence in favour of a physiological role for cytochrome c_{550} in denitrification and oxidation of methylamine or methanol has come from studies of the growth conditions that maximise expression of this protein in *P. denitrificans*. It was found that growth under denitrifying conditions and using C1 compounds as the sources of electrons substantially elevates the level of expression over that seen for aerobic growth on multicarbon compounds [55]. The molecular basis for this type of gene expression remains enigmatic.

P. denitrificans contains cytochrome *c'* [26], a protein whose function has not been firmly identified in any bacterium in which it occurs. A postulated role in *P. denitrificans* is as a sensor for nitric oxide which might in principle be linked to a chemotaxis system [20].

The role of cytochrome c_{553i} also remains to be determined. Disruption of its gene does not prevent growth on methanol or methylamine and it is not a substitute for cytochrome c_{550} because strains deleted for both cytochromes still grow on both methanol and methylamine [63]. A triple mutant in cytochromes c_{550}, c_{551i} and c_{553i} fails to grow, as expected on the basis of the requirement (see above) for cytochrome c_{551i} in methanol oxidation, but again growth on methylamine is still possible [63]. Whilst these observations do not necessarily rule out a non-obligatory role for cytochrome c_{553i} in methanol and methylamine oxidation they do tend to point towards another role for this cytochrome. The finding that the gene for cytochrome c_{553i} (*cycB*) is located adjacent to a reading frame that probably encodes a periplasmic PQQ-dehydrogenase of unknown specifity suggests that cytochrome c_{553i} may play a role as an electron acceptor from this putative enzyme [48].

A feature that distinguishes *T. pantotropha* from *P. denitrificans* is the failure of the former to grow on methanol. It was shown by Moir and Ferguson [35] that *T. pantotropha* could adapt to growth on methanol in parallel to the appearance of a 26 kDa *c*-type cytochrome. Subsequently it has been argued that this cytochrome is an acceptor from an ethanol dehydrogenase, the activity of which, rather than of methanol dehydrogenase, accounted for the appearance of a spontaneous mutant able to grow on methanol [49].

Finally, Long and Anthony [33] reported that *P. denitrificans* can also synthesise a number of other *c*-type cytochromes which have not been fully characterised.

8. Competition for Electrons Between Electron Transfer Chains

The branching of the *P. denitrificans* electron transport system raises the question of how electrons are distributed between the available pathways. The simplest mechanism would be competition for the pools of ubiquinol or reduced *c*-type cytochrome. Whether or not this is the mechanism remains to be seen, but in one case, competition between oxygen and the denitrifying electron transport chain in whole cells, it has been argued that this cannot be the whole story [1-3]. Evidence that access of nitrate across the cytoplasmic membrane is controlled has been presented [1-3]. However, even the periplasmic nitrate reductase can be inhibited by the presence of other electron acceptors. For example, hydrogen peroxide appears an effective electron acceptor [51]. In this case access of nitrate cannot be controlled and there presumably must be competition for electrons.

9. Biogenesis of Electron Transport Chain Components

The electron transport proteins of *P. denitrificans* clearly require to be assembled. The extent to which such assembly is spontaneous is not known although the finding that pseudoazurin is readily expressed in *E. coli* [66], an organism not known to synthesise this class of type 1 copper protein suggests that the copper can be inserted spontaneously. It is, however, clear that the *c*-type cytochromes require a set of many gene products for their assembly [43-45]. The same set appears to function for all the great range of *c*-type cytochromes because mutants in the assembly of this class of proteins are always pleitotropic for all the cytochromes [40-41,43-45]. A number of lines of evidence point to the polypeptide of a *c*-type cytochrome and the haem being translocated separately to the periplasm [40-41,43-45]. There is a possibility that the two thiols, destined eventually to become attached to the haem first form a disulphide bridge and are then reductively linked to the haem. Such a pathway would be consistent with the reductase role recently advocated for one of the *P. denitrificans* gene products, CcmG, that is required for *c*-type cytochrome biogenesis and is a periplasmic disulphide oxidoreductase [45].

The TTQ moiety in the periplasmic methylamine dehydrogenase requires post-translational modification of the polypeptide on which it is found. One or more of these steps requires *c*-type cytochromes because a mutant of *P. denitrificans* that is pleiotropically deficient in these molecules synthesises the polypeptides of an inactive form of the enzyme [42]. An interesting aspect of methylamine dehydrogenase is that the polypeptide bearing the TTQ group is synthesised initially carrying an unusually long target sequence with similarities to the signal sequence for the nitrous oxide reductase [42,30]. The correlation between the unusual sequence and the presence of redox centres suggests a functional significance that is discussed in a detailed hypothesis by Berks [6]. Another important aspect of biogenesis is that of control. Not all the enzymes shown in Figure 1 are synthesised together [56,64]. Amongst important recent studies are those establishing the roles of the Nnr and Fnr proteins in controlling the expression of the nitrate, nitrite and nitric oxide reductases [65], the stability of the mRNA for the enzymes of denitrification [4] and work showing that respiration on nitrous oxide results in low levels of expression of both nitrite reductase and pseudoazurin [37].

10. Acknowledgements

The work described here that comes from the author's own laboratory owes everything to the efforts of many able co-workers whose names are to be found in the references and has been supported by the BBSRC and SERC. I particularly thank Simon Baker and Alrik Koppenhöfer for their help with this manuscript. The coverage in this article is inevitably selective and the author apologies for omission of papers that have had to be omitted.

11. References

1. Alefounder, P.R. and Ferguson, S.J. (1980) The location of dissimilatory nitrite reductase and the control of dissimilatory nitrate reductase in *Paracoccus denitrificans, Biochem. J.* **192**, 231-240.
2. Alefounder, P.R., McCarthy, J.E.G. and Ferguson, S.J. (1981) The basis of the control of nitrate reduction by oxygen in *Paracoccus denitrificans, FEMS Microbiol Lett.* **12**, 321-326.
3. Alefounder, P.R., Greenfield, A.J., McCarthy, J.E.G. and Ferguson, S.J. (1983) Selection and organisation of denitrifying electron transfer pathways in *Paracoccus denitrificans, Biochim. Biophys. Acta* **724**, 20-39.
4. Bauman, B., Snozzi, M., Zehnder, A.J.B. and van der Meer, J.R. (1996) Dynamics of denitrification activity of *Paracoccus denitrificans* in continuous culture during aerobic-anaerobic changes, *J. Bacteriol.* **178**, 4367-4374.
5. Bedzyk, L.A., Escudero, K.W., Gill, R.E., Griffin, K.J. and Frerman, F.E. (1993) Cloning, sequencing and expression of the genes encoding subunits of *Paracoccus denitrificans* electron transfer flavoprotein, *J. Biol. Chem.* **268**, 20211-20217.
6. Berks, B.C. (1996) A common export pathway for proteins binding complex redox cofactors? *Mol. Microbiol.* **22**, 393-404.
7. Berks, B.C., Richardson, D.J, Reilly, A., Willis, A.C. and Ferguson, S.J. (1995) The napABC gene cluster encoding the periplasmic nitrate reductase system of *Thiosphaera pantotropha, Biochem. J.* **309**, 983-992.
8. Berks, B.C., Page, M.D., Richardson, D.J., Reilly, A., Cavill, A., Outen, F. and Ferguson, S.J. (1995) Sequence analysis of subunits of the membrane-bound nitrate reductase from a denitrifying bacterium: the integral membrane subunit provides a prototype for the electron carrying arm of a redox loop, *Mol. Microbiol.* **15**, 319-331.
9. Berks, B.C., Ferguson, S.J, Moir, J.W.B., and Richardson, D.J. (1995) Enzymes and associated electron transport systems that catalyse their respiratory reduction of nitrogen oxides and oxyanions, *Biochim. Biophys. Acta* **1232**, 97-173.
10. Berks, B.C., Richardson, D.J., Page, M.D. and Ferguson, S.J. (1996) An alternative model for haem ligation in nitrate reductase and analogous respiratory cytochrome *b* complexes (response to the Microcorrespondence by van der Oost *et al.*), *Mol. Microbiol.* **22**, 195-196.
11. Boyington, J.C., Gladyshev, V.N., Khangulov, S.V., Stadtman, T.C. and Sun, P.D. (1997) Crystal structure of formate dehydrogenase H: Catalysis involving Mo, molybdopterin, selenocysteine, and an Fe_4S_4 cluster, *Science* **275**, 1305-1308.
12. Carr, G.J. and Ferguson, S.J. (1990) The nitric oxide reductase of *Paracoccus denitrificans, Biochem. J.* **269**, 423-429.

86

13. Chen, L., Durley, R., Poliks, B.J., Hamada, K., Chen, Z., Mathews, F.S., Davidson, V.L., Satow, Y., Huizinga, E., Vellieux, F.M.D. and Hol, W.G.J. (1992) Crystal structure of an electron-transfer complex between methylamine dehydrogenase and amicyanin, *Biochemistry* **31**, 4959-4964.

14. Chen, L., Durley, R.C.E., Mathews, F.S., and Davidson, V.L. (1994) Structure of an electron-transfer complex; methylamine dehydrogenase, amicyanin and cytochrome c_{551}, *Science* **264**, 86-90.

15. De Gier, J.W.L., Lubben, M., Reijnders, W.N.M., Westerhoff, H.V., Tipker, C.A., Slotboom, D.J., and van Spanning, R.J.M. (1994) The terminal oxidases of *Paracoccus denitrificans*, *Mol. Microbiol.* **13**, 183-186.

16. De Gier, J.W.L., van der Oost, J., Harms, N., Stouthamer, A.H. and van Spanning, R.J.M. (1995) The oxidation of methylamine in *Paracoccus denitrificans*, *Eur. J. Biochem.* **229**, 148-154.

17. De Gier, J.W.L., Schepper, M., Reijnders, W.N.M., Westerhoff, H.V., Tipker, C.A., Slotboom, D.J., van Spanning, R.J.M. and van der Oost, J. (1996) Structural and functional analysis of aa_3-type and cbb_3-type cytochrome *c* oxidases of *Paracoccus denitrificans* reveals significant differences in proton-pump design, *Mol. Microbiol.* **20**, 1247-1260.

18. Dickins, M.A., Dhawan, T., Gunsalus, R.P., Schroeder, I., and Cecchini, G. (1996) Cloning, sequencing and expression of the succinate-ubiquinone oxidoreductase (SdhCDAB) operon from *Paracoccus denitrificans*. EMBL database entry PDU31902.

19. Ferguson, S.J. (1982) Aspects of the control and organization of bacterial electon transport, *Biochem. Soc. Trans.* **10**, 198-200.

20. Ferguson, S.J. (1991) The functions and synthesis of bacterial *c*-type cytochromes with particular reference to *Paracoccus denitrificans* and *Rhodobacter capsulatus*. *Biochim. Biophys. Acta* **1058**, 17-20.

21. Fujiwara, T. and Fukumori, Y. (1996) Cytochrome *cb*-type nitric oxide reductase with cytochrome *c* oxidase activity from *Paracoccus denitrificans* ATCC 35512, *J. Bacteriol.* **178**, 1866-1871.

22. Fülöp, V., Moir, J.W.B., Ferguson, S.J. and Hajdu, J. (1995) The anatomy of a bifunctional enzyme: Structural basis for reduction of oxygen to water and synthesis of nitric oxide by cytochrome cd_1, *Cell* **81**, 369-377.

23. George, C.L. and Ferguson, S.J. (1987) Immunochemical probing of the structure and cofactor of NADH dehydrogenase from *Paracoccus denitrificans*, *Biochem J.* **244**, 661-668.

24. George, C.L., Ferguson, S.J., Cleeter, M.W.J. and Ragan, C.I. (1986) Structural relationships between NADH dehydrogenases of *Paracoccus denitrificans* and bovine heart mitochondria as revealed by immunological cross-reactivities, *FEBS Lett.* **198**, 135-139.

25. George, C.L. and Ferguson, S.J. (1984) Immunochemical identification of a two-subunit NADH-ubiquinone oxidoreductase from *Paracoccus denitrificans*, *Eur. J. Biochem.* **143**, 567-573.

26. Gilmour, R., Goodhew, C.F. and Pettigrew, G.W. (1991) Cytochrome *c'* of *Paracoccus denitrificans*, *Biochim. Biophys. Acta* 1059, 233-238.

27. Goodhew, C.F., Wilson, I.B.H., Hunter, D.J.B. and Pettigrew, G.W. (1990) The cellular location and specificity of bacterial cytochrome *c* peroxidases, *Biochem. J.* **271**, 707-712.

28. Goodhew, C.F., Pettigrew, G.W., Devreese, B., van Beeumen, J., Van Spanning R.J.M., Baker, S.C., Saunders, N., Ferguson, S.J. and Thompson, I.P. (1996) The cytochromes c_{550} of *Paracoccus denitrificans* and *Thiosphaera pantotropha*: a need for the re-evaluation of the history of *Paracoccus* cultures, *FEMS Microbiol. Lett.* **137**, 95-101.

29. Grisch, P. and de Vries, S. (1997) Purification and initial kinetic and spectroscopic characterization of NO reductase from *Paracoccus denitrificans*, *Biochim. Biophys. Acta* **1318**, 202-216.

30. Hoeren, F.U., Berks, B.C., Ferguson, S.J. and McCarthy, J.E.G. (1993) Sequence and expression of the gene encoding the respiratory nitrous-oxide reductase from *Paracoccus denitrificans*. New and conserved structural and regulatory motifs, *Eur.J.Biochem.* **218**, 49-57.

31. Iwata, S., Ostermeier, C., Ludwig, B. and Michel, H. (1995) Structure at 2.8 Å resolution of cytochrome *c* oxidase from *Paracoccus denitrificans*, *Nature* **376**, 660-669.

32. John. P. and Whatley, F.R. (1975) *Paracoccus denitrificans* and the evolutionary origin of the mitochondrion. *Nature* **254**, 495-498.

33. Long, A. and Anthony, C. (1991) Characterization of the periplasmic cytochromes *c* of *Paracoccus denitrificans*: identification of the electron acceptor for methanol dehydrogenase, and description of a novel cytochrome *c* heterodimer, *J. Gen. Microbiol.* **137**, 415-425.

34. Machová, I. and Kucera, I. (1991) Evidence for the role of soluble cytochrome *c* in the dissimilatory reduction of nitrite and nitrous oxide by cells of *Paracoccus denitrificans*, *Biochim. Biophys. Acta.* 1058, 256-260.

35. Moir, J.W.B. and Ferguson, S.J. (1993) Spontaneous mutation of *Thiosphaera pantotropha* enabling growth on methanol correlates with synthesis of a 26 kDa *c*-type cytochrome, *FEMS Microbiol. Lett.* 113, 321-326.

36. Moir, J.W.B. and Ferguson, S.J. (1994) Properties of a *Paracoccus denitrificans* mutant deleted in cytochrome c_{550} indicate that a copper protein can substitute for this cytochrome in electron transport to nitrite, nitric oxide and nitrous oxide, *Microbiology* **140**, 389-397.

37. Moir, J.W.B., Richardson, D.J. and Ferguson, S.J. (1995) The expression of redox proteins of denitrification in *Thiosphaera pantotropha* grown with oxygen, nitrate and nitrous oxide as electron acceptors, *Arch. Microbiol.* **164**, 43-49.

38. Moir, J.W.B., Wehrfritz, J-M., Spiro, S. and Richardson, D.J. (1996) The biochemical characterization of a novel non-hame iron hydroxylamine oxidase from *Paracoccus denitrificans* GB17, *Biochem. J.* 319, 823-827.

39. Moir, J.W.B., Crossman, L.C., Spiro, S. and Richardson, D.J. (1996) The purification of ammonia monooxygenase from *Paracoccus denitrificans*, *FEBS Lett.* 387, 71-74.

40. Page, M.D. and Ferguson, S.J. (1989) A bacterial *c*-type cytochrome can be translocated to the periplasm as an apo form; the biosynthesis of cytochrome cd_1 (nitrite reductase) from *Paracoccus denitrificans*, *Mol. Microbiol.* **3**, 653-661.

41. Page, M.D. and Ferguson, S.J. (1990) Apo forms of cytochrome c_{550} and cytochrome cd_1 are translocated to the periplasm of *Paracoccus denitrificans* in the absence of haem incorporation caused by either mutation of inhibition of haem synthesis, *Mol. Microbiol.* **4**, 1181-1192.

42. Page, M.D. and Ferguson, S.J. (1993) Mutants of *Methylobacterium extorquens* and *Paracoccus denitrificans* deficient in *c*-type cytochrome biogenesis synthesise the methylamine dehydrogenase polypeptides but cannot assemble the tryptophan-tryptophylquinone group, *Eur. J. Biochem* **218**, 711-717.

43. Page, M.D. and Ferguson, S.J. (1995) Cloning and sequence analysis of the *cycH* gene from *Paracoccus denitrificans*; the *cycH* gene product is required for assembly of all *c*-type cytochromes, including cytochrome c_1, *Mol. Microbiol.* **4**, 1181-1192.

44. Page, M.D., Pearce, D.A., Norris, H.A.C. and Ferguson, S.J. (1997) The *Paracoccus denitrificans ccmA, B* and *C* genes: cloning and sequencing, and analysis of the potential of their products to form a haem or apo-*c*-type cytochrome transporter, *Microbiology* **143**, 563-576.

45. Page, M.D. and Ferguson, S.J. (1997) *Paracoccus denitrificans* CcmG is a periplasmic protein-disulphide oxidoreductase required for *c*- and aa_3-type cytochrome biogenesis; evidence for a reductase role *in vivo*, *Mol. Microbiol.* **24**, 977-990.

46. Parsonage, D., Greenfield, A.J. and Ferguson, S.J. (1985) The high affinity of *Paracoccus denitrificans* cells for nitrate as an electron acceptor. Analysis of possible mechanisms of nitrate and nitrite movement across the plasma membrane and the basis for inhibition by added nitrite of oxidase activity in permeabilised cells, *Biochim. Biophys. Acta* **807**, 81-95.

47. Pennoyer, J.D., Ohnishi, T. and Trumpower, B.L. (1988) Purification and properties of succinate-ubiquinone oxidoreductase complex from *Paracoccus denitrificans*, *Biochim. Biophys. Acta* **935**, 195-207.

48. Ras, J., Reijnders, W.N.M., van Spanning, R.J.M., Harms, N., Oltmann, L.F. and Stouthamer, A.H. (1991) Isolation sequencing and mutagenesis of the gene encoding cytochrome c_{553i} of *Paracoccus denitrificans* and characterization of the mutant strain, *J. Bacteriol.* **173**, 6971-6979.

49. Ras, J., Hazelaar, M.J., Robertson, L.A., Kuenen, J.G., van Spanning, R.J.M., Harms, N., Oltmann, L.F. and Stouthamer, A.H. (1995) Methanol oxidation in a spontaneous mutant of *Thiosphaera pantotropha* with a methanol-positive phenotype is catalysed by a dye-linked ethanol dehydrogenase, *FEMS Microbiol. Lett.* **127**, 159-164.

50. Richardson, D.J. and Ferguson, S.J. (1992) The influence of carbon substrate on the activity of the periplasmic nitrate reductase in aerobically grown *Thiosphaera pantotropha*, *Arch. Microbiol.* **157**, 535-537.

51. Richardson, D.J. and Ferguson, S.J. (1995) Competition between hydrogen peroxide and nitrate for electrons from the respiratory chains of *Thiosphaera pantotropha* and *Rhodobacter capsulatus*, *FEMS Microbiol. Lett.* **132**, 125-129.

52. Richter, O-M. H., Tao, J., Turba, A. and Ludwig, B. (1994) A cytochrome ba_3 oxidase functions as a quinol oxidase in *Paracoccus denitrificans*, *J. Biol. Chem.* **269**, 23079-23086.

53. Sears H.J., Ferguson, S.J., Richardson, D.J. and Spiro, S. (1993) The identification of a periplasmic nitrate reductase in *Paracoccus denitrificans*, *FEMS Microbiol. Lett.* **113**, 107-112.

54. Smith, L. and Davies, H.C. (1991) The reactions of the oxidase and reductases of *Paracoccus denitrificans* with cytochromes *c*, *J. Bioenergetics and Biomembranes* **23**, 303-319.

55. Stoll, R., Page, M.D., Sambongi, Y. and Ferguson, S.J. (1996) Cytochrome c_{550} expression in *Paracoccus denitrificans* strongly depends on growth condition: Identification of promoter region for *cycA* by transcription start analysis, *Microbiology* **142**, 2577-2585.

56. Stouthamer, A.H. (1992) Metabolic pathways in *Paracoccus denitrificans* and closely related bacteria in relation to phylogeny of prokaryotes, *Antonie van Leeuwenhoek* **61**, 1-33.

57. Takano, S., Yano, T. and Yagi, T. (1996) Structural studies of the proton translocating NADH ubiquinone oxidoreductase (NDH-1) of *Paracoccus denitrificans*: Identity. property and stoichiometry of the peripheral subunits, *Biochemistry* **35**, 9120-9127.

58. Turba, A., Jetzek, M. and Ludwig, B. (1995) Purification of *Paracoccus denitrificans* cytochrome c_{552} and sequence analysis of the gene, *Eur. J. Biochem.* **231**, 259-265.

59. van der Oost, J., Nederkoorn, P.H., Stouthamer, A.H., Westerhoff, H.V., van Spanning, R.J.M. (1996) An alternative model for haem ligation in nitrate reductase and analogous respiratory cytochrome *b* complexes, *Molecular Microbiology* **22**, 193-195.

60. van Spanning, R.J.M., Wansell, C., Harms, N., Oltman, L.F. and Stouthamer, A. (1990) Mutagenesis of the gene encoding cytochrome c_{550} of *Paracoccus denitrificans* and analysis of the resultant physiological effects, *J. Bacteriol.* **172**, 986-996. A correction to the amino acid sequence of cytochrome c_{550} reported in this paper can be found in *J. Bacteriol.* **172**, 3534.

61. van Spanning, R.J.M., Wansell, C.W., Reijnders, W.N.M., Oltmann, L.F. and Stouthamer, A.H. (1990) Mutagenesis of the gene encoding amicyanin of *Paracoccus denitrificans* and the resultant effect on methylamine oxidation, *FEBS Lett.* **275**, 217-220.

62. van Spanning, R.J.M., Wansell, C.W., de Boer, T., Hazelaar, M.J., Anazawa, H., Harms, N., Oltmann, L.F. and Stouthamer, A.H. (1991) Isolation and characterization of the *moxJ*, *moxG*, *moxI*, and *moxR* genes of *Paracoccus denitrificans*: inactivation of *moxJ*, *moxG*, and *moxR* and the resultant effect on methylotrophic growth, *J. Bacteriol.* **173**, 6948-6961.

63. van Spanning, R.J.M., Wansell, C.W., Reijnders, W.N.M., Harms, N., Ras, J., Oltmann, L.F. and Stouthamer, A.H. (1991) A method for introduction of unmarked mutations in the genome of *Paracoccus denitrificans*: Construction of strains with multiple mutations in the genes encoding periplasmic cytochromes c_{550}, c_{551i}, and c_{553i}. *J. Bacteriol.* **173**, 6962-6970.

64. van Spanning, R.J.M., de Boer, A.P.N., Reijnders, W.N.M., De Gier, J.W.L., Delorme, C.O., Stouthamer, A.H. Westerhoff, H.V., Harms, N. and van der Oost, J. (1995) Regulation of oxidative phosphorylation; the flexible respiratory network of *Paracoccus denitrificans*, *J. Bioenergetics and Biomembranes* **27**, 499-512.

65. van Spanning, R.J.M., de Boer, A.P.N., Reijnders, W.N.M., Westerhoff, H.V., Stouthamer, A.H. and van der Oost, J. (1997) FnrP and NNR of *Paracoccus denitrificans* are both members of the FNR family of transcriptional activators but have distinct roles in respiratory adaptation in response to oxygen limitation, *Mol. Microbiol.* **23**, 893-907.

66. Williams, P.A., Fülöp, V., Leung, Y-C., Chan, C., Moir, J.W.B., Howlett, G., Ferguson, S.J., Radford, S.E. and Hajdu, J. (1995) Pseudospecific docking surfaces on electon transfer proteins as illustrated by pseudoazurin, cytochrome c_{550} and cytochrome cd_1 nitrite reductase, *Nature Structural Biology* **2**, 975-982.

67. Wodra, C., Kosta, S., Egert, M., Kelly, D.P. and Friedrich, C.G. (1994) Identification and sequence analysis of the *soxB* gene essential for sulfur oxidation of *Paracoccus denitrificans* GB17. *J. Bacteriol.* **176**, 6188-6191.

68. Yano, T. Sled, V.D., Ohnishi, T. and Yagi, T. (1996) Expression and characterisation of the flavoprotein subcomplex composed of 50-kDa (NQO1) and 25-kDa (NQO2) subunits of the proton-translocating NADH-quinone oxidoreductase of *Paracoccus denitrificans*, *J. Biol. Chem.* **271**, 5907-5913.

69. Zickermann, I., Anemuller, S., Richter, O.H., Tautu, O.S., Link, T.A. and Ludwig, B. (1996) Biochemical and spectroscopic properties of the four subunit quinol oxidase (cytochrome ba_3) from *Paracoccus denitrificans*, *Biochim. Biophys. Acta* **1277**, 93-102.

70. Zickermann, I., Tautu, O.S., Link, T.A., Korn, M., Ludwig, B. and Richter, O-M.H. (1997) Expression studies on the ba_3 quinol oxidase from *Paracoccus denitrificans* - a bb_3 variant is enzymatically inactive, *Eur. J. Biochem.* **246**, 618-624.

GENETICS AND REGULATION OF C1 METABOLISM IN METHYLOTROPHS

M.E. LIDSTROM, L. CHISTOSERDOVA, S. STOLYAR,
A.L. SPRINGER
University of Washington, Box 351750, Seattle WA 98195-1750, USA

1. Introduction

Methylotrophic bacteria are capable of growth on compounds containing no carbon-carbon bonds. Although both aerobic and anaerobic methylotrophs are known, this chapter will focus on two groups of aerobic methylotrophs, the serine cycle Methylobacterium strains, and the methane-utilizing bacteria.

Methylotrophs grow on one-carbon compounds by oxidizing them to CO_2 via specialized oxidative systems (Figure 1). The methanotrophs contain monooxygenases that oxidize methane to methanol [1,2]. In the gram-negative methylotrophs, methanol is oxidized to formaldehyde by methanol dehydrogenase, a periplasmic quinoprotein that contains pyrroloquinoline quinone (PQQ) as a prosthetic group [1], and methylamine is oxidized to formaldehyde by a different periplasmic quinoprotein, the tryptophan tryptophylquinone (TTQ)-containing methylamine dehydrogenase [2]. Carbon is assimilated either at the level of formaldehyde in so-called "true" methylotrophs, or at the level of CO2 in the autotrophic methylotrophs [1]. The autotrophic methylotrophs generally use the Calvin-Benson cycle for carbon assimilation, while the true methylotrophs contain either the serine cycle (α-proteobacterial methylotrophs) or the ribulose monophosphate (RuMP) cycle (β- and γ-proteobacterial methylotrophs) [2]. Formaldehyde is oxidized to CO_2 either by a linear pathway involving a formaldehyde oxidation system, or by cyclic pathways involving ribulose monophosphate cycle intermediates [2].

In the methylotrophs, the central metabolic intermediate is formaldehyde, a toxic compound. Therefore, one of the central issues for understanding methylotrophic metabolism is to determine how these organisms handle formaldehyde to ensure sufficient flux for efficient metabolism, while providing safeguards against formaldehyde accumulation. It is clear that different groups of methylotrophs have evolved different mechanisms to handle formaldehyde. For instance, some methylotrophs oxidize formaldehyde to formate via a glutathione derivative, which minimizes the pool of free formaldehyde inside the cell [1]. In addition, the gram-negative methylotrophs have localized the formaldehyde production systems, such as MeDH and MADH, in the periplasm where they are physically separated from the cytoplasmic formaldehyde-consuming systems [1]. To pursue this issue in more detail, my group has used a

89

G.W. Canters and E. Vijgenboom (eds.),
Biological Electron Transfer Chains: Genetics, Composition and Mode of Operation, 89-97.
© 1998 *Kluwer Academic Publishers.*

molecular approach to study formaldehyde production and consumption systems in the a-proteobacterial methylotroph, *Methylobacterium extorquens* AM1.

A. Methane metabolism in methanotrophic bacteria

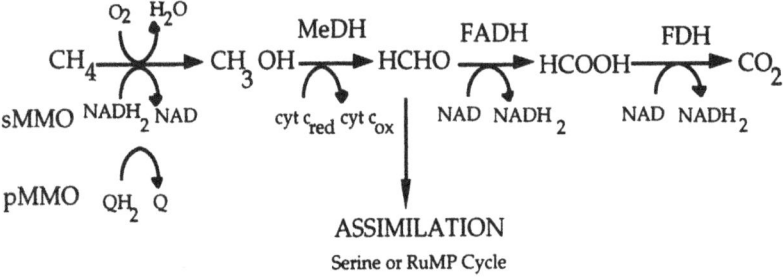

B. Methanol and methylamine metabolism in gram-negative methylotrophs.

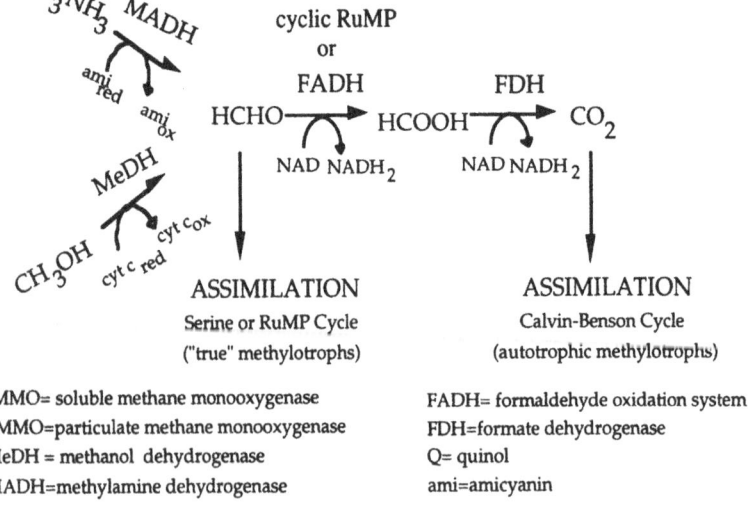

sMMO= soluble methane monooxygenase
pMMO=particulate methane monooxygenase
MeDH = methanol dehydrogenase
MADH=methylamine dehydrogenase

FADH= formaldehyde oxidation system
FDH=formate dehydrogenase
Q= quinol
ami=amicyanin

Figure 1. Methylotrophic metabolism in gram-negative methylotrophic bacteria.

2. Survey of Genes Involved in Methylotrophic Metabolism in *Methylobacterium extorquens* AM1

M. extorquens AM1 is a serine cycle methylotroph that grows on both methanol and methylamine using the enzymes shown in Figure 1B. Over the past two decades, genetic studies of this bacterium by several groups have resulted in the identification of a large number of genes involved in methylotrophic metabolism. A total of approximately 80 genes have been identified by a combination of sequencing and/or mutational analysis, and these are located in 9 different linkage groups. Fifty-three of these genes have been shown to be or are inferred to be involved in methylotrophic metabolism. Eleven of these methylotrophy genes are required for utilization of methylamine, termed *mau* genes [3], 13 are required for the serine cycle [4], 2 are required for formaldehyde oxidation, and 27 are required for methanol oxidation (*mox* genes) [5,6]. In this contribution, some details of our work on the methanol oxidation and formaldehyde oxidation systems will be presented, with emphasis on the implications of our results for understanding formaldehyde partitioning in this bacterium.

3. Genes Involved in the Methanol Oxidation (Mox) System in *M. extorquens* AM1

The genes involved in the oxidation of methanol to formaldehyde, termed *mox* genes [6], are found in five linkage groups, designated *mxa, mxb, mxc, mxd* and a group containing genes for synthesis of the prosthetic group PQQ (*pqq* genes) (Table 1) [6]. The *mxa* linkage group consists of a large 14-gene cluster (*mxaFJGIRSACKLDH1H2B*) transcribed in one direction, and a single gene (*mxaW*), adjacent to *mxaF* and transcribed in the opposite direction. This linkage group includes genes encoding the structural proteins of the methanol dehydrogenase (MeDH), the electron acceptor (cytochrome c_L), several genes involved in insertion of calcium into the enzyme, a few genes of unknown function and a transcriptional regulator [5,6,7]. The *mxb* linkage group contains two transcriptional regulators for the entire Mox system (*mxbDM*) and immediately downstream but transcribed separately, 4 *pqq* genes (*pqqABC/DE*). This linkage group also contains 30 other genes 3' to the *pqq* genes, many of which encode serine cycle enzymes [4]. The *mxc* linkage group contains 2 more Mox transcriptional regulators (*mxcQE*), and the *mxd* linkage group contains two genes required for generating active MeDH (*mxdRS*), but which have not yet been sequenced and for which the function is unknown. [5,6,7]. Two more *pqq* genes form the fifth linkage group, *pqqEF*. The 14 genes found in the large *mxa* cluster (*mxaFJGIRSACKLDEHB*) may be cotranscribed, as we have been unable to detect any promoters other than the one immediately 5' to *mxaF*. Promoters have also been identified for *pqqA, mxbM,* and *mxaW* [5,8,11].

TABLE 1. Genes involved in methanol oxidation in *M. extorquens* AM1

Gene	Function	Linkage Group	Significant Similarities*
mxaW	unknown	A	none
mxaF	MeDH large subunit	A	PQQ dehydrogenases (20-25% ID)
mxaJ	assembly, possible chaperone	A	none
mxaG	cytochrome *cL*	A	other cyts *c* (15-20% ID)
mxaI	MEDH small subunit	A	none
mxaR	unknown	A	none
mxaS	unknown	A	none
mxaA	calcium insertion in MeDH	A	none
mxaC	calcium insertion in MeDH	A	none
mxaK	calcium insertion in MeDH	A	none
mxaL	calcium insertion in MeDH	A	none
mxaD	unknown (no mutants)	A	none
mxaE	unknown (no mutants)	A	none
mxaH	unknown (no mutants)	A	none
mxaB	transcriptional regulator	A	NarL class response regulators (25-30% ID)
mxbM	transcriptional regulator	B	OmpR class response regulators (25-40% ID)
mxbD	transcriptional regulator	B	EnvZ class sensor kinases (25-40% ID)
mxcE	transcriptional regulator	C	NarL class response regulators (25-40% ID)
mxcQ	transcriptional regulator	C	NarX class sensor kinases (25-40% ID)
mxdR	unknown	D	not sequenced
mxdS	unknown	D	not sequenced

PQQ synthesis genes

Gene	Function	Linkage Group	Significant Similarities*
pqqA	putative precursor peptide	B	none
pqqB	unknown	B	none
pqqC/D	unknown	B	none
pqqE	putative cofactor syn gene	B	*M. thermoautotrophicum* MoaA (20% ID)
pqqF	putative endopeptidase	E	*Neurospora* mitochondrial peptidase MPP (22% ID)
pqqG	putative endopeptidase	E	*Mycobacterium* MPP (25% ID)

*In addition to other methanol oxidation genes [5,6,7]

We have studied the 5 regulatory genes in more detail. All of these genes are required for activity of the promoter directly upstream of *mxaF* in both *M. extorquens* AM1 and *Methylobacterium organophilum* XX [5,8,9]. It has been shown in *M. organophilum* XX that two of the regulatory genes (*mxcQE*) have similarity at the amino acid sequence level to sensor kinase/response regulator pairs of the NarX/NarL class [10], and we have confirmed that this is also true for *mxcQE* of *M. extorquens* AM1. In addition, we have sequenced the other 3 regulatory genes, and have discovered that the *mxbDM* pair show similarity to sensor kinase/response regulator pairs of the EnvZ/OmpR class [11] and that *mxaB* shows similarity to response regulators of the NarL class including MxcQ of both *M. extorquens* AM1 and *M. organophilum* XX.

The reporter gene *xylE* has been used to generate gene fusions for the promoter regions for *mxaF, mxaW, mxbD,* and *pqqA*. These have been used to assess the role of each of the regulatory genes in transcriptional control of these four genes involved in methanol oxidation [11]. The results of this work suggest that the MxcQE pair modulate expression of the MxbDM pair, which in turn are required for normal transcription from the *mxaF, mxaW,* and *pqqA* promoters, but not for the *mxbD* promoter (*mxbDM* do not appear to be autoregulated). The third regulatory system, *mxaB* is required for transcription from the *mxaF* and *pqqA* promoters, but not for the *mxaW* or *mxbD* promoters. In addition, a cytochrome of unknown function (cytochrome c_{553}) is derepressed in *mxbD, mxbM,* and *mxaB* mutants, suggesting that these genes also negatively regulate another system [5]. Presumably *mxaB* interacts with a sensor kinase, but the identity of this regulatory protein is unknown at this time. It is not yet clear why both *mxaB* and *mxbM* are required for expression of the promoters they control. Since our evidence so far suggests that *mxaB* may be transcribed from the *mxaF* promoter, it does not seem likely that *mxbM* and *mxaB* operate in a linear pathway. This 3-tiered system of transcriptional regulation probably reflects check-points for the Mox system regarding formaldehyde flux. We are currently investigating the signals that regulate each known Mox promoter. However, it seems likely that the Mox system will interact at some level with regulatory systems governing the formaldehyde consuming pathways.

4. Genes Involved in Formaldehyde Oxidation in *M. extorquens* AM1

The system involved in the oxidation of formaldehyde to formate in *M. extorquens* AM1 has not been clear. Although multiple aldehyde dehydrogenase activities can be detected in this organism, none of them seems to be involved in dissimilatory formaldehyde oxidation, and they are probably involved in aldehyde detoxification [12]. It had previously been shown that *M. extorquens* AM1 contained enzyme activities for the oxidation of methylene tetrahydrofolate (THF) to formate via tetrahydrofolate derivatives (Figure 2), and that these activities were methanol-inducible [13]. Therefore, it was suggested that formaldehyde might be oxidized to formate in *M. extorquens* AM1 by the THF-mediated pathway shown in Figure 2 [13]. Such a mechanism would minimize the amount of free formaldehyde found in the cytoplasm, since THF reacts rapidly with formaldehyde nonenzymatically to generate methylene THF. We have discovered a gene encoding methylene THF dehydrogenase (MTHFDH) in a large gene cluster that contains a number of genes encoding enzymes for the serine cycle, This gene (*mtdA*) is required for growth on C1 compounds [12]. Although we have not yet tested these mutants for methenyl THF cyclohydrolase activity, many MTHFDHs also carry out the cyclohydrolase reaction, and so this gene may be involved in mediating the first two steps shown in Figure 2. In addition, we have discovered another gene (*ffsA*) that may carry out the oxidation of the methylene moiety to the formyl level, and it is also required for growth on C1 compounds. These data suggest that formaldehyde is oxidized via folate derivatives in *M. extorquens* AM1. It is likely that other serine cycle methylotrophs contain a similar formaldehyde oxidation pathway.

5. Methane Oxidation in Methanotrophs

Methanotrophic bacteria are metabolically similar to those methylotrophs unable to grow on methane, except for the ability to utilize methane (Figure 1). These bacteria contain an enzyme, the methane monooxygenase (MMO) that oxidizes methane to methanol. Two types of MMO are known, a membrane-bound enzyme (pMMO) that is found in all known methanotrophs, and a cytoplasmic enzyme (sMMO) that is also found in a few methanotrophs as a second enzyme [2]. The sMMO is an iron enzyme, while the pMMO is a copper enzyme, and in those methanotrophs that contain both MMOs, the signal for differential expression is the amount of copper in the growth medium. At low copper levels, sMMO is expressed, at higher copper levels, pMMO is expressed [14]. Although a great deal is known about the sMMO, until recently the pMMO was less well understood. However, advances have been made in the past year in understanding the biochemistry of the pMMO, and two different groups have reported its purification from *Methylococcus capsulatus* Bath [15,16]. In both cases the pMMO was found to consist of 3 subunits, of approximately 45, 27 and 25 kDa, which have been designated PmoB, PmoA and PmoC, respectively. In both cases, the active enzyme contained approximately 15 coppers per protein [15,16].

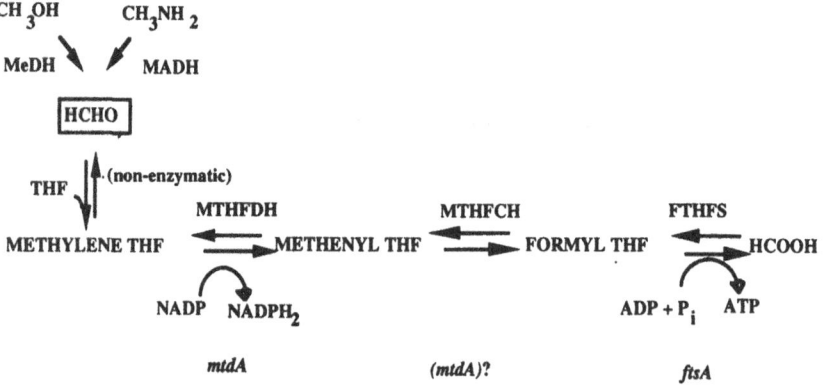

Figure 2. The oxidative THF pathway. THF, tetrahydrofolate; MTHFDH, methylene THF dehydrogenase; MTHFCH, methenyl THF cyclohydrolase; FTHFS, formyl THF synthase; FDH, formate dehydrogenase; mtdA, gene encoding MTHFDH; ffsA, gene possibly encoding FTHFS or a similar enzyme.

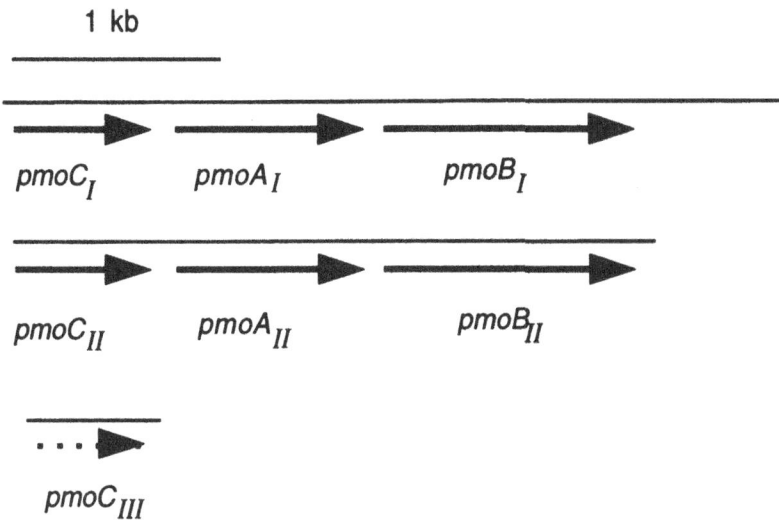

Figure 3. The *pmo* genes in *Methylococcus capsulatus* Bath.

We have previously reported that methanotrophs contain 2 copies of the genes encoding the 27 and 45 kDa subunits, respectively [17]. We have now completed the cloning and sequencing of both gene clusters, and have shown that the genes for all three subunits are present in each gene cluster in the order, *pmoCAB* (Figure 3) [18]. In addition, we have discovered a third copy of *pmoC*, which has not yet been sequenced. This third *pmoC* is not linked to other copies of *pmoAB*. In each case, the copies are almost identical, differing in only 13 positions in 3,183 bp of the *pmoCAB* region. At the amino acid level, each translated gene product contained only one differing residue in each copy. The sequences diverged 5' to *pmoC* and 3' to *pmoB*, but for the two gene clusters, the intergenic regions were almost identical. Hydropathy plots of the three translated subunits suggest that PmoA and PmoC are both integral membrane polypeptides, with 6 transmembrane segments each, while PmoB is predicted to contain two short transmembrane segments with two large periplasmic soluble portions [17,18]. The 27 kDa subunit (PmoA) is known to bind the irreversible inhibitor, acetylene, and therefore it is likely that this subunit contains the methane binding site. However, the location of the coppers and the oxygen activation site(s) are not yet known.

The only entries in the database showing similarity to any of the *pmo* gene products are the corresponding subunits of the ammonia monooxygenase (AMO) [18,19]. A comparison of *pmoA/amoA* sequences suggested that these enzymes are evolutionarily related, presumably evolving from a common ancestor [19]. It is not yet known whether any other homologs exist in prokaryotes, but so far the genome sequences available do not show any genes with substantial similarities at the amino acid level to either *pmo* or *amo* genes.

We have generated insertion mutations in each gene except *pmoC3* using a kanamycin-resistance cassette, and in each case have replaced the wild-type gene in the chromosome with the inserted gene by allelic exchange techniques. All of these mutants grew normally on methane, showing that each of the gene clusters was functional, and that each alone was sufficient to provide normal growth on methane. It has not yet been possible to obtain double mutants that are pMMO-negative, so we do not yet know whether pMMO is required for growth on methane, even under conditions in which sMMO is expressed. We are currently studying the expression of each set of *pmo* genes under a variety of growth conditions, to determine whether they are differentially expressed.

6. Summary

Methylotrophs have an unusual mode of metabolism, in that it involves production and consumption of a toxic compound, formaldehyde, as a central intermediate. These bacteria have developed a number of strategies for handling formaldehyde. In the gram-negative methylotrophs, part of this strategy has involved localizing the formaldehyde-producing systems (such as MeDH and MADH) in the periplasm, separating them from the cytoplasmic formaldehyde-consuming systems. In *M. extorquens* AM1, our results suggest that free formaldehyde may not exist to an appreciable extent in the cytoplasm. It appears instead that methylene THF is the cytoplasmic intermediate for both assimilation by the serine cycle (which has been known for many years) and for dissimilatory metabolism to formate. In the methanotrophs, methanol is produced from methane either in the cytoplasm, by the sMMO, or by the membrane-bound pMMO. In the latter case it is possible that the predicted periplasmic portions of PmoB may serve to provide methanol to MeDH. However, in the case of the sMMO, growth on methane is not impaired compared to cells expressing pMMO, and therefore directed transfer of methanol to the periplasm is probably not necessary for optimal growth on methane.

7. References

1. Anthony, C. (1982) *The Biochemistry of Methylotrophs*, Academic Press, London.
2. Lidstrom, M.E. (1991) The methylotrophic bacteria, in A. Balows, H. G. Truper, M. Dworkin, W. Harder, and K-H. Schleifer (eds.), *The Prokaryotes Second Edition*. Springer-Verlag, New York, pp. 431-445.
3. Chistoserdov, A.Y., Chistoserdova, L.V., McIntire, W.S. and Lidstrom, M.E. (1994) Genetic organization of the *mau* gene cluster in *Methylobacterium extorquens* AM1: Complete nucleotide sequence and generation and characterization of *mau* mutants, *J. Bacteriol.* **176**, 4052-4065.
4. Chistoserdova, L. (1996) Metabolism of formaldehyde in *Methylobacterium extorquens* AM1, in M.E. Lidstrom and F.R. Tabita, (eds.), *Microbial Growth on C1 Compounds*, Kluwer Academic Publishers, Dordrecht, pp. 16-24.

5. Springer, A.L., Chou, H.-H., Fan, W.-H., Lee, E. and Lidstrom, M.E. (1995) Methanol oxidation mutants in *Methylobacterium extorquens* AM1: identification of new genetic complementation groups, *Microbiol.* **141**, 2985-2993.

6. Lidstrom, M.E., Anthony, C., Biville, F., Gasser, F., Goodwin,. P., Hanson, R.S. and Harms, N. (1994) New unified nomenclature for genes involved in the oxidation of methanol in Gram-negative bacteria, *FEMS Microbiol. Lett.* **117**, 103-106.

7. Amaratunga, K., Goodwin, P.M., O'Connor, C.D. and Anthony, C. (1997). The methanol oxidation genes *mxaFJGIR(S)ACKLD* in *Methylobacterium extorquens*, *FEMS Microbiol. Lett.* **146**, 31-38.

8. Morris, C.J. and Lidstrom, M.E. (1992) Cloning of a methanol-inducible *moxF* promoter and its analysis in *moxB* mutants of *Methylobacterium extorquens* AM1, *J. Bacteriol.* **174**, 4444-4449.

9. Xu, H.H., Viebahn, M. and Hanson, R.S. (1993) Identification of methanol-regulated promoter sequences from the facultative methylotrophic bacterium *Methylobacterium organophilum* XX, *J. Gen. Microbiol.* **139**, 743-752.

10. Xu, H.H., Janka, J.J., Viebahn, M. and Hanson, R.S. (1995) Nucleotide sequence of the *mxcQ* and *mxcE* genes, required for methanol dehydrogenase synthesis in *Methylobacterium organophilum* XX: a two-component regulatory system, *Microbiology* **141**, 2543-2551.

11. Springer, A.L., Morris, C.J., and Lidstrom, M.E. (1997) Molecular analysis of *mxbD* and *mxbM*, a putative sensor-regulator pair required for oxidation of methanol in *Methylobacterium extorquens* AM1, *Microbiology* **143**, 1737-1744.

12. Chistoserdova, L.V. and Lidstrom, M.E. (1994) Genetics of the serine cycle in *Methylobacterium extorquens* AM1: Identification of *sgaA* and *mtdA* and sequences of *sgaA*, *hprA* and *mtdA*, *J. Bacteriol.* **176**, 1957-1968.

13. Marison, I.W., and Attwood, M.M. (1982) A possible mechanism for the oxidation of formaldehyde to formate, *J. Gen. Microbiol.* **128**, 1441-1446.

14. Stanley, S.H., Prior, S.D. , Leak, D.J. and Dalton, H. (1983) Copper stress underlies the fundamental change in intracellular location of methane monooxygenase in methane-oxidizing organisms: studies in batch and continuous culture, *Biotechnol. Lett.* **5**, 487-492.

15. Zahn, J.A. and DiSpirito, A.A. (1996) Membrane associated methane monooxygenase from *Methylococcus capsulatus* (Bath), *J. Bacteriol.* **178**, 1018-1029.

16. Nguyen, H.-H. A.T., Elliot, S.J., Kent, B.H., Nakagawa, H., Costello, A.M., Peeples, T.L., Wilkinson, B., Marimoto, H., Williams, P.G., Floss, H.G., Lidstrom, M.E., Hodgson, K.O., and Chan, S.I. (1996) The biochemistry of the particulate methane monooxygenase, in M.E. Lidstrom and F.R. Tabita (ed.), *Microbial Growth on C1 Compounds*, Kluwer Academic Publishers, Dordrecht, pp. 150-158.

17. Semrau, J.D., Chistoserdov, A., Lebron, J., Costello, A., Davagnino, J., Kenna, E., Holmes, A.J., Finch, R., Murrell, J.C. and Lidstrom, M.E. (1995) Particulate methane monooxygenase genes in methanotrophs, *J. Bacteriol.* **177**, 3071-3079.

18. Stolyar, S., Costello, A.M., Peeples, T.L., and Lidstrom, M.E., Genetic organization of genes encoding the particulate methane monooxygenase from the methane oxidizing bacterium *Methylococcus capsulatus* Bath, in preparation.

19. Holmes, A.J., Costello, A., Lidstrom, M.E. and Murrell, J.C. (1995) Evidence that particulate methane monooxygenase and ammonia monooxygenase are homologous enzymes, *FEMS Microbiol. Lett.* **13**, 202-208.

HIERARCHICAL CONTROL OF ELECTRON-TRANSFER

H. V. WESTERHOFF[1,2], P. R. JENSEN[3], L. EGGER[1], W. C. VAN HEESWIJK[1], R. VAN SPANNING[1], B. N. KHOLODENKO[4,5], J. L. SNOEP[1]

[1]*MicroPhysiology, BioCentrum Amsterdam, Dep. of Biology, Vrije Universiteit, De Boelelaan 1087, 1081 HV Amsterdam, The Netherlands*
[2]*Mathematical Biochemistry, ECSI, BioCentrum Amsterdam, University of Amsterdam*
[3]*Dep. of Microbiology, Technical University of Denmark, Lyngby Denmark*
[4]*A.N. Belozersky Institute, Moscow State University*
[5]*Department of Pathology, Anatomy and Cell Biology Thomas Jefferson University, Philadelphia, USA*

1. Summary

In this chapter the role of electron-transfer in determining the behaviour of the ATP synthesising enzyme in *E. coli* is analysed. It is concluded that the latter enzyme lacks control because of special properties of the electron-transfer components. These properties range from absence of a strong back pressure by the protonmotive force on the rate of electron-transfer to hierarchical regulation of the expression of the genes that encode the electron-transfer proteins as a response to changes in the bioenergetic properties of the cell. The discussion uses Hierarchical Control Analysis as a paradigm. This allows one to analyse a complex system of regulatory interactions in terms of the importance of the contributing factors.

2. The Complex Live Cell

Biology is a science, but a special one. It studies systems that completely adhere to the physical and chemical principles, but are nonetheless special in that they result from a long selection process. In all cases this selection has been for functionality, be it that the type of functionality may have differed widely between cases and is often unclear to us. In physics and chemistry, explanations in terms of function (e.g., this chemical bond is weak because then the molecule is hydrolysed more readily) are not acceptable. In biology they are, be it that they should not displace the physical chemical explanation. In other words, biological phenomena should be explained in terms of their mechanism (how they work, based on underlying physical chemical properties) as well as in terms of function (why a phenomenon contributes to functions of the

G.W. Canters and E. Vijgenboom (eds.),
Biological Electron Transfer Chains: Genetics, Composition and Mode of Operation, 99-114.

organism that enhance its fitness). The former type of explanation may be called mechanistic, the latter teleological.

This chapter does *not* deal with the latter special nature of biology, biophysics and biochemistry. It only discusses mechanistic aspects of biology. Yet it addresses another aspect that makes biology special with respect to much of physics and chemistry. Biological systems have been selected for fitness under a variety of conditions. Therewith they have developed a large number of response systems and regulatory mechanisms. The reason why these systems exist in biology is not because they are simple (they are not), but because they led to fitness. Accordingly, biophysicists/biochemists seeking mechanistic explanations should be prepared to accept that such explanations may not be simple but complex. By contrast, for a long time the paradigm for physics and chemistry has been that of any two explanations of a phenomenon, the simple one is to be preferred. Accordingly, systems are dissected into elements until the latter are simple enough to treat by simple, preferably linear methods. Subsequently, the functioning of the system as a whole should be understandable in terms of all its components functioning simultaneously in the same way as they would function individually. (Only recently physicists and chemists have begun to consider non-linear systems to be somewhat more than a bother to the beauty of simple physics. They have begun to realise that such systems can give rise to properties that are entirely different from the properties of the linear, near equilibrium world. Examples are found in hydrodynamics, aerodynamics, superconduction and electronics.)

Being biochemists by training we tend to ask: What then is the very essence of the complexness of biological systems? What is the simplest system that exhibits such complexity? What are the minimum properties required for a biological system to be called BioComplex? Isn't the complexness just the phenomenon that biological systems contain many components? Well, it is not. BioComplexity is defined as new functional properties arising from the combination of biological components. Accordingly, two components can already give rise to complexity. One example is the two components cytochrome oxidase and H^+-ATPase, which only when organised properly together (i.e., in a proton impermeable membrane and in proper relative orientation), give rise to ATP synthesis driven by electron-transfer. Another example is that of the lactose transporter and the phosphotransferase system in *E. coli*. If these two transport systems were active individually, *E. coli* would simultaneously utilise lactose and glucose. However, the regulatory organisation is such that glucose is used preferentially [1,2].

BioComplexity can therefore be defined as the difference between the functional whole and its parts. The essence of BioComplexity is therefore not *'difficult'* or *'multitude'*, but *'organisation'*. Non-linear interactions give rise to properties that are not exhibited by the components in isolation. In living cells such non-linear interactions arise already in enzyme kinetics when the substrate concentration exceeds the Michaelis constant. For instance, when a step in a metabolic pathway is insensitive to products, whereas its immediate product lies above the K_M of the subsequent step, then the former step can exert an extremely strong control on the concentration of its immediate product. Feedback inhibition, induction and repression of gene expression all tend to lead to truly new properties, such as well-controlled homeostatic regulation.

BioComplexity can be studied by looking both at the intact system of components and at all its isolated components. An implication of the phenomenon of BioComplexity is that a component behaves differently when inside the intact system than when in isolation, i.e., in the intact system the component does not control its own behaviour. In this chapter we shall discuss an experimental example of this case and address the question where in the rest of the system the control may reside.

3. The H$^+$-ATPase in Intact *E. coli* Behaves Differently From the Isolated Enzyme.

The example is that of the H$^+$-ATPase in *E. coli*. The question we shall ask here is whether the H$^+$-ATPase in *E. coli* controls its own flux. This flux is the synthesis of ATP driven by the electrochemical potential difference for protons across the plasma membrane ($\Delta\tilde{\mu}_{H^+}$):

$$ADP + phosphate + nH^+_{out} \Longleftrightarrow ATP + nH^+_{in} \tag{1}$$

The reaction is catalysed by the H$^+$-ATPase. When examined in isolation, the rate of an enzyme catalysed reaction should be proportional to the concentration of the enzyme. Accordingly, if the H$^+$-ATPase determined its own rate (the flux of ATP synthesis), then elimination of 50 % of the enzymes from *E. coli* should reduce the ATP synthetic flux by 50 %. Under appropriate conditions the flux of ATP synthesis is approximately proportional to the growth rate of the cells and the implication should therefore translate to a 50 % reduction in growth rate, resulting from a 50 % reduction in the

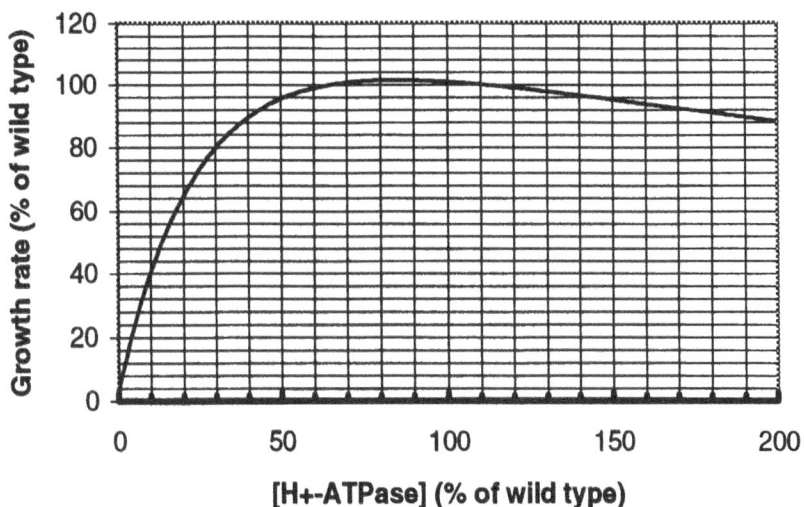

Figure 1. Dependence of growth rate on amount of H$^+$ATPase. Based on [3].

number of H⁺-ATPases.

There is no reason for research into complexity to be vague. In fact, because quite often a number of positive and negative factors are active simultaneously, it is important to quantify effects. The control by the H⁺-ATPase on its own flux is quantified by its flux control coefficient, defined by:

$$C_{ATPase}^{J_p} = \frac{dJ_p / J_p}{d[ATPase]/[ATPase]} \qquad (2)$$

Here J_p refers to the steady state flux of ATP synthesis and $[ATPase]$ to the concentration of the H⁺-ATPase. This control coefficient is the percentage change in flux divided by the percentage change in enzyme activity, for small changes. If equal to 1, the enzyme is in full control of the flux. The advantage of this precise definition over the argument around the effect of a 50 % reduction is that the relative effect on the flux is likely to depend on the magnitude of the modulation of the H⁺-ATPase concentration. The choice of a 50 % modulation then becomes rather arbitrary. In addition, at strong modulations, the cells physiology may be disturbed.

The question to what extent the H⁺-ATPase behaves differently in the intact *in vivo* system, as compared to *in vitro*, may now be rephrased quantitatively as the extent to which its flux control coefficient with respect to growth rate deviates from 1. The answer is: pretty much, the control coefficient is minus 0.06, i.e., virtually zero [3,4] (cf. Fig. 1).

4. Mechanisms of Complexity

Observing that in the intact cell an enzyme behaves differently when compared to its behaviour in isolation, merely points out that that enzyme is engaged in some of the complex behaviour of the living cell. The real challenge is of course to examine the mechanism by which the complex behaviour of the enzyme *in vivo* arises, as this should shed some light on the complex behaviour of the cell. Mechanisms of BioComplexity consist of the types of regulatory interaction between molecules and the corresponding regulation properties of those molecules. For metabolic networks at steady state, Metabolic Control Analysis (MCA) has shown that the latter consist solely of the kinetic acceleration effects the various metabolite concentrations have on the various reaction rates. These acceleration effects are called the elasticity coefficients and denoted by Greek lower case epsilons:

$$\varepsilon_X^i = \left(\frac{\partial \ln v_i}{\partial \ln[X]}\right)_{\text{other variables constant}} \qquad (3)$$

Here v_i represent the rate of process (enzyme) i and [X] the concentration of the metabolite X. These elasticity coefficients present the properties of the enzymes (processes) themselves, as exhibited in isolation of the rest of the system, yet under the same conditions. For instance, the elasticity coefficients of an intracellular enzyme for its substrate depends on the concentration of its substrate and product. When the enzyme is saturated with its substrate, its elasticity for that substrate is close to zero. When the product concentration is high, then the apparent K_M for its substrate is increased and its elasticity for its substrate is increased towards the value of 1. That value of 1 is attained when the substrate concentration is far below the corresponding Michaelis constant. When the enzyme is cooperative with respect to its substrate, the elasticity coefficient approaches its Hill coefficient. Yet, although the elasticity coefficient reflects the enzyme kinetic properties under the *in vivo* conditions, it does not fully describe its behaviour *in vivo* because it does not contain the information on how the substrate and product concentrations change *in vivo* between steady states.

The case can be illustrated for the dependence of the flux through an enzyme on the concentration of the enzyme. In terms of the enzyme kinetic property under fixed *in vivo* conditions, this dependence is equal to the elasticity coefficient of the enzyme with respect to its own concentration, which equals 1:

$$\varepsilon^i_{e_i} = \left(\frac{\partial \ln v_i}{\partial \ln[e_i]} \right)_{\text{variables constant}} = 1 \qquad (4)$$

In terms of the change in rate in the actual system when the enzyme concentration is changed, this is equal to the flux-control coefficient of that enzyme:

$$C^{v_i}_{e_i} \equiv \frac{d \ln v_i}{d \ln[e_i]} = \frac{\partial \ln v_i}{\partial \ln[e_i]} + \frac{\partial \ln v_i}{\partial \ln[X]} \bullet \frac{d \ln[X]}{d \ln[e_i]} + \frac{\partial \ln v_i}{\partial \ln[Y]} \bullet \frac{d \ln[Y]}{d \ln[e_i]} =$$
$$= 1 + \varepsilon^i_X \bullet C^X_{e_i} + \varepsilon^i_Y \bullet C^Y_{e_i} \qquad (5)$$

This flux control coefficient is the sum of the direct effect of the change in enzyme concentration and the effects of the changes in the metabolite concentrations to which the enzyme is sensitive (in this equation these are denoted by [X] and [Y]; there may be more than 2.

We may exemplify this for the simple metabolic pathway of Figure 2, where there is only a single metabolic intermediate X. Using Metabolic Control Analysis one can obtain an expression of the flux control coefficient of the rate of reaction 2 with respect to the concentration of the enzyme that catalyses that reaction (enzyme e_2):

$$C^{v_2}_{e_2} = 1 + \varepsilon^2_X \bullet C^X_{e_1} = 1 - \frac{\varepsilon^2_X}{\varepsilon^2_X - \varepsilon^1_X} = \frac{-\varepsilon^1_X}{\varepsilon^2_X - \varepsilon^1_X} \qquad (6)$$

This equation shows that the behaviour of the enzyme deviates more and more from its behaviour in isolation as it is more sensitive (elastic) to changes in the concentration of the metabolite X (i.e., as ε^2_X increases).

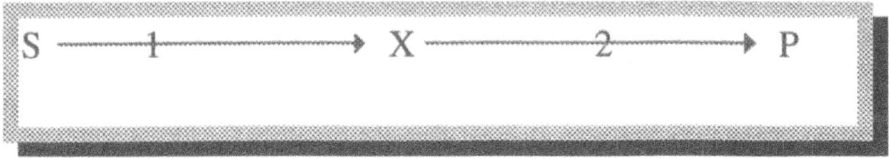

Figure 2. Metabolic control in a two step pathway. Enzyme 2 symbolises the H⁺ATPase, enzyme 1 the electron-transfer chain. Both enzymes 1 and 2 may be responsive ('elastic') towards the concentration of metabolite X (e.g., $\Delta\tilde{\mu}_{H^+}$, the protonmotive force).

Returning to the observation that the ATP synthesis flux in *E. coli* is not proportional to the concentration of the H⁺-ATPases, and that the control coefficient of the H⁺-ATPase with respect to its own flux is so far from 1 that it is even close to zero, we may consult the above equation. It suggests that the H⁺-ATPase is much more sensitive (elastic) towards an intermediate in the pathway than one of the other enzymes, i.e., ε_X^2 is very large. Consequently reduction of the number of H⁺-ATPases will lead to an increase in the metabolite X ($\Delta\tilde{\mu}_{H^+}$). This increase will then in turn accelerate the flux through the remaining H⁺-ATPases [3]. That the H⁺-ATPase is more elastic towards $\Delta\tilde{\mu}_{H^+}$ than the electron-transfer chain is in keeping with the fact that *E. coli* exhibits little respiratory control; when a protonophore is added respiration is hardly accelerated , especially after prolonged incubations [5,6].

The interesting corollary is that the flux through the H⁺-ATPase is largely controlled by an other component in the system, perhaps the electron-transfer chain. This suggests that regulation of the flux of electron-transfer (and of course of its coupling to proton pumping) may be much more important for *E. coli* physiology than regulation of the ATP synthesis in the biochemical sense, i.e., through the enzyme catalysing that reaction.

5. Hierarchical Control Analysis

The ATP synthesis we just discussed takes place in an intact cell and this implies that the enzymes that are involved have not just been added by the experimenter, but have resulted from the expression of corresponding *operons* (cf. Fig. 3). Consequently, one may speak of a hierarchy of control, the direct flux control residing with the enzymes involved in the metabolic pathway, whereas the ultimate control of the flux of ATP synthesis (and of growth rate) resides in transcription of these operons. This at least is true in the case of a dictatorial hierarchy, where the rate of transcription is not affected by what happens at the level of metabolism (Fig. 3).

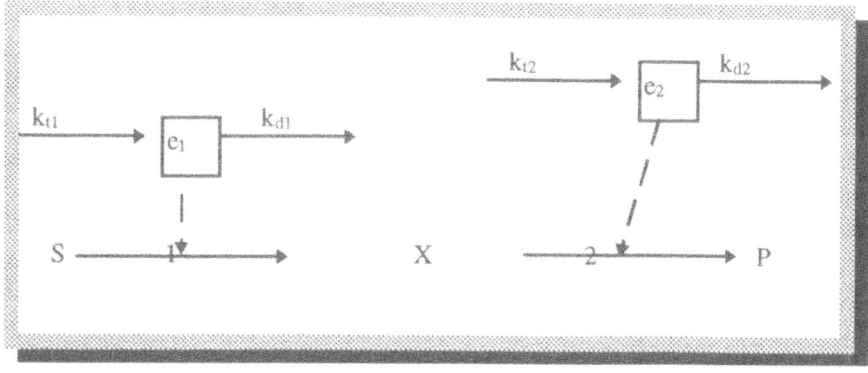

Figure 3. The two-enzyme pathway in the cell, where the enzyme concentrations are controlled by gene expression. The dictatorial case; no retroeffect from metabolism to gene expression. k_{ti} is the rate constant of transcription of operon i, and k_{di} is the rate constant of degradation of enzyme i.

In the case of a dictatorial hierarchy, the control exerted by enzymes 1 and 2 on the flux is not different from that in a simple metabolic pathway. The only difference is that on top of that metabolic control, transcription, mRNA degradation and translation, protein stability and topogenesis control the flux. Interestingly, the effect is not that control resides in transcription *rather* than in metabolism; it resides both in transcription and in metabolism; from this perspective molecular genetics and metabolic biochemistry appear equally important for the understanding of cell function. At any rate, if control of gene expression is dictatorial, then the presence of gene expression does not add any new perspective to the above analysis of the complexity around the H^+-ATPase.

However, control of gene expression is often not dictatorial, but sensitive to external signals and to metabolism itself. At times the product of a metabolic pathway represses an operon that encodes enzymes in the pathway. In other cases, the pathway substrate, or the product of the first reaction (e.g. the transport reaction) induces the enzymes. These cases have been called 'democratic control hierarchies' [7,8]. Figure 4 presents a scheme for the democratic hierarchy that may be involved in a two enzyme metabolic pathway.

When defining the control of steady-state phenomena by a property, that property should itself not be subject to variation by the system. For instance, for the case of Figure 2, it makes no sense to discuss the control exerted by the concentration [X], as that concentration is itself controlled by other properties of the system; it is in fact freely *variable* and only bound to a certain value by the condition of steady state. In Metabolic Control Analysis, one reserves the word 'variable' (short for dependent variable) for this type of property. On the other hand, again in the case of Figure 2, one may discuss the control exerted by the concentration of enzyme 1. One may modulate its concentration and examine the effect on the steady-state flux (or on the concentration of X); the system will make no attempt to redress the concentration change effected in enzyme 1. In Metabolic Control Analysis terms the enzyme concentration is a '*parameter*' in the case of Figure 2 (but not in the case of Fig. 4). In the dictatorial control hierarchy of Figure 3, this remains true, as gene expression will not respond to the modulation of the concentration of enzyme 1. However, in a democratic hierarchy (Fig. 4), the enzyme concentration is not fixed, but is itself subject to variation as the system attempts to attain a new steady state. How can one define and measure the control coefficient by an enzyme in such a system? How can one define the control coefficient by the H$^+$-ATPase on the growth rate of *E. coli* and measure it?

We shall first consider an operational approach. The best way of modulating the activity of an enzyme may be the modulation of the concentration of the enzyme and this can be done by tuning the promoter of the gene encoding it. The corresponding control is that exerted by the transcription rate constant k_t. Accordingly the control coefficient of enzyme 2 of Figure 4 with respect to the metabolic flux J is defined by:

$$C^J_{k_{t2}} \equiv \frac{d \ln J}{d \ln k_{t2}} = \frac{-\varepsilon^1_X - \varepsilon^{t_1}_X}{\varepsilon^2_X + \varepsilon^{t_2}_X - \varepsilon^1_X - \varepsilon^{t_1}_X} \tag{7}$$

The right-hand side of this equation is an elaboration for the scheme of Figure 4, with some special assumptions [9]. One deals with the metabolic level and the other with the level of 'gene expression' (a contraction of the mRNA and the protein level). The metabolic level is assumed to consist of a simple metabolic pathway (two enzymes in series) and one metabolic intermediate involved in the regulation (In Figure 2 this metabolite is written as X and is the intermediate between the two enzyme reactions). Enzymes e_1 and e_2 catalyse the two metabolic conversions. Each of these enzymes is subject to control at the transcription level: each is synthesised in a process characterised by the transcription rate constant (k_{t1} for enzyme e_1 and k_{t2} for enzyme e_2). The enzymes are considered to be degraded in monomolecular reactions (hence with elasticities of 1). The metabolic reactions rates are considered to be proportional

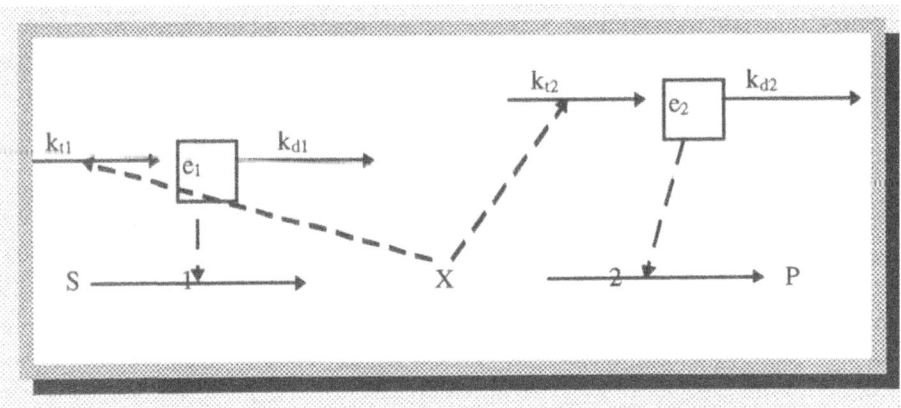

Figure 4. Scheme of hierarchical control system. Enzymes 1 and 2 catalysing reactions 1 and 2 of the bottom metabolic pathway are each produced by transcription and subject to degradation. Flows are indicated by full arrows, influences by dashed arrows.

to the concentration of the enzyme catalysing them *ceteris paribus*. The transcription rate is assumed to be influenced by the metabolite concentration [X], as quantified by the elasticity coefficient ε^t_X. An objection to the use of equation 7 as definition for the control exerted by enzyme 1 on the metabolic flux, is that the definition really describes the control exerted by the transcription rate constant and not the control by a property of the enzyme itself. How then can one define the control by an enzyme without implicating its concentration?

Kacser & Burns [10] have defined the control exerted by an enzyme in terms of the effect of a modulation of its concentration. Because it corresponds to a rather direct, molecular genetic, operational definition, we have greatly favoured this definition [11]. Heinrich and co-workers [12] have emphasised that the more fundamental definition of control should be based on a modulation of the activity of the enzyme. They advocated comparison of the effect on a steady-state system variable to the effect on the rate of the reaction catalysed by the enzyme under fixed conditions of substrates, products and modifiers:

$$C^J_{e_i} \equiv \frac{d\ln J}{\partial \ln v_i} \equiv \left.\left\{\frac{d\ln J}{dp}\right\}_{steadystate} \middle/ \left\{\frac{\partial \ln v_i}{\partial p}\right\}_{\text{all metabolic variables constant}} \right. \tag{8}$$

Different symbols are used for two different types of derivative: "∂" refers to the partial derivative of a reaction rate v with respect to any of the properties that affect it directly, at constant magnitudes of all other properties. Typically v may be written as v([S], [X], [e], T, K_M, V_{max}). This type of derivative of v with respect to any of these is taken at constant magnitudes of the others, including the system's (dependent) variables such as the metabolite concentrations [X]. By contrast, "d" refers to the derivative of a steady state property such as a flux with respect to any of the parameters (independent variables) that affect it also indirectly. This derivative is taken at constant magnitudes of all other parameters of the system, but allowing all system variables to adjust to the new steady state. Typically a steady-state flux J can be written as J([e_1], [e_2], ..., [e_n], T, V_{max1}, V_{max2}, ..., V_{maxn}, K_M, ...) {note that [X] is not in this list} and the derivative corresponds to the partial derivative of J with respect to any of the parameters in the list between parentheses [11]. Here *p* is any parameter that affects only enzyme e_i. When the parameter is the logarithm of the concentration of the enzyme, then this definition reduces to the above definition by of the flux control coefficient (see Eq. 2). Until recently it was thought that this definition produced a control coefficient that was independent of which parameter was used to vary the activity of the enzyme (for as long as that parameter only affected that enzyme) [13], but this is not always the case [14]. The important property of being parameter independent is only obtained when the process v_i is independent of all the other processes in the system. This is not the case when enzyme e_i is involved in metabolic channelling. Then also a second reaction depends on the concentration of the enzyme. The property is also absent when the enzyme concentration is not a parameter but itself subject to control. In these cases, one should retreat to an identification of the independent processes and define control coefficient in terms of a parameter affecting each 'elemental' process [15]. The definition in terms of the dependent properties should then be called 'response coefficients' [16].

The independent process of which we here wish to discuss the control, is process 2 in Figure 4. The rate equation of this enzyme-catalysed reaction contains a forward and a reverse V_{max}. We now multiply each of these V_{max}'s by a parameter λ and set this parameter equal to 1 in the steady state of interest. The definition of the control coefficient is now based on the modulation of λ without modulating the concentration of the enzyme, i.e., on a simultaneous proportional modulation of the forward and the reverse V_{max}:

$$C_i^J = \left\{ \frac{d \ln J}{d \ln \lambda} \right\}_{steady\ state} \tag{9}$$

For the simple case of democratic hierarchical control discussed above, one can show that [9]:

$$C_i^J = C_{k_{ti}}^J \tag{10}$$

Consequently, the operational definition based on the modulation of the transcription rate of the operon encoding an enzyme, may then be used to measure the control exerted by the process catalysed by the enzyme, on any steady state property in the system.

In combination with the above Equation 7 for the control coefficient in terms of a modulation of the transcription rate constant, this allows us to analyse the experimental result of Figure 1, whilst taking the hierarchical nature of the control system into account. We first note that the experimental measurement corresponded to a modulation of the transcription rate constant and a measurement of the flux of ATP synthesis. Hence it seems to correspond to the flux control coefficient by transcription (but see below). By the above analysis this should be equal numerically to the hierarchical control by the ATP synthesis process on the ATP synthesis flux. Experimentally $\epsilon^J_{ktATPase}$ was found to be very low. Using Equation 10 it can be concluded that also the value of C^J_{ATPase}, the control of the H^+-ATPase on ATP synthesis. is small. This absence of control can be understood (from Equation 7):

$$\epsilon_X^2 + \epsilon_X^{t_2} \gg -\epsilon_X^1 - \epsilon_X^{t_1} \tag{11}$$

i.e., if enzyme 1 is barely or inversely sensitive to the metabolic intermediate (the protonmotive force), neither directly, nor through regulated gene expression.

6. The Actual Measurement Was a Co-response.

Above we suggested that Jensen et al. [3,4] measured the control coefficient by the transcription rate constant of the *atp* operon. This implies that they measured the change in steady-state flux and divided that by the fractional change in transcription rate constant they brought about. It is however difficult to determine *with precision* to what extent one actually modulates transcription activity whilst modulating the *lac* promoter in front of the *atp* operon. Therefore Jensen and colleagues determined to what extent the concentration of the c-subunit of the H^+-ATPase had changed and related the change in flux to this change in enzyme concentration. This was done in an experiment where transcription was modulated by varying the concentration of the inducer IPTG. Both the flux and the concentration of the H^+-ATPase are variables in this hierarchical control system. What then does this ratio of changes in properties correspond to? Does it correspond to a control coefficient at all?

Hofmeyr and co-workers [17,18] have defined the concept of co-response (coefficient) and this concept can help us out here: a co-response is the relative change in one variable, divided by the relative change in a second variable by a certain modulation of a parameter of the system. Herewith it is accepted that the magnitude of the co-response may depend on which parameter is modulated. For the present situation, the property that is directly modulated is the rate constant of transcription of the *atp* operon (k_{tp}). The measured property [3,4] was the co-response of growth rate with ATPase concentration under modulation of the transcription rate constant:

$$'O^{J_p}_{ATPase} \equiv \frac{d \ln J_p \Big/ d \ln k_{tp}}{d \ln[ATPase] \Big/ d \ln k_{tp}} \equiv \frac{C^{J_p}_{k_{tp}}}{C^{ATPase}_{k_{tp}}} \tag{12}$$

Here the derivatives are *partial* in the sense that all other *parameters* are held constant, but *total* in the sense that all other *variables* adjust to a new steady state. J_p refers to the rate of ATP synthesis by the H^+-ATPase (the net flux through the enzyme), k_{tp} to the rate constant of transcription of the *atp* operon, and the subscript 'p' to ATPase. In a dictatorially organised system, there is no feedback from the metabolic level to the gene expression level. Consequently, the concentration of H^+-ATPase becomes a parameter. The only way in which k_{tp} affects the metabolic level of the system (which is where the rate of ATP synthesis resides), is through the H^+-ATPase concentration. Hence in dictatorial systems the above definition reduces to:

$$'O^{J_p}_{ATPase} = \frac{d \ln J_p}{d \ln[ATPase]} = C^{J_p}_{ATPase} \tag{13}$$

We conclude that if the system in *E. coli* were dictatorial, i.e., if gene expression of the bioenergetic enzymes were insensitive to cellular energetics, then the property measured by Jensen *et al.*, was equal to the control exerted by the H$^+$-ATPase on its own ATP synthesis flux. However, if the control system were democratic, then the measured property was a co-response coefficient.

To what extent does this co-response differ from the control coefficient? In a complete Hierarchical Control Analysis of the scheme of Figure 4 the co-response coefficient has been expressed in terms of all the elasticity coefficients [9]. This leads to the following comparison between the co-response coefficient $^{k_{12}}O_{e_2}^J$, the hierarchical control coefficient C_2^J, and the metabolic control coefficient c_2^J. The latter is here defined as the control coefficient in which only regulatory interactions at the metabolic level are taken into account (corresponding to a rapid experiment with an inhibitor of the H$^+$-ATPase) C_2^J is the control coefficient that also includes the effects of the adjustment of gene expression. To effect the comparison, we implement Equation 10 with i=2:

$$^{k_{12}}O_{e_2}^J = C_2^J / C_{k_{12}}^{e_2} = C_2^J \bullet \frac{\varepsilon_X^2 - \varepsilon_X^1 + \varepsilon_X^{t_2} - \varepsilon_X^{t_1}}{\varepsilon_X^2 - \varepsilon_X^1 - \varepsilon_X^{t_1}} \tag{14}$$

$$^{k_{12}}O_{e_2}^J = \frac{C_{k_{12}}^J}{C_{k_{12}}^{e_2}} = c_2^J \bullet (1 + \frac{\varepsilon_X^2}{-\varepsilon_X^1} \bullet \frac{-\varepsilon_X^{t_1}}{\varepsilon_X^2 - \varepsilon_X^1 - \varepsilon_X^{t_1}}) \tag{15}$$

$$c_2^J = \frac{\varepsilon_X^2}{\varepsilon_X^2 - \varepsilon_X^1} \tag{16}$$

Equation 14 shows that the measured co-response coefficient becomes equal to the hierarchical control coefficient, whenever the operon of enzyme 2 (the H$^+$-ATPase) is hardly elastic (sensitive) to what happens in free-energy metabolism ($\varepsilon_X^{t_2} \cong 0$). This is certainly so for the *E. coli* strain used by Jensen *et al.*, as they substituted the *lac* type promoter for the endogenous *atp* promoters. Equation 15 shows that the measured co-response coefficient can differ greatly from the metabolic control coefficient, if the electron-transfer chain activity depends strongly on gene expression being regulated by the energy state of the cell and if the electron-transfer chain is hardly sensitive to changes in the protonmotive force. The latter of these two conditions appears to be met, the former is uncertain for the moment.

7. Control of Electron Flow Involved in the Control by the H$^+$-ATPase?

One aspect of the complexity of the energy metabolism of an organism as "simple" as *E. coli*, is that when growing on glucose, *E. coli* can choose between various routes of glucose consumption. In one of these glucose is oxidised to CO_2 by NAD, the NADH being reoxidised by the electron-transfer chain. The latter process is coupled to proton pumping by the NADH dehydrogenase and by the quinol oxidase. Also here the organism can choose between routes: There are two quinol oxidases, one bd-type with a high affinity for oxygen, and a bo3-type with lower affinity for oxygen. The bo3-type may have a higher proton per electron stoichiometry than the bd-type and also the two NADH dehydrogenases may differ in H^+/e^- ratio [19].

In functional terms it may be expected that when meeting with a free energy challenge, resulting either from an increased demand on ATP or $\Delta\tilde{\mu}_{H^+}$, or from a reduced supply of electron donors or oxygen, E. coli re-routes its electron fluxes. Indeed, under anaerobic conditions the organism hardly expresses the bo3-type quinol oxidase. Adding protonophores to dissipate $\Delta\tilde{\mu}_{H^+}$, Skulachev and co-workers [20] have observed that respiratory chain components were increased in expression and concluded that the decrease in $\Delta\tilde{\mu}_{H^+}$ served as the signal to the genome.

The issue whether the concentration of the electron-transfer chain components is subject to regulation by cellular free energy properties, is also of importance for the above discussion of the control by the H^+-ATPase. As an addition to the observation stressed there, the reduction in concentration of H^+-ATPases was observed to lead to an increase in the rate of respiration [3,4], rather than the decrease that is observed in rat-liver mitochondria upon inhibition of the H^+-ATPases. Because the latter is called respiratory control, the phenomenon in E. coli was called 'inverse respiratory control'.

Analysing by spectroscopy the cytochrome content of the membrane of E. coli, we have indeed observed changes in content of cytochromes d and o (data not shown), suggesting that there is an induction of the genes encoding respiratory chain components, as the concentration of the H^+-ATPases is decreased. In terms of hierarchical control analysis, this may mean that:

$$\varepsilon_X^{t_i} < 0 \tag{17}$$

or at least that the metabolic change induced by a decrease in the number of H^+-ATPases (which may be a decrease in ATP/ADP ratio) enhances gene expression of the respiratory-chain encoding genes. This then has the effect that Equation 11 is met and hence explains the observed lack of control by the ATP synthesis process.

The actual nature of the changes in the electron-transfer chain still have to be analysed. Is it just an increase in all the cytochromes, or is it a selective increase in some components over other components? Or, is the composition of the electron-transfer chain a complex function of the concentration of the H^+-ATPases. Another issue is what the signal is for the change in gene expression witnessed here. Is it $\Delta\tilde{\mu}_{H^+}$, as proposed by Skulachev and co-workers, or is it the ATP/ADP ratio? If similar changes are seen in the case of reduced concentrations of H^+-ATPases and added protonophore, then the latter is the more likely mechanism. Then perhaps the control of the structure of the intracellar DNA by the ATP/ADP ratio through DNA gyrase [21] may be involved.

8. Electron-Transfer and Complexity, Exception or the Rule?

Above we have given an example of complexity of the living cell. We have shown that electron-transfer reactions may be regulated at the level of gene expression. This phenomenon and the degree to which electron-transfer reactions do, or do not respond to changes in the protonmotive force have vast implications for the behaviour of other components of the cell, including the enzyme that makes ATP.

The example was that of *E. coli*. However, the electron-transfer chain of *P. denitrificans* is even more versatile. Deletions of some components affect the concentrations of other components through regulated gene expression [22,23,24]. Also for that system however, the signalling molecules have not been identified and much excitement lies ahead.

In *E. coli*, the *arc* regulon is involved in regulation of the concentration of certain cytochromes. Arc responds to the oxygen concentration, probably in an indirect way. Also FNR responds to the oxygen concentration, possibly in a more direct way. In either case it is suspected that the regulation by oxygen is modulated by the magnitude of $\Delta\tilde{\mu}_{H^+}$ and by the redox potential of certain redox shuttles (quinones?). This suggests that regulation by various signals may overlap. One mechanism may be that of transfer of phosphoryl groups between two-component and other regulatory systems [25].

We are revisiting the regulation of ammonia assimilation of *E. coli*, asking the question whether that may also be inherently more complex than hitherto assumed. Here we determined the control exerted by the pivotal protein PII on the adenylylation of glutamine synthetase. We observed that there is little such control, neither on the steady-state level of adenylylation of glutamine synthetase, nor on the rate of de-adenylylation after an incubation with ammonia. As in the case of the H^+-ATPase, an important protein, PII, seemed to lack control. Then however, the cells were grown under conditions of high ammonia concentration. When these cells were deprived of ammonia PII did exert a control on the rate of de-adenylylation. Also here there appeared to be a complex regulatory system, exhibiting memory of previous ammonia conditions. A new protein, GlnK appeared to play as a substitute for or competitor with PII [26,27].

9. Concluding Remarks

In this chapter we have shown that with respect to cell function there may be more to electron-transfer proteins than the mechanism by which they transfer electrons or protons. Depending on how they respond to changes in $\Delta\tilde{\mu}_{H^+}$ and on how their relative concentrations are regulated, these proteins may regulate electron-transfer and proton pumping in a subtle manner. The effect of this regulation then reaches far beyond electron-transfer itself: the behaviour of other important intracellular enzymes may be greatly altered (here from a control of 1 to a control near 0). More generally, the living cell seems to exhibit many complex properties, especially in terms of its regulation. Perhaps this serves to make the cell versatile *vis-à-vis* the various challenges it has to withstand. At any rate, the complex regulation pattern gives the human observer the impression of a quasi-intelligent behaviour, as if the living cell does things on purpose.

10. Acknowledgement

This work was supported by the Netherlands Organisation for Scientific Research and by the European Union.

11. References

1. Monod, J. (1942). Recherches sur la croissance des cultures bacteriennes, Hermann et Cie, Paris.
2. Saier, M.H., Jr. (1989) Proteinphosphorylation and allosteric control of inducer exclusion and catabolite repression by the bacterial phosphoenolpyruvate:sugar phosphotransferase system, *Microbiol. Rev.* **53**, 109-120.
3. Jensen, P. R., Westerhoff, H. V., Michelsen, O. (1993) Excess capacity of H+-ATPase and inverse respiratory control in *Escherichia coli*, *EMBO J* **12**, 1277-1282
4. Jensen, P. R., Michelsen, O., Westerhoff, H. V. (1993) Control analysis of the dependence of *E. coli* physiology on the H+-ATPase, *Proc. Natl. Acad. Sci. USA* **90**, 8068-8072.
5. Burstein, C., Tiankova, L. and Kepes, A. (1979) Respiratory control in *Escherichia coli* K12, *Eur. J. Biochem.* **94**, 387-392.
6. Tsuchiya, T. and Rosen, B.P. (1980) Respiratory control in *Escherichia coli*, *FEBS Lett.* **120**, 128-130.
7. Westerhoff, H. V., Koster, J. G., van Workum, M., Rudd, K. E. (1990) On the Control of Gene Expression, A. Cornish-Bowden and Cardenas, M.L. Ed., *Control of Metabolic Processes* Plenum Press: New York, pp 399-412.
8. Wijker, J. E., Jensen, P. R., Snoep, J. L., Vaz Gomez, A., Guiral, M., Jongsma, A. P. M., de Waal, A., Hoving, S., Van Dooren, S., Van der Weijden, C. C., Van Workum, M., Van Heeswijk, W. C., Molenaar, D., Wielinga, P., Richard, P., Diderich, J., Bakker, B. M., Teusink, B., Hemker, M., Rohwer, J. M., Van der Gugten, A. A., Kholodenko, B. N., Westerhoff, H. V. (1995) Energy, control and DNA structure in the living cell, *Biophys Chem* **55**, 153-165.
9. Westerhoff, H.V., Jensen, P.R., Snoep, J.L. and Kholodenko, B.N. (1998) Thermodynamics of complexity. The live cell *Thermochim Acta* (in press).
10. Kacser, H., Burns, J. A. (1973) The control of flux. D.D. Davies Ed., *Rate control of biological processes*, Cambridge Univ. Press, London, pp 65-104.
11. Westerhoff, H. V., van Dam, K. (1987) *Thermodynamics and control of biological free energy transduction*, Elsevier: Amsterdam.
12. Heinrich, R., Rapoport, S. M., Rapoport, T. A. (1977) Metabolic regulation and mathematical models, *Prog Biophys Molec Biol*, **32**, 1-82.
13. Schuster, S., Heinrich, R. (1992) The definition of metabolic control analysis revisited, *Biosystems*, **27**, 1-15.
14. Kholodenko, B. N., Westerhoff, H. V., Puigjaner, J., Cascante, M. (1995) Control in channelled pathways. A matrix method calculating the enzyme control coefficients, *Biophys Chem* **53**, 247-258.
15. Kholodenko, B. N., Westerhoff, H. V. (1994) Control theory of one enzyme, *Biochim Biophys Acta*, **1208**, 294-305.

16. Rohwer, J.M. (1997) Interaction of functional units in metabolism. Control and regulation of the bacterial phosphoenolpyruvate-dependent phosphotrasferase system. PhD thesis, University of Amsterdam, Amsterdam.

17. Hofmeyr, J.-H.S., Cornish-Bowden, A., and Rohwer, J.M. (1993) Taking enzyme kinetics out of control, putting control into regulation, *Eur J Biochem.* **212**, 833-837.

18. Hofmeyr, J.-H.S., and Cornish-Bowden, A. (1996) Co-response analysis: a new experimental strategy for metabolic control analysis, *J. Theor. Biol.* **182**, 371-380.

19. Gennis R.B. and Stewart V. (1996) Neidhart F.C., et al., eds *E. coli and S. typhimurium* pp. 217 - 261. ASM Press, Washington D.C.

20. Bogachev A V. Murtazina R A. Shestopalov A I. Skulachev V P. (1995) Induction of the *Escherichia coli* cytochrome d by low DELTA-mu-H+ and by sodium ions, *Eur. J. Biochem* **232**, 304-308.

21. Van Workum, M., Van Dooren, S., Oldenburg, N., Molenaar, D., Jensen, P., Snoep, J., Westerhoff, H. (1996) DNA supercoiling depends on the phosphorylation potential in *Escherichia coli*, *Mol Microbiol*, **20**, 351-360.

22. Van Spanning, R.J.M., Wansell, C., Harms, N., Oltmann, L.F., Stouthamer, A.H. (1990) Mutagenesis of the gene encoding cytochrome-c550 of *Paracoccus denitrificans* and analysis of the resultant physiological effects, *J. Bacteriol.* **172**, 986-996.

23. Van Spanning, R. J .M., De Boer, A. P. N., Reijnders, W. N. M., De Gier, J. W. L., Delorme, C. O., Stouthamer, A. H., Westerhoff, H. V., Harms, N., and Van der Oost, J. (1995) Regulation of oxidative phosphorylation: the flexible respiratory network of *Paracoccus denitrificans*, *J. Bioenerg. Biomembr.* **27**: 499-512.

24. Van Spanning, R. J. M., De Boer, A. P. N., Reijnders, W. N. M., Westerhoff, H. V., Stouthamer, A. H., and Van der Oost, J. (1997). FnrP and NNR of *Paracoccus denitrificans* are both members of the FNR family of transcription activators but have distinct control on respiratory adaptation in response to oxygen limitation, *Mol Microbiol* **23**, 893-907.

25. Hellingwerf, K.J., Postma, P.W., Tommassen, J., and Westerhoff, H.V. (1995) Signal transduction in bacteria: phospho-neural network(s) in *Escherichia coli*?, *FEMS Microbiol.Rev.* **16**,309-321.

26. Van Heeswijk, W.C., Stegeman, B., Hoving, S., Molenaar, D., Kahn, D., and Westerhoff, H.V. (1995) An additional PII in *Escherichia coli*: a new regulatory protein in the glutamine synthetase cascade, *FEMS Microbiol.Lett.* **132**,153-157.

27. Van Heeswijk, W.C., Hoving, S., Molenaar, D., Stegeman, B., Kahn, D., and Westerhoff, H.V. (1996) An alternative PII protein in the regulation of glutamine synthetase in *Escherichia coli*, *Mol.Microbiol.* **21**,133-146.

ON THE MECHANISM OF NITRITE REDUCTASE: COMPLEX BETWEEN PSEUDOAZURIN AND NITRITE REDUCTASE FROM *A. CYCLOCLASTES*

MICHAEL E.P. MURPHY[¶], STEWART TURLEY,
and ELINOR T. ADMAN
Department of Biological Structure, Box 357420,
School of Medicine, University of Washington, Seattle, WA 98195-7420.
[¶]Present address: Department of Biochemistry and Molecular Biology,
2146 Health Sciences Mall, University of British Columbia,
Vancouver, BC V6T 1Z3.

1. Abstract

Dissimilatory nitrite reductase is a trimeric copper-containing protein in *Alcaligenes faecalis* and *Achromobacter cycloclastes*. NO is the immediate product and is an intermediate in the dissimilatory denitrification pathway, in which nitrate is completely reduced to N_2. Data from crystals of *Alcaligenes faecalis* Nir at -160°C in its oxidized and reduced forms, and with nitrite soaked into crystals, show that the reduced form binds ligands much less tightly than the oxidized form. Previouly published results have shown that the electron transfer partner of nitrite reductase, pseudoazurin, donates electrons to Nir via the Type I Cu site. Electrons are then transferred internally to the Type II Cu site, normally ligated by three histidines and a solvent. More weakly bound ligand in the reduced form suggests that nitrite first replaces the solvent ligand in the oxidized form, modifying the redox potential of the active site Type II copper so that, upon reduction, electrons are then transferred from the Type I site.

A specific complex between the physiological electron donor, pseudoazurin, and Nir has been characterized by site-directed mutagenesis studies. Preliminary data on two crystal forms (I: space group I222, a=114.5Å, b=148.3Å, c=171.4Å; II: $P2_12_12_1$ a=106.5Å b=116.2Å c=116.6 Å) of a putative complex between AcNir and pseudoazurin show that Form I binds one molecule of pseudoazurin per trimer, and this at half occupancy in the present crystals. The other potential binding sites are occluded by packing interactions. Form II crystals, while having enough space to accomodate a fully loaded trimer, do not show interpretable electron density for pseudoazurin in difference maps.

G.W. Canters and E. Vijgenboom (eds.),
Biological Electron Transfer Chains: Genetics, Composition and Mode of Operation, 115-128.
© 1998 Kluwer Academic Publishers.

2. Introduction

The copper containing nitrite reductase (Nir) requires a specific protein, reduced pseudoazurin (Paz), when inactivated aerobically [1]. In the denitrification pathway of some soil bacteria, nitrite reductase normally reduces nitrite to NO, but in the presence of oxygen, peroxide is formed [2]. This finding suggested to us that this would be a good system in which to study specificity of electron transfer.

We initially determined the structure of pseudoazurin [3] learning that it had a Greek-key β barrel fold with a Type I copper ligated by Cys78, His40, His81 and Met 86. We have studied the structures of two mutants found to increase the redox potential of pseudoazurin by 150 mV or more [4]. The structure of the mutant P80A revealed that a water replaced the surface proline atoms, and electrostatic calculations showed that the dipole associated with this water was largely responsible for the increase in redox potential. The structure of the P80I mutant showed that the slightly larger side chain forced a change in the copper site geometry toward the reduced geometry, thereby increasing the redox potential. A distinct conformational change at Met7 and Pro35 occurs in the reduced forms of the native and mutant pseudoazurins, a consequence of the slight expansion at the copper site of the reduced form.

We have studied nitrite reductase from two bacteria, *Achromobacter cycloclastes* [5,6] (AcNir) and *Alcaligenes faecalis* [7,8] AfNir. The structure of AcNir revealed that it consists of a trimer of two-domain monomers, each domain having a cupredoxin-like fold, the N-terminal domain containing a Type I copper (ligated by Cys136, His 145, Met150, and His95), and the C-terminal domain not. A Type II copper resides at the interface between two monomers, and is ligated by two histidine ligands (His100 and His135) from domain 1 of one monomer, and one histidine (His306) from domain 2 of another monomer [5]. The structure of the Type-II depleted AcNir showed that the Type II copper is replaced by solvent, and the trimeric structure remains intact. Nitrite-soaked crystals showed that a half-occupancy nitrite binds with its oxygens toward the Type II copper, replacing the water, or hydroxyl, normally bound in the oxidized form. There are no detectable changes in structure from pH 5.0 to pH 6.8 [6]. Figure 1 is a view into the active site from the solvent, and shows only domain 1 of one monomer (to the right) and domain 2 of the other (to the left).

Structural and site-directed mutagenesis studies on AfNir have shown that electrons must go through the Type I site to the Type II site: a M150E mutant cannot be reduced by pseudoazurin, and in fact contains Zn in the Type I Cu site [8]. Recent structural studies at low temperature [9,10] show that native AfNir reduced by ascorbate does not have any water or hydroxyl ligand bound to the Type II copper, and that fully occupied nitrite binds to the oxidized form also with oxygens toward the copper albeit with unequal Cu-O distances. Two nearby residues, His255 and Asp98, each from different monomers, are important in the mechanism. Asp 98 hydrogen bonds to the copper solvent ligand, or to the oxygen of the bound nitrite with the shorter Cu-O distance. It also hydrogen bonds to another solvent that in turn is hydrogen bonded to His255. It is

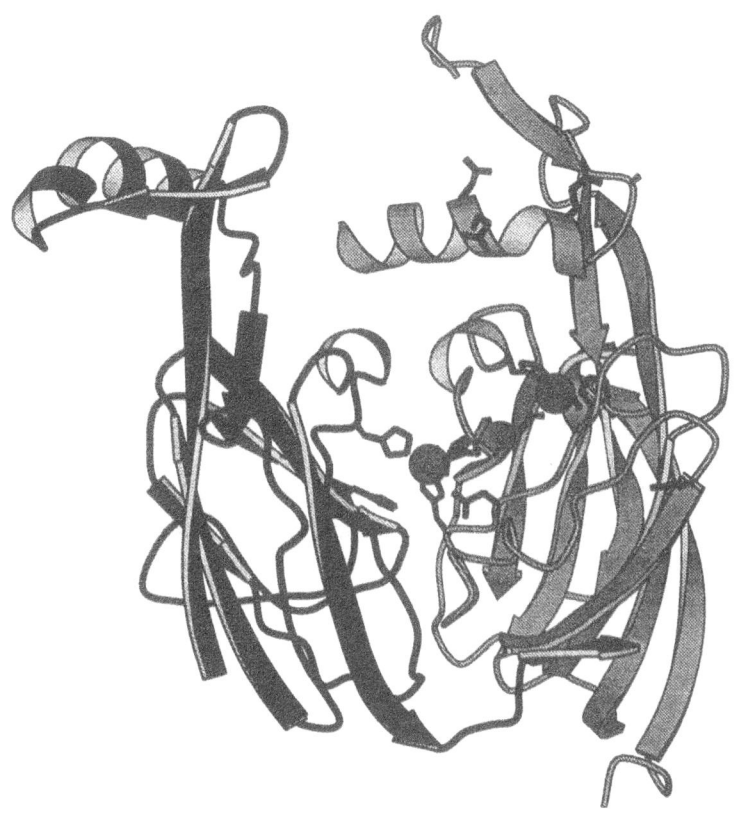

Figure 1. View into active site of Nir. Type I Cu is in upper right, Type II Cu in center. Ligands, H255 and D98 are shown and some negatively charged residues on the tower helix belonging to domain 2 of the right hand monomer.

our view that nitrite displaces the solvent ligand of the Type II Cu in its oxidized form, the protein is reduced by pseudoazurin, electrons are transferred through the Type I Cu to the Type II Cu and then to the bound nitrite, the N-O bond breaks between the oxygen bound to the copper and the nitrogen, and NO diffuses away. A proton residing on Asp98 returns to the bound oxygen returning the site to its original form. The water that is also a product of the reaction is produced upon the initial displacement of the copper ligand. It is not yet clear where the proton at this step comes from.

The focus of this paper is, however, the nature of the specific interaction between Nir and pseudoazurin (Paz). Based on our proposed model of the interaction of Nir with pseudoazurin, that apposes the surface of pseudoazurin closest to its copper (a surface that is circumscribed with a number of lysine residues) with the surface of Nir that affords the shortest distance of its Type I copper to the surface (and has several negatively charged residues around it), we carried out site directed mutagenesis studies on each partner. We demonstrated that upon individual modification of nearly all the lysines on the surface of the molecule, those closest to the copper site (K10, K38, K77, K57) increased the Km for interaction with AfNir the most [11]. Similarly, site directed mutation of the negatively charged E113, E118, E197, D201, E204 and D209 residues of AfNir and measurement of their Km with some of the mutant pseudoazurins allowed us to propose a specific model of interaction [12], shown in Figure 2.

This model, although generally like our original model in that positive charges were apposed to negative charges, differed in that the surface copper histidine ligands of each Type I copper did not directly interact, and indeed it appeared there might be an an interaction of the ligand histidine of Paz with a carbonyl main-chain oxygen of Nir, much like has been seen in the interaction of subunit II of cytochrome oxidase with the heme containing subunits [13].

While much evidence points to our original hypothesis being generally correct the final "proof" would be a structure of a complex. Here we report that we have indeed obtained the structure of a complex, and that in general it too is consistent with the original hypothesis, but differs in interesting details.

3. Experimental Procedures

3.1 PREPARATION OF CRYSTALS AND DATA COLLECTION.

Purified AcNir and pseudoazurin were supplied by J. LeGall and Ming Liu prepared as described previously [14]. Crystals were obtained using the hanging drop method with a reservoir solution containing 55% methylpentanediol, 80mM potassium phosphate, pH 5.8-6.2 and 0.04% β-octyl glucoside. The drop contained 5 μl of protein solution and 3 μl of reservoir solution equilibrated at 12°C. The protein solution contained approximately 20mg/ml of Nir and approximately 1:1 monomer to pseudoazurin. An important step in obtaining crystals of the complex was passing the 1:1 solutions through a Centricon filter which retains material of >100kD. Data was collected from crystals at -160 °C on an Raxis-II using copper K_α radiation produced from a Rigaku RU-200 rotating anode generator operated at 50 kV and 100 mA and focussed with mirrors. The data sets were processed with DENZO (15). Data from six crystals in all were collected; two from Form I, and four from Form II. Only data from the first of the two Form I crystals was used, and data from three of the Form II crystals were merged yielding the statistics shown in Table 1.

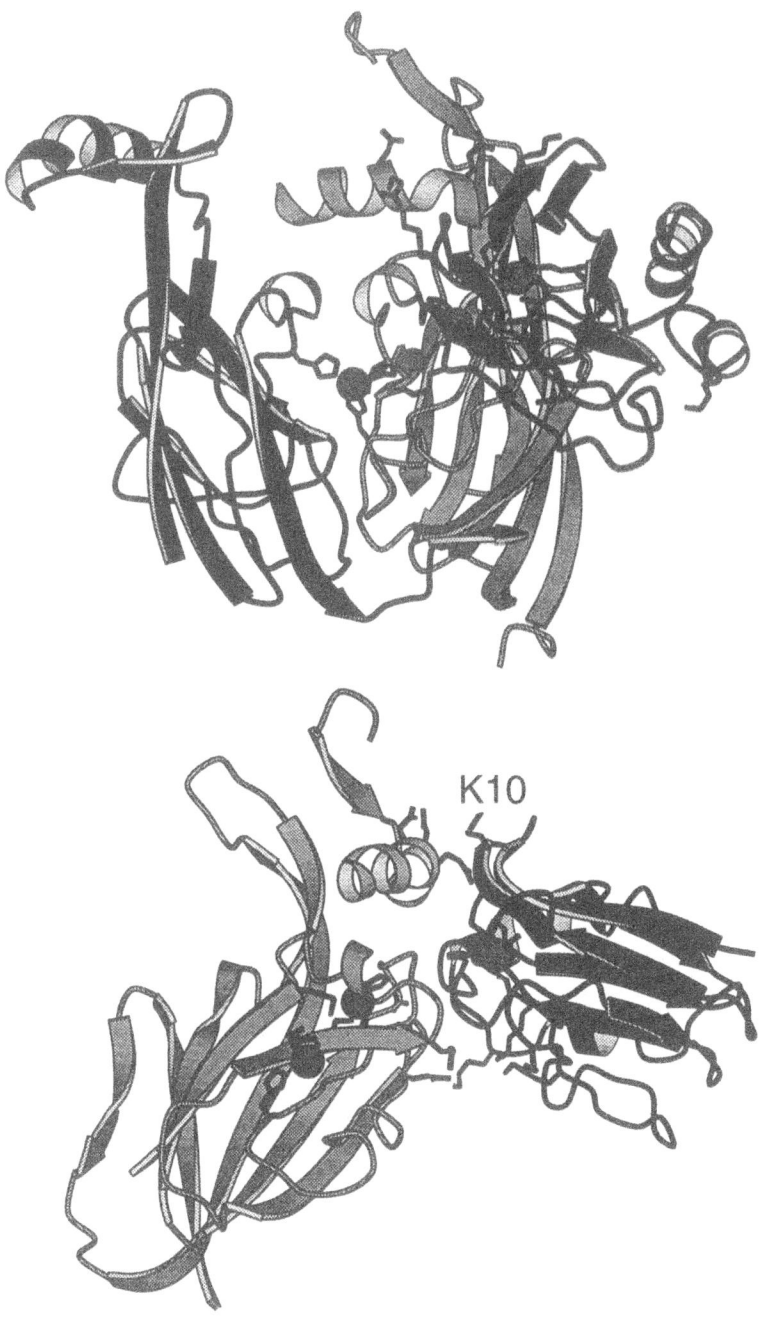

Figure 2. Complex of Nir and Paz, from site directed mutagenesis studies. Upper: view into active site Lower: view rotated 90° around vertical axis, showing domain 1 of Nir and Paz. In this model K10 interacts with E197 of Nir.

Table 1: Data collection statistics.

Crystal	Form I	Form II
Cell Dimensions (Å)	a=114.5, b=148.3, c=171.5	a=106.5, b=116.2, c=116.6
Space group	I222	$P2_12_12_1$
Resolution (Å)	3.2 (3.4-3.2)[a]	2.4 (2.7-2.4)
R-merge[b] on I	0.154 (0.315)	0.072 (0.181)
<I> / <σ(I)>	5.6 (2.3)	8.1 (3.9)
Completeness	0.78 (0.78)	0.97 (0.97)
Unique Reflections	18384 (2258)	52415 (12627)
Redundancy	2.0 (1.9)	1.9 (1.8)

[a]Values in parentheses are for the highest resolution shell.

b
$$R-merge = \frac{\sum_{hkl}\sum_{i=1}^{n}|I_i(h,k,l) - \langle I(h,k,l)\rangle|}{\sum_{hkl}\sum_{i=1}^{n} I_i(h,k,l)}$$

3.2 STRUCTURE SOLUTION AND REFINEMENT

The structures of each form were solved using the program Amore, implemented in the CCP4 package [16] using the entire trimer of Nir (pH6.2 model, PDB code 1NIA) as the search model. SigmaA difference maps [17] (in CCP4) were computed to search for the complexed pseudoazurin. For refinement a randomly selected portion (10%) of the data was reserved for computation of the free R-factor. Refinement was carried out with the program X-PLOR [18] using the parameter set parhcsdx.pro [19]. Manual adjustment and modeling was done using the program O [20]. Figures were made using Molscript [21].

4. Results

4.1 FORM I CRYSTALS

The V_m, the ratio of crystal volume to molecular weight, for this crystal form assuming a complete complex (three pseudoazurins plus the Nir trimer) would be 2.4, quite possible for 364 kD of protein. The space group for the orthorhombic form of Nir [6] is $P2_12_12_1$, a=99.3 Å, b= 115.2 Å, c = 116.0 Å, with V_m = 3.1 for the 108kD trimer, while the original $P2_13$ form a=b=c=98.6 Å has the 36kD monomer in the asymmetric unit [5] and V_m=2.2.

Inasmuch as there could have been as many as nine domains with the characteristic cupredoxin fold, using the trimer as the search model seemed most appropriate. Using data from 20-3.5Å, a peak twelve times the rms value of the rotation function was found, and using data from 8-3.5Å, a peak 28.5 times the rms value of the translation function was found. Rigid body refinement of this model in Xplor yielded an R-value of 0.33 for data from 10 to 3.2Å. Using this model to compute a SigmaA weighted difference map (a map using coefficients that take account of missing scattering matter, and reduce the bias of the known model), the density was explored by superposing the Nir of our predicted complex model onto the Nir in the I222 crystal found from the rotation function, and searching in the expected regions for pseudoazurin. Packing of the Nir excluded two of the three possible sites, but one remained possible. This region exhibited density that could be interpreted as the pseudoazurin molecule. Significantly, it could be seen that it had to be rotated 90° from its expected position, about an axis roughly parallel to the Cu-Cu vector between the Type I centers of each protein. The most telling feature of the difference map was tetrahedral density around one of the larger difference peaks, which we interpreted as the copper and its ligands, and which allowed us to manually rotate and translate the pseudoazurin so that we could begin to see reasonable density for the beta strands of the molecule. Limited attempts to find its position using the rotation function failed. While the model had been generated using our well refined pseudoazurin from *Alcalgenes faecalis* we had a partially refined model from unpublished data collected on a pseudoazurin from *Achromobacter cycloclastes,* which we superposed on our construct to initiate refinement.

Refinement of the trimer of Nir and the pseudoazurin monomer first as rigid bodies resulted in an R-factor of 0.31 for both the working and free R. Refinement proceeded, periodically examining difference maps, finding new density for the C-terminal tail of chain A of AcNir which had different packing interactions in the pH6.2 form used as the starting model, and some new density associated with C-terminal residues of pseudoazurin. The average B-value for the pseudoazurin was nearly twice that of the AcNir; setting the occupancy of pseudoazurin to 0.5 resulted in average B values near those of the Nir. For 18488 reflections in the range of 10-3.2 Å the working R-factor is 0.245, and the Rfree 0.297, with reasonable geometry. No solvent is included.

Since the occupancy of the pseudoazurin is low, we tried two experiments to convince ourselves that it indeed contributed to the intensities. Removing it from the model results in an R-factor of 0.272 (R-free = 0.306); rotating it by 90° and refining results in R of 0.303 (R-free = 0.326). This model appears sufficient to locate side chains and the relative orientations of the molecules, but certainly is not one from which accurate small distances can be obtained.

The resultant model of the complex is shown in Figure 3.

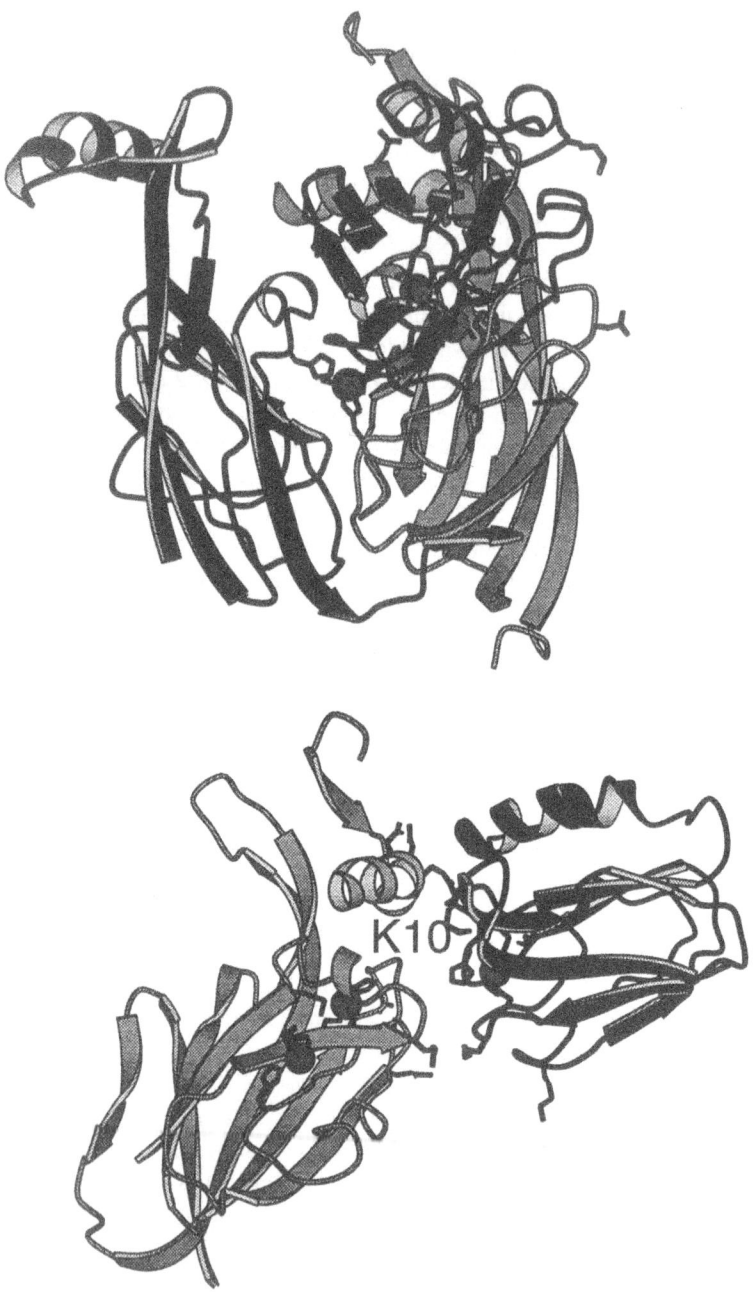

Figure 3. Complex of monomer of Nir and Paz, model from Form I crystal structure. Upper: view into active site. Lower: rotated 90° as in Figure 2. Only K10 of the charged residues is labelled

4.2 FORM II CRYSTALS

A second crystal form which diffracted to at least 2.6Å has $P2_12_12_1$ crystal symmetry and cell dimensions quite similar to the orthorhombic native AcNir. Nevertheless, the cell volume is such that the Vm is 2.44 for three pseudoazurin molecules plus an Nir trimer. This structure was also solved using Amore, and a search model consisting of the Nir trimer, yielded an R-factor of 0.31. Examination of a packing model in which the model complex was superimposed on the Nir found in the $P2_12_12_1$ cell suggested that there indeed was room for all of the pseudoazurin molecules to fit into the unit cell as well. Nevertheless, examination of a SigmaA difference map (2.8Å) phased with the Nir trimer failed to reveal any density interpretable as pseudoazurin. Mapping the orientation of the complex from the I222 cell into this cell showed that only one of the three possible sites for pseudoazurin would accomodate this orientation. The crystal contacts for theNir trimer in the pH6.2 orthorhombic cell are quite different than for the Nir trimer in this new cell.

5. Discussion

It can been seen in Figures 2 and 3 that the model found in the crystal has the proposed hydrophobic face of pseudoazurin interacting with the hydrophobic face of Nir. Our mutagenesis studies with AfNir indicated that mutating residues E118, E197, D201, E204and D205 to Ala caused an increase in Km, and that it was likely that pseudoazurin residues K10 interacted with Nir:E197; K77 and K57 interacted with Nir:E113 and E118 (towards the bottom of the lower part of Figure 2) , and that K38 interacted with Nir:E204.

  While the fundamental premise that charged residues interact across an otherwise hydrophobic interface, the specific interactions differ from the proposed model, probably because of sequence differences between *Achromobacter cycloclastes* and *Alcaligenes faecalis* Nir and Paz. While all of the lysine residues in AfPaz are also lysines in AcPaz, there is an additional lysine, 110, in AcPaz. Similarly, all the Asp and Glu residues examined in AfNIR are Asp and Glu in AcNir, except for E113 which is Q in AcNir.

  However a new negatively charged residue, E139 provides the dominant interaction actually seen in the crystal model, between Nir E139 and pseudoazurin K38. K10 is in the vicinity of E204. Another new interaction appears between Nir:E201 and AcPaz:K110. In AfNir residue 139 is a proline and in AfPaz, residue 110 is an isoleucine, so that these particular interactions are not likely in the AfNir system. AcPaz:R114 also interacts with AcNir: E204.

  Two views of the interface between AcNir and AcPaz are shown in Figure 4. The upper view is in the same orientation as Figure 3b. Aside from the charge-charge interactions at the outside of the interaction surface, the interactions are largely hydrophobic. The nearest Paz residue to the Nir Cu ligand His145 is His81, but Met141 of Nir lies between them.

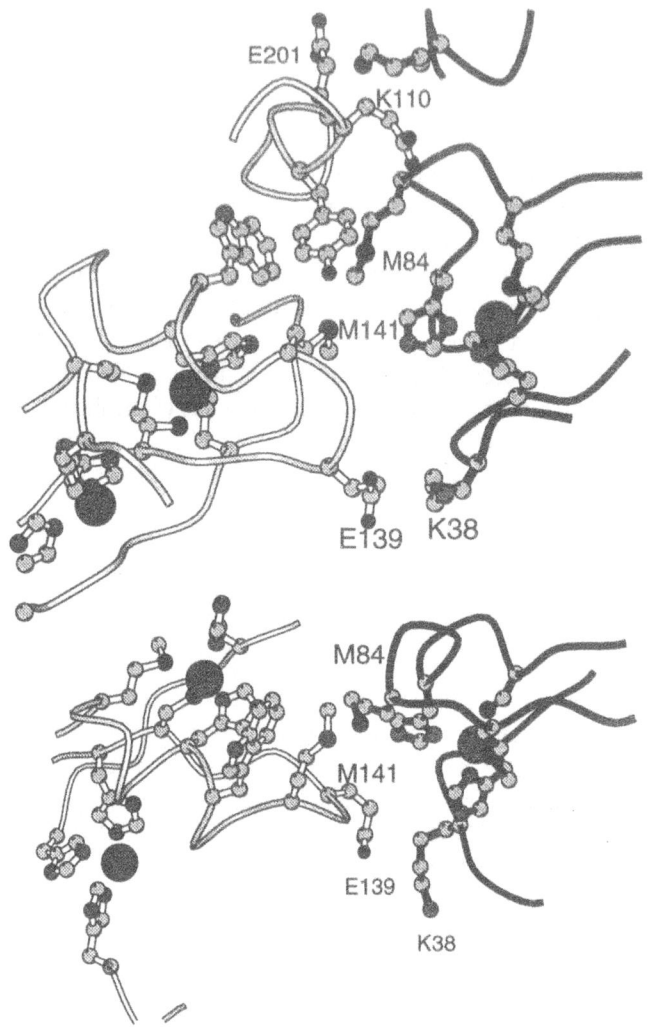

Figure 4. Interface of Nir Paz complex. Upper: close up of lower half of Figure 3. Lower: rotated 90° around horizontal axis from upper, omitting E201-K110 interaction for clarity.

The Cu-Cu distance is about 16Å; the NE2-SD distances 4-5Å. There does not appear to be a continuous path of connected bonds/hydrogen bonds between the copper atoms; rather the shortest path includes van der Waals interactions passing through the methionines. There is, however, room for a solvent molecule between His81 of Paz, and main chain atoms of the loop 139-141 of Nir, which might allow for a covalent/hydrogen bonded pathway between the Type I copper centers. The relative orientations of various side chains enclosing this region are intriguing: His81 is parallel to Nir Y203 which in turn is perpendicular to W144. Paz Met84 is packed up against Nir Y203, and Paz Met16 (not shown) is similarly packed up against His81. The internal Met7 and external Pro35 of Paz, seen to undergo small but significant conformational changes upon change in oxidation state [4] are far from the interaction surface with Nir, so it is unlikely that those conformational changes serve to signal oxidation state change as we once supposed.

K57 and K77 in Paz do not appear to interact with AcNir in this crystal complex: K77 is likely to have a reduced charge because of its interactions with surrounding acidic residues on Paz, and in any case faces solvent in Form I crystals. There is a crystal contact that involves a symmetry related molecule of AcNIR and the region near K57, although there are no visible specific interactions with it. Surface residues are frequently quite mobile so that it is difficult to judge exactly where they are. This problem is compounded here with having only a partially occupied pseudoazurin, and only a 3.2Å structure.

Figure 5 illustrates the overall complex seen in the Form I crystals. Clearly the presence of pseudoazurin had some effect on the crystallization of Nir, although crystallization from a relatively low ionic strength buffer may also have contributed. Low occupancy may have contributed to the absence of interpretable density for the Form II crystals: it may be that there are multiple orientations of Paz at a given site within the crystal, although this seems unlikely given the relatively high resolution of the crystals. There is precedent for rotationally disordered electron transfer partners in a crystal, in the early work of a complex of cytochrome c and cytochrome c peroxidase [22], but those crystals diffracted only to 3.3Å. Subsequent efforts with a recombinant cytochrome c peroxidase with two amino acid substitutions that should not have affected complex formation yielded well ordered crystals of the desired complex [23]. Further work will have to be done on our Form II crystals. Superposition of the Nir of the Form I complex onto the Nir found in the Form II crystals shows that if Paz is present, its orientation will have to be slightly different than in Form I crystals. Nevertheless it is significant that the surface on Nir that is proposed to interact with Paz *is* exposed to solvent in these crystals: Paz would just have to tilt somewhat differently.

It is our view that the oppositely charged surfaces provide a target for matching the two interacting surfaces, but that it may not be important that they have a specific relative orientation for electron transfer. It appears that if the complex we have described is a functional complex, a tunnel of relatively low dielectric constant with many polarizable electrons (from aromatic and sulfur-containing residues) lies between the coppers. On the other hand if this orientation is unique and essential, perhaps it provides a medium through which well extended wave functions afford the opportunity for quantum mechanical tunneling.

6. Acknowledgments

This work has been supported by NIH grant GM31770 and the Medical Research Council of Canada. We are extremely grateful to Dr. Jean Legall and Ming Liu, and to Professor T. Beppu, and Drs Makoto Nishiyama and Mutsuko Kukimoto for their generous collaborations. We also thank Bruce Averill, Gerard Canters and Antonio Xavier for their generosity to ETA.

126

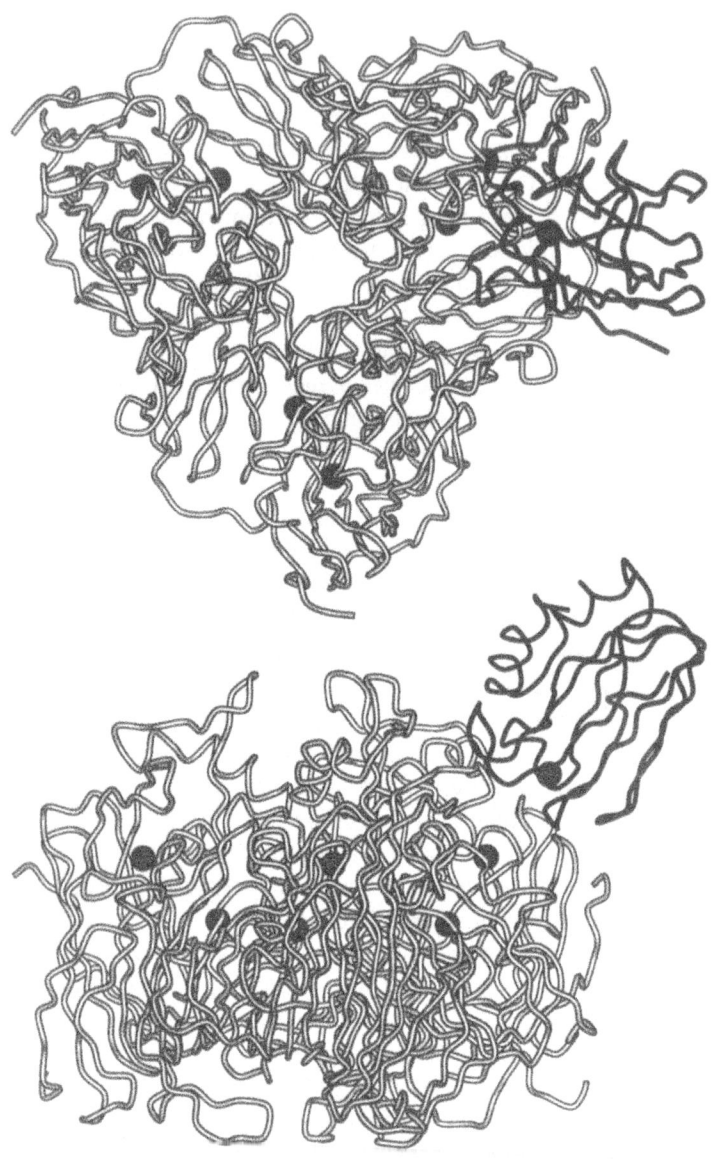

Figure 5. Complete complex of Nir and Paz as seen in Form I crystals. Upper: view along trimer axis, Lower: rotated 90° around horizontal axis

7. References

1. Kakutani, T., Watanabe, H., Arima, K., & Beppu, T. (1981) A Blue Protein as an Inactivating Factor for Nitrite Reductase from *Alcaligenes faecalis* Strain S-6. *J. Biochem.* **89**, 463–472.

2. Kakutani, T., Watanabe, H., Arima, K., & Beppu, T. (1981) Purification and Properties of a Copper-containing Nitrite Reductase from a Denitrifying Bacterium *Alcaligenes faecalis* Strain S-6. *J. Biochem.* **89**, 453–461.

3. Adman, E. T., Turley, S., Bramson, R., Petratos, K., Banner, D., Tsernoglu, D., Beppu, T. and Watanabe, H. (1989) A 2.0Å structure of the blue copper protein (cupredoxin) from *Alcaligenes faecalis* S-6. *J. Biol. Chem.* **294**, 87-99..

4. Peters-Libeu, C., Kukimoto, M., Nishiyama, M. and Adman, E.T. (1997) Site Directed Mutants of Pseudoazurin: Explanation of Increased Redox Potentials from Xray Structures and from Calculation of Redox Potential Differences, submitted to Biochemistry.

5. Godden, J.W., Turley, S., Teller, D.C., Adman, E.T., Liu, M.Y., Payne, W.J., and LeGall, J. (1991) The 2.3 Å structure of nitrite reductase from *Achromobacter cycloclastes*, *Science* **253**:,438-442., 1991.

6. Adman, E.T., Godden, J.W., and Turley, S. (1995) The Structure of Copper-nitrite Reductase from *Achromobacter cycloclastes* at Five pH values, with NO_2^- Bound and with Type II Copper Depleted, *J. Biol. Chem.* **270**, 27458-27474.

7. Kukimoto,M., Nishiyama, M., Murphy, M.E.P., Turley, S., Adman, E.T., Horinouchi, S., and Beppu, T.. (1994) ray Structure and site-directed mutagenesis of a nitrite reductase from *Alcaligenes faecalis* S-6: roles of two copper atoms in nitrite reduction, *Biochemistry* **33**, 5246-5252.

8. Murphy, M.E.P., Turley, S., Adman, E.T., Kukimoto, M., Nishiyama, M., Sasaki, H., and Tanokura, M. (1995) The Structure of *Alcaligenes faecalis* Nitrite Reductase, and a Copper Site Mutant M150E, That Contains Zinc, *Biochemistry* **34**, 12107-12117.

9. Work presented at American Chemical Society Meeting, March, 1996, and International Union of Crystallography, August 1996.

10. Murphy, E.P., Turley, S., and Adman, E.T. (1997) Binding of Nitrite to Oxidized and Reduced Forms of *A. faecalis* Nitrite Reductase, Submitted.

11. Kukimoto, M., Nishiyama, M., Ohnuki, T., Turley, S., Adman, E.T., Horinouchi, S., and Beppu, T. (1995) Identification of interaction site of pseudoazurin with its redox partner, copper-containing nitrite reductase from *Alcaligenes faecalis* S-6, *Protein Engineering* **8**, 153-158.

12. Kukimoto, M., Nishiyama, M., Tanokura, M. Adman, E.T., Horinouchi, S. (1996) Studies on Protein-Protein Interaction between Copper-containing Nitrite Reductase and Pseudoazurin from *Alcaligenes faecalis* S-6, *J. Biol. Chem.* **271**, 13680-13683.

13. Tsukihara, T., Aoyama, H., Yamashita, E., Tomizaki, T., Yamaguchi, H., Shinzawa-itoh, K., Nakashima, R., Yaono, R., and Yoshikawa, S., (1996) The Whole Structure of the 13-Subunit Oxidized Cytochrome c Oxidase at 2.8 Å, *Science* **272**, 1136-1144.

14. Liu-M-Y., Liu, M.-C., Payne, W.J. and Legall, J. (1986) Properties and electron transfer specificity of copper proteins from the denitrifier *Achromobacter cycloclastes J. Bacteriol.* **166**, 604-608.

15. Otwinowski, Z. (1993) Oscillation data reduction program In Data Collection and Processing, Sawye,r L., Evans, P. R., Leslie, A. G. W. eds Proceedings of the CCP4 Study weekend. UK SERC Daresbury Laboratory pp 87-91.

16. Collaborative Computational Project, Number 4 (1994) *Acta Crystallogr.* **D50**, 760–763

17. Brünger, A.T. (1990) *X-plor, Version 3.1* Yale University, New Haven.

18. Read-R-J. (1986) Improved Fourier coefficients for maps using phases from partial structures with errors, *Acta Crystallogr* **A42**, 140-149.

128

19. Engh, R.A. & Huber, R. (1991) Accurate bond and angle parameters for X-ray protein structure refinements, *Acta Crystallogr.* **A47**, 392–400.

20. Jones, T. A., Zou, J.-Y., Cowan, S. W.,& Kjeldgaard, M. (1991) Improved methods for building protein models in electron density maps and the location of errors in these models, *Acta Crystallogr.* **A47**, 110–119.

21. Kraulis, P. (1991) MOLSCRIPT: a program to produce both detailed and schematicplots of protein structures, *J. Appl. Cryst.* **24**, 946–950.

22. Poulos, T, L. , Sheriff, S., Howard, A. J. (1987) Cocrystals of yeast cytochrome *c* peroxidase and horse heart cytochrome *c*, *J. Biol. Chem.* **262**, 13881-4.

23. Pelletier, H., and Kraut, J. (1992) Crystal structure of a complex between electron transfer partners, cytochrome *c* peroxidase and cytochrome *c*, *Science* **258**, 1748-1755.

STRUCTURAL RESEARCH ON THE METHYLAMINE DEHYDROGENASE REDOX CHAIN OF *PARACOCCUS DENITRIFICANS*

F. S. MATHEWS, Z.-W. CHEN, R.C.E. DURLEY
Department of Biochemistry and Molecular Biophysics, Washington University Medical School
St. Louis, MO 63110 USA
V.L. DAVIDSON, L.H. JONES, M.E. GRAICHEN AND J.P. HOSLER
Department of Biochemistry, University of Mississippi Medical Center, Jackson, MS 39216, USA
A. MERLI, D.E. BRODERSEN, G.L. ROSSI
Istituto di Scienze Biochimiche, Universita di Parma
43100 Parma, Italy

1. The Methylamine Dehydrogenase Enzyme System.

Methylotrophic bacteria can utilize C_1 compounds as the sole source of carbon and energy. Oxidation of methylamine in these organisms is catalyzed by methylamine dehydrogenase (MADH), an inducible, periplasmic enzyme. This enzyme is a quinoprotein and contains the novel cofactor tryptophan tryptophylquinone (TTQ) (Fig. 1), which is derived from two tryptophan side chains [1], one of which has been modified to contain an orthoquinone. The reaction catalyzed by MADH is

$$CH_3NH_3^+ + H_2O \rightarrow NH_4^+ + CH_2O + 2H^+ + 2e^-.$$

Subsequently, electrons are transferred to the membrane bound terminal oxidase, cytochrome aa_3 via a series of soluble electron carrier proteins. In facultative autotrophs such as *Paracoccus denitrificans* the initial electron acceptor in this chain is amicyanin, a blue copper protein [2]. The next acceptor along the chain is cytochrome c_{551i} [3]. All three

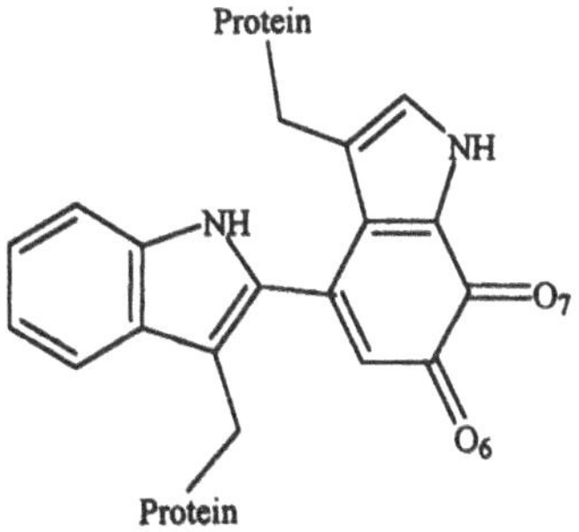

Figure 1. Tryptophan tryptophylquinone (TTQ), the cofactor of MADH.

129

G.W. Canters and E. Vijgenboom (eds.),
Biological Electron Transfer Chains: Genetics, Composition and Mode of Operation, 129-146.
© 1998 *Kluwer Academic Publishers.*

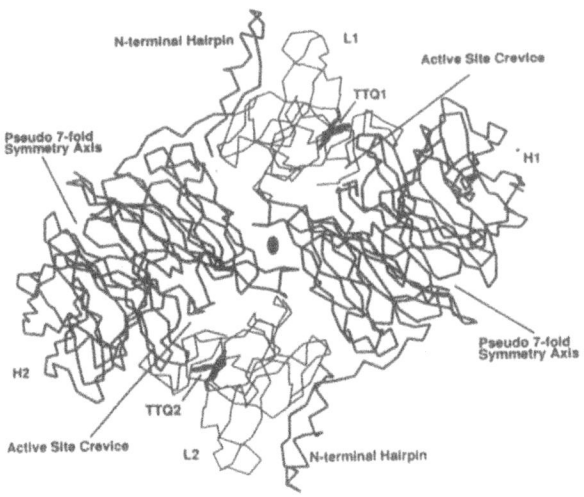

Figure 2. Cα diagram of the H₂L₂ tetramer of MADH viewed along the molecular 2-fold axis. Reproduced with permission [31].

proteins are induced when the bacterium is grown on methylamine as the sole carbon source. A possible sequence of electron carriers for this organism [4] is

$$MADH \rightarrow Amicyanin \rightarrow Cyt\ c_{551i} \rightarrow Cyt\ c_{550} \rightarrow Cyt\ aa_3$$

MADH is a hetero-tetramer consisting of two identical pairs of heavy (H) and light (L) subunits (Fig. 2) of 47 kDa and 15 kDa molecular mass, respectively. The crystal structure of MADH from *P. denitrificans* (PD-MADH) is known at 1.75 Å resolution [5]. The H subunit consists of 7 antiparallel 4-stranded β-sheets [Fig. 3] of the type first observed in influenza neuraminidase [6]. The L subunit consists of five β-strands (Fig. 3) which form two antiparallel β-sheets and contains the TTQ prosthetic group. This subunit is crosslinked by 6 disulfide bridges and an additional covalent bond between tryptophylquinone 57 (Trq) and Trp 108 which together form the TTQ cofactor.

Anaerobic reductive titration of the TTQ cofactor in MADH by dithionite proceeds through a semiquinone intermediate with distinct spectral properties [7]. When reduced by methylamine, the enzyme is converted to the fully reduced form with no production of a semiquinone intermediate. The 2-electron redox potential for the TTQ cofactor is +100 mV. Steady state kinetic studies of MADH using phenazine ethosulfate as electron acceptor support a ping-pong mechanism [8]. Using methylamine as substrate, the deuterium kinetic isotope effect is 3.0, indicating that proton abstraction from substrate is partially rate limiting. A variety of primary amines can be oxidized by MADH, but no secondary or tertiary amines.

Amicyanin has a molecular mass of about 12.5 kDa [9]. It is a one-electron carrier with a redox potential of +294 mV and is an obligatory intermediate in the transfer of electrons from methylamine dehydrogenase to cytochrome c_{551i} [10]. The amicyanin gene is located immediately downstream of that for the small subunit of MADH; inactivation of the former

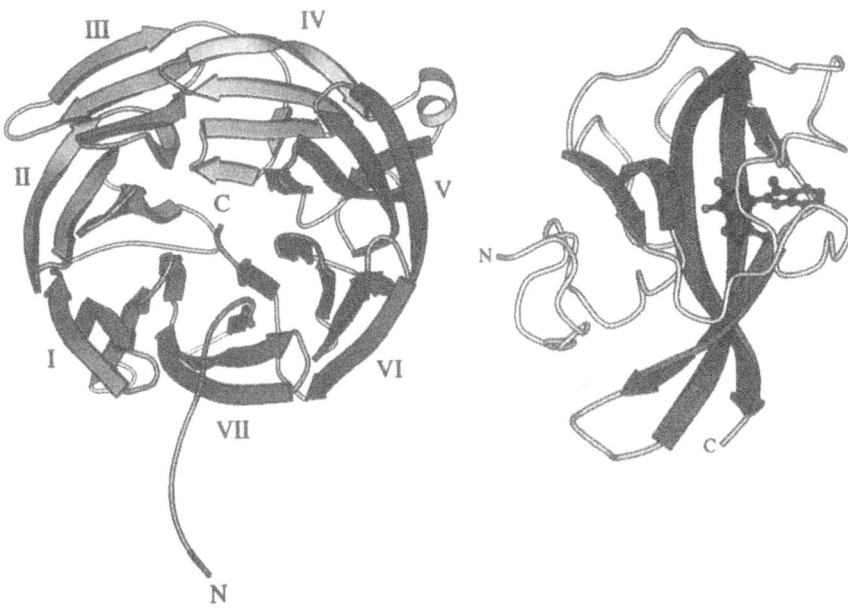

Figure 3. Schematic ribbon drawings of the H subunit (left) and of the L subunit (right) of PD-MADH.

by means of gene replacement results in complete loss of the ability of the bacterium to grow on methylamine [9]. The crystal structure of amicyanin is known at 1.3 Å resolution [11]. The protein has a β sandwich topology with 9 β strands forming two mixed β sheets (Fig. 4). The copper atom is located in a pocket between the β sheets and has four coordinating ligands, two histidine nitrogens, one cysteine sulfur and one methionine sulfur.

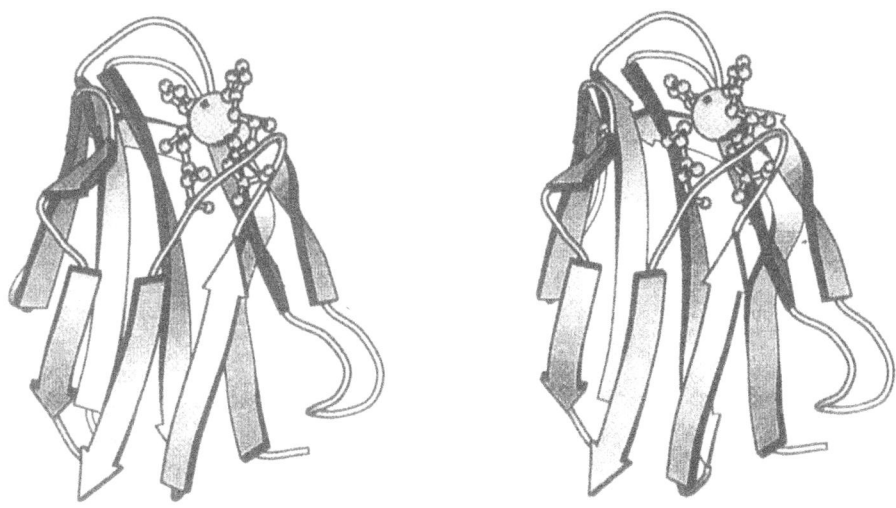

Figure 4. Stereo diagram of amicyanin. The copper ion and its four ligands, His53, Cys92, His95 and Met98 are shown.

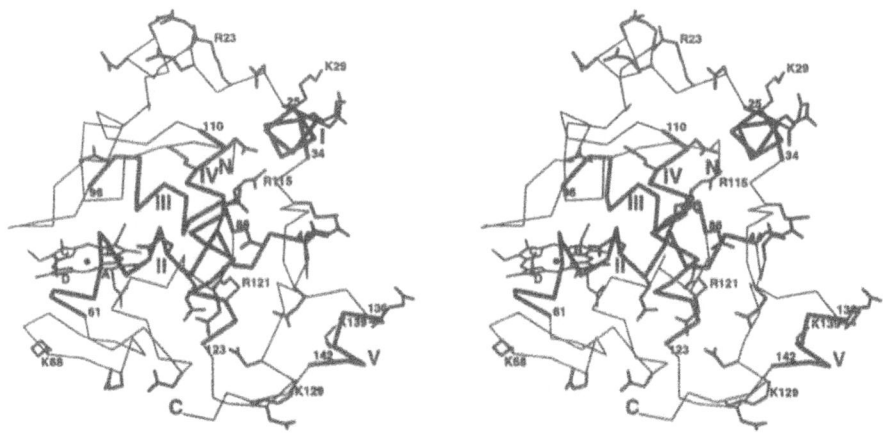

Figure 5. Stereo diagram of cytochrome c_{551i}. The Cα-atoms plus acidic and basic side chains are shown. The 5 α-helices are labeled I-V. The starting and ending Cα atoms of each helix and the basic side chains, as well as the heme pyrrole rings are labeled.

Its amino acid sequence and secondary and tertiary structures are very similar to those of plastocyanin, but with an additional 20 residues at the N terminus with no counterparts in the other blue copper proteins; a portion of these residues forms the first β strand of the structure.

Cytochrome c_{551i} is located in the periplasm of *P. denitrificans* and is induced during growth on methylamine or methanol [3]. Its gene is a member of the methanol oxidation (*mox*) gene cluster [12] and is believed to be the primary acceptor of electrons from methanol dehydrogenase. It is the most efficient electron acceptor from amicyanin *in vitro* when compared to two other periplasmic cytochromes, cytochrome c_{553i} and cytochrome c_{550} [3]. Cytochrome c_{551i} has a molecular mass of 17.5 kDa, a pI of 3.3 and a redox potential of +190 mV [13]. The heme group is partially exposed to solvent, at the vinyl and

Figure 6. Stereo view of a ribbon diagram of the ternary complex. Only half the hetero-octamer, HLAmiCyt is shown. Reproduced by permission [32].

TABLE 1. MADH-amicyanin interactions. The atom involved follows the residue; (H) and (L) refer to the H or L subunit, respectively.

Amicyanin	MADH	Distance (Å)
Van der Waals (≤ 3.5 Å)		
Ala50 O	Leu71 $C^{\delta 2}$ (L)	3.4
Met71 C^{γ}	Thr54 $O^{\gamma 1}$ (L)	3.5
Lys73 C^{γ}	Val127 O (L)	3.3
His91 $N^{\epsilon 2}$	Pro145 O (H)	3.3
His95 $N^{\epsilon 2}$	Glu101 C^{β} (L)	3.1
Phe97 C^{α}	Asp167 $O^{\delta 2}$ (H)	3.3
Phe97 $C^{\epsilon 2}$	Pro100 C^{γ} (L)	3.5
Phe97 C^{ζ}	Arg184 $N^{\epsilon 2}$ (H)	3.4
Salt Bridge		
Arg99 $N^{\eta 2}$	Asp167 $O^{\delta 1}$ (H)	2.7
Arg99 $N^{\eta 2}$	Asp167 $O^{\delta 2}$ (H)	2.7
Potential Salt Bridge		
Lys68 N^{ζ}	Asp115 $O^{\delta 2}$ (L)	4.4

Additional Interface residues (> 3.5 Å separation)

Amicyanin	MADH	
Met28	Phe43 (H)	Ala55 (L)
Met51	Ser144 (H)	Ser56 (L)
Pro52	Ala146 (H)	Val58 (L)
Asn54	Tyr169 (H)	Phe102 (L)
His56		Trp108 (L)
Val58		Phe110 (L)
Lys68		Gly111 (L)
Gly69		Asp115 (L)
Thr93		Lys129 (L)
Pro94		
Pro96		
Arg99		

methylene carbon atoms on pyrrole ring C, and is surrounded by a very hydrophobic patch of residues (Fig. 5) [14]. In the crystal lattice of the ternary complex, two cytochrome molecules pack together about a crystallographic 2-fold axis in such a way that this patch is buried. The protein is highly acidic, having 27 acidic residues and eight basic residues in the molecule. Most of the charged residues are located on the surface of the molecule. However, these charged residues are not distributed evenly over the surface, but are located mostly on the sides and back of the molecule, away from the heme group and its hydrophobic surroundings (Fig. 5).

2. Binary and Ternary Complexes

Complex formation between MADH and amicyanin in solution causes perturbation of the absorption spectrum of TTQ and a 73 mV decrease in the redox potential of the copper center of amicyanin [15]. This decrease in potential facilitates an otherwise thermodynamically unfavorable electron transfer from amicyanin (Em = 295 mV) to cytochrome c_{551i} (Em = 190 Mv), to which it subsequently donates electrons. Kinetic [16] and chemical crosslinking [17] studies suggest a role for both electrostatic and hydrophobic interactions in stabilizing complex formation. Resonance Raman studies [18] indicate no structural change of either the protein-bound TTQ or copper on complex formation, implying that the observed spectral perturbations of TTQ and redox potential change of amicyanin were probably due to inductive effects.

TABLE 2. Amicyanin-cytochrome c_{551i} interactions.
The atom involved in the interaction follows the residue.

Amicyanin	Cyt c_{551i}	Distance (Å)
Van der Waals ≤ 3.5 Å		
Asp24C$^\beta$	Tyr77C$^{\epsilon2}$	3.4
Glu31C$^\delta$	Asp75O$^{\delta1}$	3.5
Glu31O	Pro71C$^\alpha$	3.2
Thr32C$^{\gamma2}$	Ile69O	3.4
Glu34O$^{\epsilon1}$	Ile69C$^\delta$	3.4
Arg48$^{N\epsilon}$	Tyr77O$^\eta$	3.4
Hydrogen Bonds		
Asp24O$^{\delta1}$	Thr79O$^{\gamma1}$	3.4
Glu31O$^{\epsilon2}$	Asp75O$^{\delta1}$	2.7
Glu31O	Gly72N	3.1
Glu34N	Lys68O	3.4
Salt Bridge		
Lys29N$^\zeta$	Asp75O$^{\delta1}$	2.9
Potential Salt Bridge		
Glu34O$^{\epsilon1}$	Lys68N$^\zeta$	5.9
Bridging Solvent Molecules		
Sol1	Glu31O$^{\epsilon1}$ (Ami)	2.7
Sol1	Gly72O (Cyt)	2.6
Sol1	Asp75O$^{\delta2}$ (Cyt)	2.7
Sol1	Tyr77O (Cyt)	2.5
Sol88	His36N (Ami)	2.9
Sol88	Ile69O (Cyt)	3.0

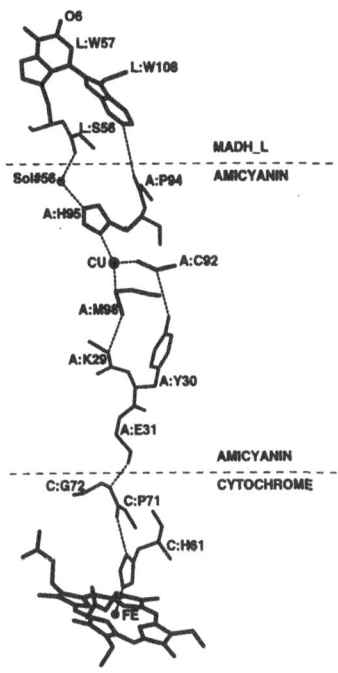

Figure 7. Pathways calculated for electron transfer in the ternary complex. Two paths of roughly similar coupling link O6 of TTQ and the copper of amicyanin. Between the copper and the heme iron atom, there are two branches in amicyanin, one through Cys92 and the other through Met98.

Crystals of the binary and ternary complexes, grown from 2.4 M Na/K-phosphate buffer, pH 6.5, are distinct from one another. The former have a tetragonal unit cell containing one heterohexamer, $(HLAmi)_2$ in the asymmetric unit [19] while the latter have a C-centered orthorhombic cell containing one heterotetramer, HLAmiCyt, in the asymmetric unit [14]. Both complexes will form crystals which are isomorphous with the native when apo-amicyanin is substituted for amicyanin. The amicyanin/MADH interface is very similar in the three crystallographic environments in which it is found in the binary [19] and ternary [14] complexes. For brevity, only the ternary complex will be discussed, except where differences exist.

The arrangement of the cytochrome, amicyanin and one HL dimer of MADH is shown in Fig. 6. The TTQ, copper and heme groups are arranged in a linear fashion, placing the cytochrome and MADH on opposite sides of the amicyanin. The distance from O6 of TTQ to the copper atom is 16.8 Å while the atom on TTQ closest to copper, CH2 of Trp 108, is 9.3 Å away (see Fig. 7). The copper to iron distance is 24.8 Å and the distance from copper to the nearest atom of the heme is about 21 Å. The MADH-amicyanin interface (Table 1), has a surface area of about 750 $Å^2$. Twenty four residues of MADH (2/3 from the L subunit) are associated with the interface, of which 15 are hydrophobic. Likewise, 10 out of 18 residues of amicyanin within the interface are hydrophobic. The interface is quite hydrophobic, despite the presence of the 8 charged residues and 9 neutral hydrophilic residues since most of their side chains are directed into solution at the edge of the interface.

136

There is one strong salt bridge (2.7 Å separation, partially buried) and one weak salt bridge (4.4 A separation, solvent accessible) connecting the H and L subunits to amicyanin, respectively. In addition, 3 water molecules bridge MADH and amicyanin.

The amicyanin-cytochrome c_{551i} interface (Table 2) is considerably different from the MADH-amicyanin interface. It is smaller, about 425 Å^2 in area and more polar. Only 2 of the 10 amicyanin residues in the interface are hydrophobic and 6 of the 12 interface residues of the cytochrome are hydrophobic. Thus, approximately 65% of the residues in the amicyanin-cytochrome c_{551i} interface are hydrophilic compared with about 40% in the MADH-amicyanin interface. The number of connections between amicyanin and the cytochrome is also greater than between amicyanin and MADH, with 1 salt bridge, 3 main chain to main chain hydrogen bonds and one Glu-Asp interaction, possibly stabilized by protonation of one of the acidic side chains or neutralized by a cation. Two solvent molecules are also involved in bridging the subunits. In addition to the direct connections between amicyanin and cytochrome c_{551i}, there is a potential salt bridge between Glu34 of amicyanin and Lys68 of the cytochrome which, in the crystal structure, point away from the interface into solution. An interaction between them can be modeled and might form at lower ionic strength. The remaining charged or neutral groups point into solution, away from the interface, except for Thr32 which forms a hydrogen bond to a solvent molecule within the interface.

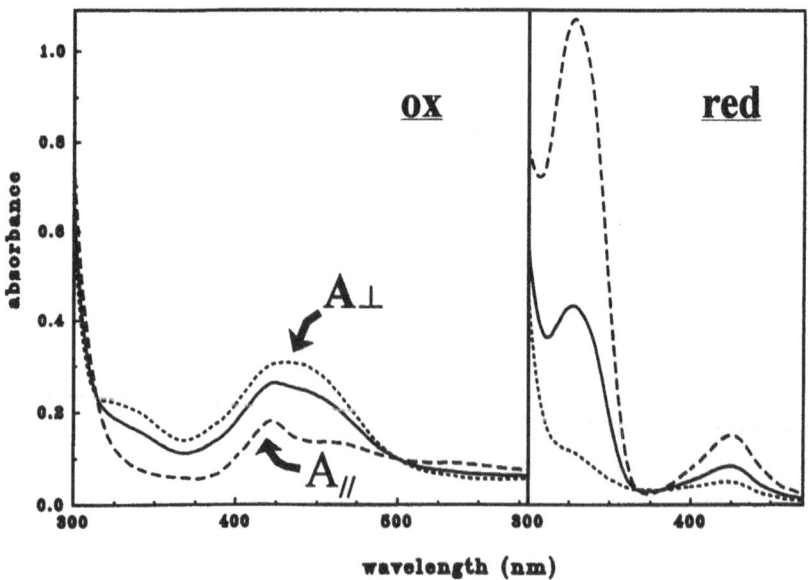

Figure 8. Polarized absorption spectra of a single crystal of the binary complex between MADH and apo-amicyanin recorded at pH 7.5. On the left is the oxidized complex and on the right is the complex reduced by 0.2 mM methylamine. The solid lines show the isotropic equivalent spectra computed from the two polarized spectra shown in dashed and dotted lines. Reproduced by permission from [21].

Two prominent electron transfer paths, suggested by the program PATHWAYS-II [20], connect the copper atom to the indole ring of Trp57 (Fig. 7). One utilizes Trp108 of TTQ while the other follows two main chain residues of the L subunit. The computed relative efficiencies of these two paths indicates that in the latter the coupling is about 3-fold more efficient than the former, but depends on the presence of a water molecule which might not always be occupied, thereby reducing the relative efficiency of that pathway. In the case of the electron transfer from copper to iron, the two most prominent paths follow either Cys92 or Met98 (Fig. 7), both of which are copper ligands, to a common point and continue through three residues of amicyanin and two of the cytochrome. Both branches appear to have approximately equal efficiency.

3. Activity in the Crystals

Although the binary and ternary complexes provide much detailed structural information about the interacting surfaces and arrangement of cofactors among the partners, and suggest potential pathways for electrons to flow during transfer, questions do arise about their physiological relevance and catalytic competence. To address these questions, we have undertaken a single crystal polarized absorption study of the reactivity of the complexes in their crystalline states [21]. This technique is useful for probing the redox properties of proteins in the solid state [22]. Spectra were recorded of crystals of the MADH binary and ternary complexes prepared using either copper-containing amicyanin or copper-free apo-

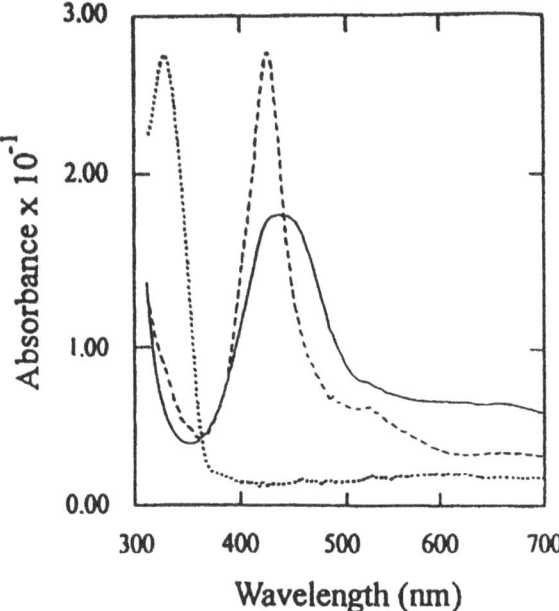

Figure 9. Solution spectra of the three redox forms of MADH at pH 7.5. The spectrum of the oxidized TTQ is solid, that of the semiquinone is dashed and that of the fully reduced is dotted. Reproduced by permission from [23].

138

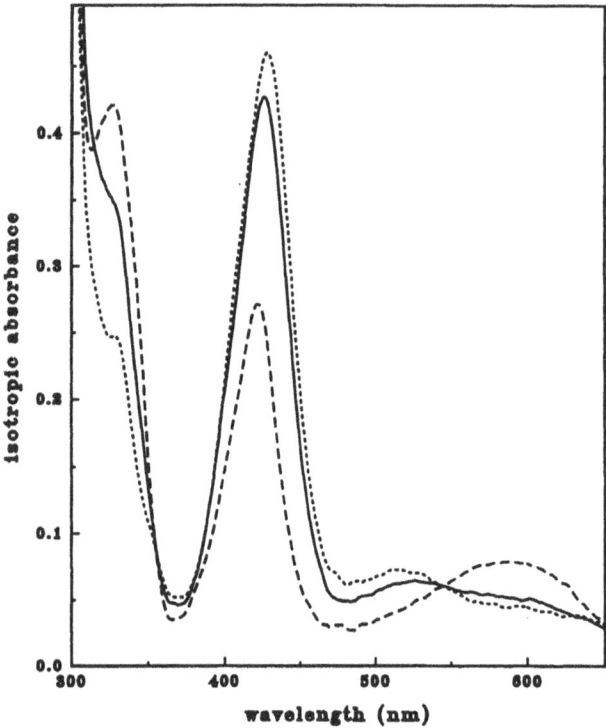

Figure 10. Isotropic spectrum of a single crystal of the MADH-amicyanin binary complex after reduction by 0.2 mM methylamine, observed at pH 5.7 (dashed line), pH 7.5 (solid line) and pH 9.0 (dotted line). The three absorption maxima, at 330 nm, 420 nm and 590 nm correspond to the reduced TTQ, TTQ semiquinone and Cu(II), respectively. Reproduced by permission from [21].

amicyanin. The crystals of these holo and apo complexes are isomorphous, which is very advantageous for these microspectrophotometry studies since it provides an internal control, allowing studies of the reactions of MADH in these crystalline complexes with and without the possibility of electron transfer through the copper atom.

Polarized absorption spectra of crystals of the apo-binary complex (Fig. 8) show that they are reduced fully by methylamine. The isotropic equivalent spectra of both oxidation states, computed from the two polarization directions, are similar to those observed in solution (Fig. 9) [23]. Minor variations can be attributed to differences in ionic strength, composition of the medium and pH [24]. Thus, the crystal lattice and the presence of apo-amicyanin in complex with MADH has only a minor effect on the spectra. The relative absorbance in the two directions of polarization differs markedly in the two oxidation states. In the oxidized crystal, $A_\perp$ is generally 2-fold larger than $A_\parallel$ while the reverse is true for the reduced crystal, indicating that the direction of the transition dipole moment of TTQ changes when the enzyme is reduced. These changes may reflect changes in the electronic structure and/or orientation of the TTQ.

When methylamine is added to crystals of the holo-binary complex, spectral changes

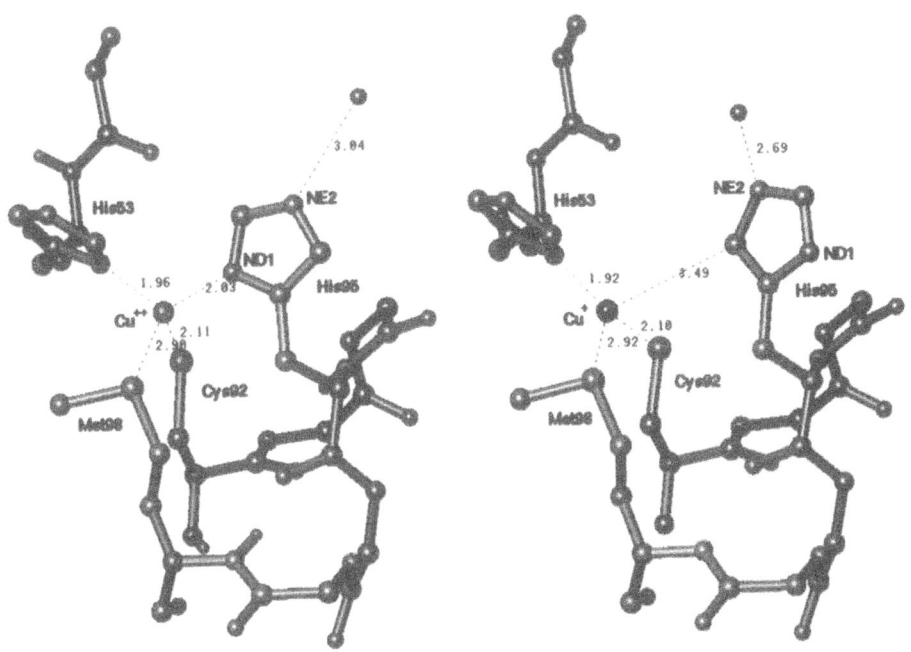

Figure 11. Copper coordination of oxidized amicyanin (left), pH ~5.5, and of reduced amicyanin (right), pH ~5.5. In the latter, His 95 has rotated 180° about the Cα-Cβ bond so that atom ND1 is no longer coordinated to the copper atom.

indicate formation of significant amounts of the semiquinone form of TTQ (Fig. 10). The only way the semiquinone can be formed in these crystals is by transfer of one electron from TTQ to the copper of amicyanin after the TTQ has first been reduced fully to the hydroquinone form by substrate . This demonstrates that MADH in the crystalline holo-binary complex is competent both in catalysis and electron transfer. The amount of semiquinone formed is dependent upon pH. At pH 5.7 a large fraction of the TTQ remains reduced and a significant absorbance by Cu(II) at about 600 nM can be observed, whereas at pH 9.0 the TTQ is mostly in the semiquinone form. Furthermore, after the reaction of the crystal with methylamine is complete, the ratio of semiquinone to reduced TTQ can be shifted reversibly by altering the pH, suggesting that the difference between the redox potentials for the TTQ semiquinone/reduced couple and the Cu++/Cu+ couple is pH dependent.

Two factors could cause the electron distribution between TTQ and the copper in the binary crystals to be pH dependent. One is stabilization of the TTQ semiquinone at high pH. This could arise, for example, by dissociation of a proton from reduced TTQ but not from the semiquinone form at high pH. Redistribution of electrons at high pH between reduced and oxidized MADH to form semiquinone has been observed at high pH during titration of MADH with substoichiometric amounts of methylamine at low ionic strength [25]. The other factor could be a pH dependence of the redox potential of amicyanin when it is complexed with MADH. Reduction of amicyanin in the crystalline state produces a conformational change below pH 6 resulting from protonation of the exposed histidine

140

ligand of copper followed by rotation of 180° about the Cα-Cβ bond and a movement away from the copper into solution (Fig. 11) [26]. In the binary complex, such a histidine flip would move the imidazole ring about 0.7 Å closer to MADH, according to model building studies, promoting disruption of the MADH-amicyanin interface. This would destabilize the reduced form of amicyanin in the complex and lower its redox potential with respect to TTQ at lower pH.

Within the holo-ternary complex, heme reduction occurs when crystals are treated with methylamine at pH 7.5 (Fig. 12). The rate of heme reduction is dependent on pH. Even after several days there had been little reduction of heme when substrate was added at pH 5.7. This is consistent with the observation that electron transfer from reduced TTQ to copper in the crystals of the binary complex is much less favorable at that pH. At pH 6.5, many hours are required to obtain even partial heme reduction while at pH 9.0 reduction is nearly complete within 35 minutes. When methylamine is added to the apo-ternary complex crystal at pH 7.5, very little heme reduction occurs after 1 hour, but is nearly complete after four days. Conversely, no reduction of the heme occurs at pH 5.7. Since the overall distance from TTQ to heme is the same in the holo-ternary and apo-ternary complexes and pathways for electron transfer from TTQ to heme are undoubtedly available within the apo-ternary

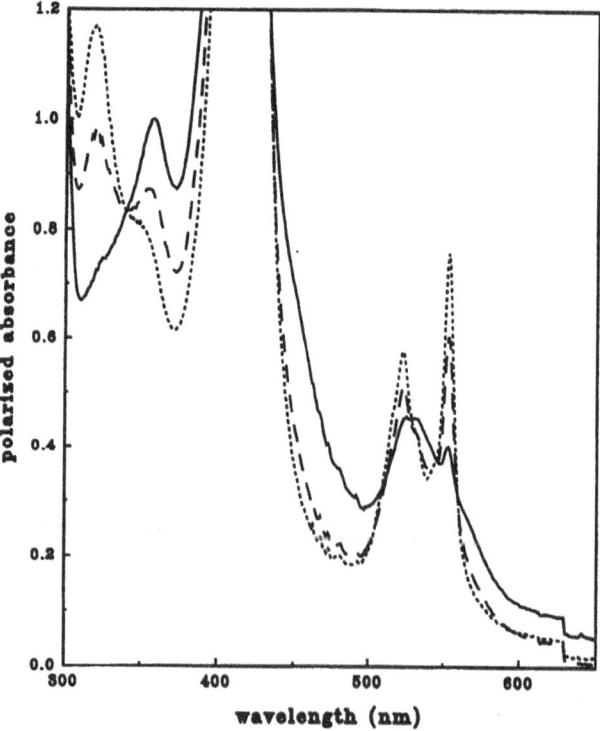

Figure 12. Polarization absorption spectrum of a single crystal of the ternary complex between MADH, amicyanin and cytochrome c_{551i} recorded at pH 7.5. Spectra are recorded for the native complex (solid line) and at 13 minutes (dashed line) and 60 minutes (dotted line) after addition of 0.2 mM methylamine. Reproduced by permission from [21].

TABLE 3. Analysis of Site-Directed Mutants of Amicyanin

	Wild Type	R99D	R99L	K68A	R99L/K68A	M71R	F97E
Steady State K_m[1]							
10 mM buffer	1 µM	76 µM	31 µM	3 µM	98 µM	2 µM	151 µM
+200 mM NaCl	8 µM	64 µM	59 µM	7 µM	60 µM	23 µM	ND
K_d[2]							
10 mM buffer	<5 µM	90 µM	59 µM	<5 µM	85 µM	<5 µM	228 µM
+200 mM NaCl	15 µM	179 µM	158 µM	5 µM	107 µM	22 µM	ND

1. K_m is the Michaelis constant for the substrate-dependent reduction of oxidized amicyanin by MADH under steady-state conditions.
2. K_d is the equiliburium constant for complex formation between fully reduced MADH and oxidized amicyanin determined from transient kinetic experiments.

complex, the fundamental question arises as to how the pH and the presence of copper in the ternary complex can enhance, by some orders of magnitude, the rate of the electron transfer reaction from TTQ to the heme in the crystalline state.

The results of the single crystal microspectrophotometric studies show that the holo-binary and the holo-ternary complexes are competent both for substrate oxidation and for electron transfer. They do not prove that the proteins are oriented in the crystallized complexes exactly as they would be in solution or that it is the only possible orientation for these proteins. The studies do, however, clearly demonstrate that catalysis and long range electron transfer from TTQ to copper and from TTQ to heme via copper can and do occur in a predictable manner when the proteins are present in this orientation.

4. Mutational Studies

Molecular biology provides an important tool for studying the electron transfer properties of the MADH system. Using site-directed mutagenesis, it is possible to study the effects of changes at specific amino acid sites or combinations of sites on the electron transfer process.

The amicyanin molecule is the simplest target for mutagenic studies. It has been expressed heterologously in E. coli, in the case of the *Thiobacillus* protein [9], as have other type I blue copper proteins. Expression of the MADH gene will be much more difficult since a modified form of *Paracoccus* or a related methylotroph is probably needed for the host system since post translational modification of the tryptophan precursor of TTQ of the L subunit requires some products of the *mau* gene cluster of the organism [27]. Expression of cytochrome c_{551i} is possible in principle, since covalent attachment of the heme in c-type cytochromes has been achieved [28], but considerable effort will be required to determine the optimal conditions for such expression.

TABLE 4. Ionic strength and pH effects on Amicyanin-MADH complex formation

Spectral Perturbations: Features in Difference Spectra

Present at low ionic strength (10 mM phosphate)
Absent at high ionic strength (200 mM NaCl)

Membrane Retention, Centricon 30: 2:1 Mixture of MADH and amicyanin

Amicyanin retained at low ionic strength
Amicyanin released at 200 mM NaCl

Redox Potential of Amicyanin in the presence of MADH

Low ionic strength: 221 mV
200 mM NaCl: 278 mV
(Amicyanin alone): 294 mV

Crosslinking Studies with EDC[1]

At pH 7.5, low ionic strength:
A·H and A·H·L bands present
At pH 7.5, 200 mM NaCl:
A·H band, A·H·L band less pronounced
At pH 6.5, low ionic strength:
A·H and A·H·L bands present
At pH 6.5, 200 mM NaCl:
only A·H band present

1. EDC: 1-ethyl-3-[3-(dimethylamino)propyl)carbodiimide.

Recently, the *mau*C gene for amicyanin from *P. denitrificans*, originally prepared in pUC18 by Andrei Chistoserdov (SUNY, Stoney Brook), has been expressed in *E. coli* using a pUC19 plasmid and the BL21 expression system [29]. This has enabled several amicyanin mutants to be designed and overexpressed on the basis of the structure of the binary and ternary complexes. Fortunately, amicyanin is central to the ternary complex, since it interacts with both MADH and cytochrome c_{551i} and contains the conduits for electron transfer between them.

So far, 6 mutant forms of amicyanin have been prepared and analyzed in a preliminary fashion (Table 3) [29]. All of these mutations at the MADH/amicyanin interface.

The interaction between MADH and amicyanin appears to be stabilized by both electrostatic and hydrophobic forces. The importance of the electrostatic interactions was demonstrated by membrane retention, TTQ spectral perturbation and reduction of copper redox potential in the complex, all of which were abolished by high ionic strength (Table 4) [30]. On the other hand, crosslinking between MADH and amicyanin, although somewhat reduced, was maintained under conditions of high ionic strength [17].

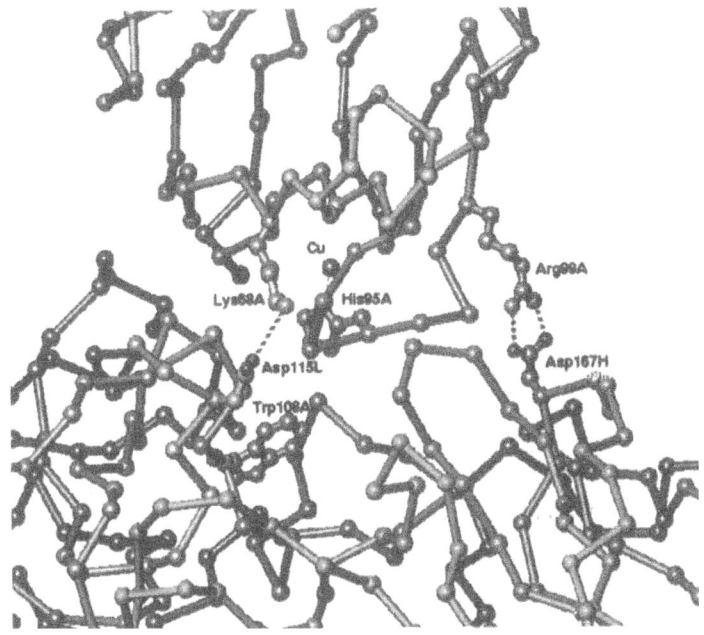

Figure 13. MADH-amicyanin interface showing two salt bridges, a strong one between Arg99A and Asp167H and a weak one between Lys68A and Asp115L. The copper, His95A and Trp108L are also indicated.

There is one salt bridge, connecting Arg99 of amicyanin with Asp167 of the H subunit, and one "potential" salt bridge connecting Lys68 of amicyanin with Asp115 of the L subunit (Table 1, Fig. 13). To test whether either of these might contribute to the electrostatic stability of the MADH/amicyanin complex, R99 was mutated to D and L and K68 was mutated to A (Table 3). A double mutation, R99L/K68A was also made. The R99 mutations resulted in a marked decrease in the affinity of amicyanin for MADH and made the affinity insensitive to ionic strength. In contrast, the K68A mutation had almost no effect at low salt, but does influence the ionic strength dependence, as the binding is still tight at high salt. The R99L/K68A double mutant showed roughly 2-fold higher K_d indicating contributions of both pairs of interactions to the stability of the complex.

Amicyanin contains 7 hydrophobic groups surrounding His95 which are buried in the interface, including Met71 and F97 (Table 1, Fig. 14). When Met71 is changed to R, there is little effect on the binding affinity of amicyanin for MADH. This result is surprising. It may be that the long aliphatic side chain of arginine is able to replace the methionine side chain in the hydrophobic interface and still maintain the guanidinium group on the periphery of the interface. Modeling of this mutation lends support to this possibility. Replacement of F97 by E, on the other hand, destabilizes the complex to the greatest extent, at least at low ionic strength, indicating that the glutamic acid side chain disrupts the hydrophobic interface.

A complete interpretation of these kinetic results with the mutants of amicyanin requires structural data from their complexes. So far, crystallization trials have been carried out with the R99D, K68A and F97E mutants. Crystals of both the binary and ternary

144

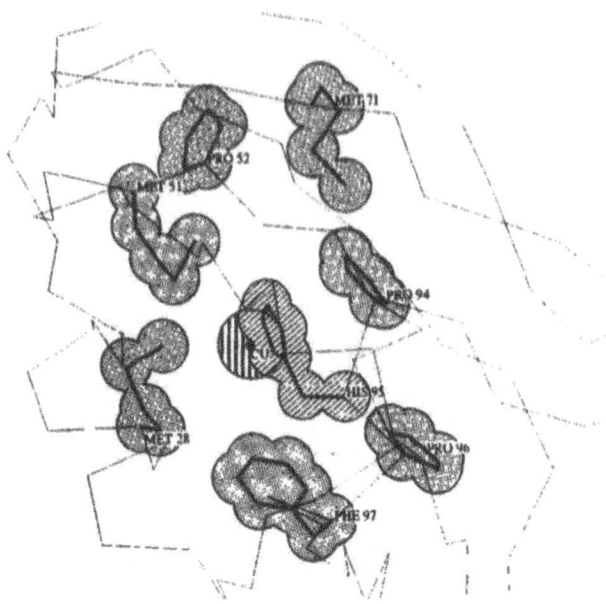

Figure 14. The hydrophobic surface of amicyanin surrounding His95. This surface is buried in the complex with MADH.

complexes of R99D and K68A have been obtained, although the K68A binary complex crystals are still very small. Data have been recorded from the R99D ternary complex and are currently being analyzed.

Acknowledgements

This work was supported by NSF Grant No. MCB-9419899 (F.S.M.), by USPHS Grant No. GM41574 (V.L.D.) and by CNR Grant No. 93.01080.PF70 (G.L.R) and by Nato Grant No. 940187 (F.S.M. and G.L.R.).

References

1. McIntire, W. S., Wemmer, D. E., Chistosedov, A. and Lidstrom, M. E. (1991) A new cofactor in a prokaryotic enzyme: tryptophan tryptophylquinone as the redox prosthetic group in methylamine dehydrogenase. *Science* **252**, 817-824.
2. Husain, M. & Davidson, V. L. (1985). An inducible periplasmic blue copper protein from *Paracoccus denitrificans. J. Biol. Chem.* **260**, 14626-14629.
3. Davidson, V. L. and Kumar, M. A. (1989) Cytochrome c-550 mediates electron transfer from inducible periplasmic c-type cytochromes to the cytoplasmic membrane of *Paracoccus denitrificans. FEBS Lett.* **245**, 271-273.
4. Anthony, C. (1982) *The Biochemistry of Methylotrophs*, Academic Press, New York.
5. Chen, L., Doi, M., Durley, R.C.E, Chistoserdov, A.Y., Lidstrom, M.E., Davidson, V.L. and Mathews, F. S. (1997) Refined crystal structure of methylamine dehydrogenase from *Paracoccus denitrificans* at 1.75 Å resolution, manuscript in preparation.
6. Varghese, J. N., Laver, W. G. and Colman, P. W. (1983) Structure of the influenza virus glycoprotein antigen neuraminidase at 2.9 Å resolution. *Nature* **303**, 35-40.
7. Davidson, V. L. (1989) Steady-state kinetic analysis of the quinoprotein methylamine dehydrogenase from

Paracoccus denitrificans. Biochem. J. **261**, 107-111.

8. Davidson, V.L. and Jones, H. L. (1991) Inhibition by cyclopropylamine of the quinoprotein methyamine dehydrogenase is mechanism-based and causes covalent cross-linking of the α and β subunits. *Biochemistry* **30**, 1924-1928.

9. van Spanning, R. M. J., Wansell, C. W., Reijnders, W. N. M., Oltmann, L. F. and Stouthamer, A. H. (1990) Mutagenesis of the gene encoding amicyanin of *Paracoccus denitrificans* and the resultant effect on methylamine oxidation. *FEBS Lett.* **275**, 217-220.

10. Husain, M. and Davidson, V. L. and Smith, A. J. (1986) Properties of *Paracoccus denitrificans* amicyanin. (1985) *Biochemistry* **25**, 2431-2436.

11. Cunane, L.M., Chen, Z.-w., Durley, R.C.E. and Mathews, F.S. (1996) X-ray structure of the cupredoxin amicyanin, from *Paracoccus denitrificans,* refined at 1.31 Å resolution. *Acta Cryst.* **D52**, 676-686.

12. van Spanning, R. M. J., Wansell, C. W., de Boer, T., Hazelaar, M. J., Anazawa, H., Harms, N., Oltmann, L. F. and Stouthamer, A. H. (1991). Isolation and characterization of the *moxJ, moxG, moxI,* and *moxR* genes of *Paracoccus denitrificans*: inactivation of moxJ, moxG, and moxR and the resultant effect on methylotrophic growth. *J. Bacteriol.* **173**, 6948-6961.

13. Gray, K. A., Knaff, D. B., Husain, M. and Davidson, V. L. (1986) Measurement of the oxidation-reduction potentials of amicyanin and c-type cytochromes from *Paracoccus denitrificans. FEBS Lett.* **207**, 239-242.

14. Chen, L., Durley, R., Mathews, F. S and Davidson, V. L. (1994). Structure of an electron transfer complex: methylamine dehydrogenase, amicyanin cytochrome c_{551i}. *Science* **264**, 86-90.

15. Gray, K. A., Davidson, V. L. and Knaff, D. B. (1988) Complex formation between methylamine dehydrogenase and amicyanin from *Paracoccus denitrificans. J. Biol. Chem.* **263**, 13987-13990.

16. Davidson, V.L. and Jones, L.H. (1991) Intermolecular electron transfer from quinoproteins and its relevance to biosensor technology. *Anal. Chim. Acta.* **249**, 235-240.

17. Kumar, M. A. and Davidson, V. L. (1990) Chemical cross-linking study of complex formation between methylamine dehydrogenase and amicyanin from *Paracoccus denitrificans. Biochemistry* **29**, 5299-5304.

18. Backes, G., Davidson, V.L., Huitman, F., Duine, J.A. and Sanders-Loehr, J. (1991) Characterization of the tryptophan-derived quinone cofactor of methylamine dehydrogenase by resonance Raman spectroscopy. *Biochemistry* **30**, 9201-9210.

19. Chen, L., Durley, R., Poliks, B.J., Hamada, K., Chen, Z., Mathews, F.S., Davidson, V.L., Satow, Y., Huizinga, E., Vellieux, F.M.D. and Hol, W.G.J. (1992). Crystal structure of an electron-transfer complex between methylamine dehydrogenase and amicyanin. *Biochemistry* **31**, 4959-4964.

20. J. J. Regan, J.J. (1993). *Pathways II software v2.01*, San Diego, personal communication.

21. Merli, A., Broderson, D.E., Morini, B., Chen, Z.-w., Durely, R.C.E., Mathews, F.S., Davidson, V.L. and Rossi, G.L. (1996) Enzymatic and electron transfer activities in crystalline protein complexes. *J. Biol. Chem.* **271**, 9177-9180.

22. Rivetti, C., Mozzarelli, A., Rossi, G. L, Henry, E. R. and Eaton, W. A. (1993) Oxygen binding by single crystals of hemoglobin. *Biochemistry* **32**, 2888-2906

23. Bishop, G.R., Brooks, H.B. and Davidson, V.L. (1996) Evidence for a tryptophan tryptophylquinone aminosemiquinone intermediate in the physiologic reaction between methylamine dehydrogenase and amicyanin. *Biochemistry* **35**, 8948-8954.

24. Kuusk, V. and McIntire, W. S. (1994)Influence of monovalent cations on the ultraviolet-visible spectrum of tryptophan tryptophylquinone-containing methylamine dehydrogenase from bacterium W3A1. *J. Biol. Chem.* **269**, 26136-26143.

25. Davidson, V.L., Jones, L.H. & Kumar, M.A. pH-dependent semiquinone formation by methylamine dehydrogenase from *Paracoccus denitrificans*. Evidence for intermolecular electron transfer. *Biochemistry* **29**, 10786-10791 (1990).

26. Cunane, L. M., Durley, R. C. E., Chen, Z., Tarng, W. and Mathews, F. S. (1997) unpublished results.

27. van der Palen, C.J., Slotboom, D.J., Jongejan, L., Reijnders, W.N., Harms, N., Duine, J.A. and van Spanning, R.J. (1995) Mutational analysis of mau genes involved in methylamine metabolism in *Paracoccus denitrificans. Eur. J. Biochem.* **230**, 860-871.

28. Kim, J., Fuller, J.H., Kuusk, V., Cunane, L., Chen, Z.-w., Mathews, F.S. and McIntire, W.S. (1995) The cytochrome subunit is necessary for covalent FAD attachment to the flavoprotein subunit of p-cresol methylhydroxylase. *J. Biol. Chem.* **270**, 31202-31209.

29. Davidson, V.L., Jones, L.H., Graichen, M.E., Mathews, F.S. and Hosler, J.P. (1997) Factors which stabilize the methylamine dehydrogenase-amicyanin electron transfer protein complex revealed by site-directed mutagenesis. *Biochemistry* **36**, 12733-12738.

146

30. Gray, K.A., Davidson, V.L. and Knaff, D.B. (1988) Complex formation between methylamine dehydrogenase and amicyanin from *Paracoccus denitrificans. J. Biol. Chem.* **263,** 13987-13990.
31. Mathews, F.S. (1995) X-ray studies of quinoproteins. *Methods in Enzymology* **258**, 191-216.
32. Mathews, F.S., Chen, L. , Durley, R.C.E., Chen, Z.-w. and McIntire, W.S. (1996) Structural Studies of Methylamine Dehydrogenase, in M.E. Lidstrom and R. Tabita (eds), *Proceedings of the 8th International C_1 Symposium*, Kluwar Press, Amsterdam, pp. 205-212.

Chapter 3
Oxido-reductases: Structure and Function

MICROBIAL AMINE OXIDOREDUCTASES
Their diversity, role, structure and mechanism

J. A. DUINE and A. HACISALIHOGLU
Department of Microbiology & Enzymology
Delft University of Technology
Julianalaan 67
2628 BC Delft, The Netherlands.

1. Introduction

The well known enzyme involved in oxidation of amines is monoamine oxidase (MAO, EC 1.1.3.4). The reason for this familiarity is the fact that this flavoprotein oxidizes amine hormones and neurotransmitters, compounds regulating the behaviour of mammalian organisms. So far, less well known mammalian amine oxidases (with respect to physiological role and mechanism of action) were: histaminase, diamine oxidase, semicarbazide-sensitive amine oxidase, etc., enzymes rubricated under EC 1.4.3.6. Since it was only recently established for some of the latter that they have topaquinone (TPQ, Fig. 1) as cofactor [1], it can be expected that knowledge on these enzymes will rapidly increase the coming years. A quinone like TPQ appears to be well suited as participant in the conversion of an amine by an oxidase since protein-lysine 6-oxidase (EC 1.4.3.13), oxidizing the lysyl residues in collagen so that cross-links can be formed, has the related LTQ (Fig. 1) as cofactor [2].

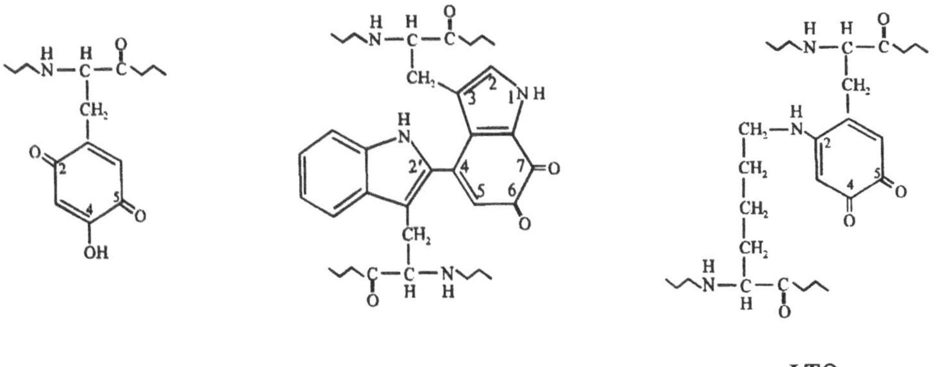

TPQ   TTQ   LTQ

Figure 1.. Structures of quinone cofactors

149

G.W. Canters and E. Vijgenboom (eds.),
Biological Electron Transfer Chains: Genetics, Composition and Mode of Operation, 149-164.
© *1998 Kluwer Academic Publishers.*

In contrast to mammalian organisms, microbes can utilize amines as sole C- (and energy-) source, as sole N-source, or as both. Amines occur as such in nature and they are formed in the pathway of amino acid degradation *via* oxidative decarboxylation or in that of degradation of substituted amines, e.g. the degradation of choline or the osmoprotectant betaine, yielding methylamine. A large diversity of unrelated microbial amine oxidoreductases appears to be involved in amine oxidation, the unrelatedness as presented in Table 1 referring to the type of cofactor participating in the reaction (in analogy with the situation for e.g. alcohol oxidoreductases, it is reasonable to assume that not only the cofactor but also the protein structures of these enzymes will be completely different). This review summarizes the properties of microbial amine oxidoreductases and, based on a comparison, tries to give an explanation for the diversity. Since progress in knowledge and the interest of our research group mainly concerns the quinoprotein amine oxidoreductases, some aspects of the structure and mechanism of these types of enzymes will be discussed.

2. General Aspects

2.1. BIOENERGETICS

Formally, the conversion of primary amines into products by amine oxidoreductases can be presented in the following way:

$$R\text{-}CH_2NH_2 + H_2O \to R\text{-}HCO + NH_3 + 2[H]$$

However, based on what happens with the reducing equivalents, indicated here with 2[H], two fundamentally different classes of oxidoreductase with respect to functioning can be distinguished, the oxidases and the dehydrogenases. In case of an oxidase, the reducing equivalents are converted into H_2O_2 whereas in the case of a dehydrogenase, they are transferred directly to the respiratory chain (indirect transfer does not occur since NAD(P)-dependent amine dehydrogenases have not been found sofar). Generation of useful energy from the amine oxidation step only occurs when the dehydrogenase is used as catalyst, not in the case of the oxidase. Related to the aspect of energy provision by the oxidation step and the toxicity of the H_2O_2 and aldehyde formed, the localization of the amine dehydrogenases and oxidases is of importance.

2.2. LOCALIZATION

In view of the toxicity of the products of the amine conversion, it is not unexpected that amine oxidation by yeasts and fungi, always found to be catalyzed by amine oxidases, has been reported [3-5] to take place in peroxisomes, these organelles containing not only the amine oxidase but also catalase. This conclusion has been derived from electron microscopy pictures [3,4] as well as from the presence of a specific leader sequence in the gene [5]. However, a contradictory conclusion was reported for the case of *Aspergillus niger* when grown on n-butylamine as a N-source, the enzyme being localized in the cell wall [6]. Since depending on the nature of the amine used as N-source, different quinoprotein amine oxidases are induced in yeasts and fungi (see 2.3.), the discrepancy may be explained by assuming that the "methylamine oxidase" of these organisms is localized in the peroxisomes whereas their "benzylamine oxidase" is localized in the outer side of the cell wall, perhaps the conversion of the different aldehyde products requiring different sites of production.

The problem of local build up of H_2O_2, which can even damage the (quinoprotein) amine oxidase itself [7], may be solved in a simpler way in other cases. Amine oxidases have been described where removal of catalase (observed as a "contaminant absorbing at 410 nm") by purification methods was rather difficult (e.g. see the purification of methylamine oxidase from the Gram-positive bacterium *Arthrobacter* P1 [8]). On the other hand, such an association does not always occur, perhaps because generation of H_2O_2 is only harmful when it occurs in the cytoplasm. For instance, quinoprotein arylalkylamine oxidase of *Klebsiella oxytoca* [9] (and most probably also that of *Escherichia coli* [10]) is found in the periplasm of this Gram-negative bacterium whereas catalase and the NAD-dependent aldehyde dehydrogenase oxidizing the corresponding aldehyde product, are found in the cytoplasm [9]. A gene encoding an NAD-dependent aldehyde dehydrogenase is located upstream of that encoding the amine oxidase in the *E. coli* chromosome [11, and J. van Beenen, unpublished results]. The enzyme, for which phenylacetaldehyde is a good substrate, has been characterized and appears to be an NAD-dependent aldehyde dehydrogenase in the case of *E. coli* [D. Luykx, unpublished results] (which might be similar to EC 1.2.1.28/39/53) and an NADP-dependent aldehyde dehydrogenase in the case of *Arthrobacter globiformis* [12] (which might be similar to EC 1.2.1.7). In view of the type of enzymes involved and their localization, *Enterobacteriaceae* may produce H_2O_2 when they are present in an aerobic environment where substantial amounts of tyrosine and/or phenylalanine are present. However, phenylacetic acid is a growth substrate for *E. coli* but not all strains can metabolize its product, 4-hydroxyphenylacetic acid. Therefore, perhaps tyrosine can only be used as nitrogen-source and not as a C-source in the latter cases [11].

Conversion of methylamine appears to be a special case since formaldehyde is extremely toxic to the cell. Perhaps for that reason, nucleophilic compounds are present in the cytoplasm which form an adduct with formaldehyde after which a specific dehydrogenase oxidizes the adduct to the corresponding formate ester. Although it is generally believed that the nucleophile is GSH in all cases, it was recently demonstrated that many Gram-positive bacteria (in which GSH is lacking) use mycothiol (Fig. 2) for that purpose [13] and the Gram-negative *Methylobacterium extorquens* THF [14]. Although many oxidoreductases are able to oxidize formaldehyde *in vitro* [15], it seems that for most organisms only the way *via* adduct formation with a nucleophile and subsequent oxidation with a specific dehydrogenase has physiological significance, as was demonstrated for *Paracoccus denitrificans* (using GSH for that purpose [16]).

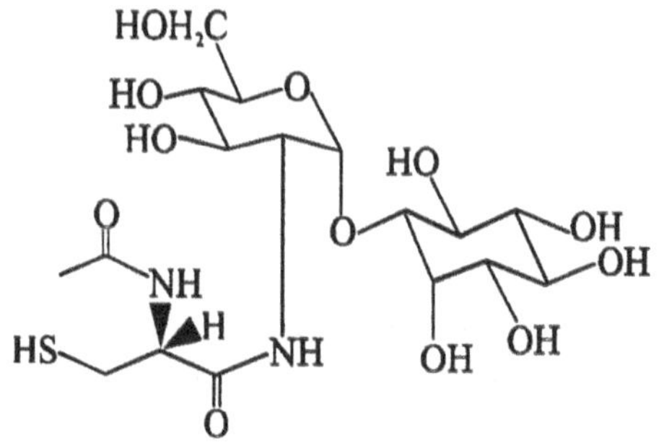

Figure 2. Structure of mycothiol, 1-O-2'[N-L-cysteinyl]amido-2'-deoxy-α-D-glucopyranosyl)-D-*myo*-inositol

2.3. TYPES AND THEIR DISTRIBUTION

2.3.1. *Flavoprotein Amine Oxidases*

Flavin-containing amine oxidase oxidizing primary amines has been purified from *Aspergillus niger* [17]. The enzyme consists of 4 subunits of 55.6 kDa and its flavin is non-covalently bound. Some similarity exists with human MAO-A and -B with respect to typical inhibitors and amino acid sequence, but the flavin is bound in a different way [18]. Tyramine oxidase from the bacterium *Micrococcus luteus* (formerly *Sarcina lutea*) is also a flavoprotein with a molecular mass of 129 kDa and FAD as cofactor [19]. Polyamine oxidase, only oxidizing spermine at the secondary N, has been isolated from *Penicillium chrysogenum* and *Aspergillus terreus* [20]. The dimeric flavoprotein has subunits of 80 kDa and contains FAD. Putrescine oxidase (EC 1.4.3.10) has been isolated from the bacterium *Micrococcus rubens*. The monomeric enzyme has a molecular mass of 80 to 88 kDa and contains 1 FAD per enzyme molecule [21,22]. Cyclohexylamine oxidase (EC 1.4.3.12) has been isolated from a *Pseudomonas* species [23]. The enzyme has a narrow substrate specificity as it oxidizes only alicyclic primary amines but not primary or secondary aliphatic or aromatic amines. It has a molecular mass of 80 kDa and contains FAD.

Summarizing, many amine oxidases appear to be of the flavoprotein type but in lack of further structural information, it cannot be decided whether they are mutually related to each other or to MAO.

2.3.2. *Quinoprotein (TPQ-Containing) Amine Oxidases*

The distribution of this type of amine oxidoreductase appears to be broad since it occurs in animals, plants, and microbes (yeasts, fungi, Gram-negative as well as Gram-positive bacteria). The microbial members of this type of enzyme have a restricted substrate specificity, but all together they cover a wide spectrum of substrates, from arylalkyl amines to methylamine.

2.3.3. Flavohemoprotein Amine Dehydrogenase
Spermidine dehydrogenase from *Serratia marcescens* has a molecular mass of 76 kDa, and contains 1 FAD and 1 protoporphyrin per enzyme molecule. The enzyme attacks at the secondary N position in spermidine, yielding 1,3-diaminopropane and γ-aminobutyraldehyde as products [24].

2.3.4. Quinoprotein (TTQ-Containing) Amine Dehydrogenases
The first discovered TTQ-containing enzyme, methylamine dehydrogenase (EC 1.4.99.3), has meanwhile been detected in many Gram-negative, methylamine-utilizing, methylotrophic bacteria. In addition, a TTQ-containing arylalkylamine dehydrogenase (EC 1.4.99.4) has been isolated from *Alcaligenes faecalis* [25].

2.3.5. Quinohemoprotein Amine Dehydrogenases
An arylalkylamine dehydrogenase, isolated from *Pseudomonas putida*, has been described consisting of subunits of 58 and 42 kDa, the heavy subunit containing a heme *c*, and the enzyme being inhibited by carbonyl group reagents [26,27]. Further characterization in two research groups (A. Hacisalihoglu, J.A. Jongejan, J.A. Duine; O. Adachi, c.s.) has revealed the following: although not observed in earlier studies, the enzyme appears to contain an additional 20 kDa subunit; when the small subunit was separated from the large ones by gel filtration under denaturing conditions, the former appeared to contain a chromophore having an absorption maximum at 390 nm; when the enzyme was first derivatized, with concomitant inhibition, by p-nitrophenylhydrazine, the isolated subunit showed a maximum at 468 nm (shoulder at 520 nm). Since the subunit also showed a positive reaction in the redox cycling assay, it seems higly likely that a quinone cofactor is present in the small subunit. Resonance Raman spectroscopy confirmed this but the cofactor identity could not be established [N. Nakamura and J. Sanders-Loehr, unpublished results]. Work is in progress to identify the quinone cofactor.

A spermidine dehydrogenase has been isolated from *Citrobacter freundii* [28]. It appeared to be localized in the periplasm, was induced by spermidine, converted spermidine into 1,3-diaminopropane and γ-aminobutyraldehyde, had a molecular mass of 63 kDa, and was inhibited by carbonyl group reagents. The authors concluded that the enzyme is a quinohemoprotein because it contains a heme (responsible for the cytochrome *b*-like spectrum of the enzyme) and PQQ, the latter being detected only after the enzyme was hydrolyzed. Two spermidine dehydrogenases were found in *C. freundii* but they showed identical N-terminal amino acid sequences. A constitutive spermidine dehydrogenase with the same molecular mass was found in *Pseudomonas aeruginosa*, but whether this enzyme is also of the quinohemoprotein type is unknown [29]. Finally, although it has been claimed that the spermidine dehydrogenases oxidize at the secondary N position and contain PQQ, a discrepancy exists between these

TABLE 1. Types of microbial amine oxidoreductases

Type	Cofactor	Occurance	Remarks
Oxidases			
Flavoprotein	FAD	Eukaryotes, prokaryotes	Related to MAO; mutually related?
Quinoprotein	Cu^{2+}/TPQ		
Dehydrogenases			
Quinoprotein	TTQ	Gram-neg. bacteria	
Quinohemoprotein	Quinone/Heme c	*P. putida*	Structure of the quinone cofactor to be elucidated
Quinohemoprotein	Quinone/Heme	Gram-neg. bacteria	Presence of quinone cofactor, oxidation at secondary nitrogen to be established
Flavohemoprotein	Flavin/Heme	*S. marcescens*	Oxidizes at secondary nitrogen
Others	Unidentified	Several bacteria	

claims. According to the transaminase mechanism attributed to the quinoprotein amine oxidoreductases [30], only primary Ns can be oxidized. This is illustrated by the fact that quinoprotein bovine plasma amine oxidase oxidizes only the primary N in spermine and spermidine [31] and addition of 3-pyrrolines, only containing a secondary N, leads to irreversible inactivation of this enzyme because these amines are converted into a compound which forms an adduct with the cofactor [32]. Thus two possibilites exist which can explain the discrepancy: 1. In view of the difficulties signalized in ref. [32] to determine the reaction products of spermidine conversion, the dehydrogenases discussed here contain a quinone cofactor but they oxidize the primary N of their substrates; 2. These dehydrogenases are quinoproteins oxidizing spermidine at the secondary N position but with an as yet unknown mechanism.

A number of amine dehydrogenases have been described for which the cofactor identity is still unknown. An amine dehydrogenase oxidizing lower aliphatic amines (but not methylamine) has been found in the extract of *Mycobacterium convolutum*. The activity corresponded with the remarkably low molecular mass of 38.5 kDa (determined by gel filtration) [33]. Similar low values have been found in the case of amine dehydrogenase from *Pseudomonas* K95 [34]. A membrane-integrated tyramine dehydrogenase with unidentided cofactor has been detected in *P. aeruginosa* [35].

2.4. DIVERSITY IN ONE AND THE SAME ORGANISM

Our studies on the distribution of some of the amine oxidoreductase types among a limited number of bacteria revealed [9] that always only one type was present in a certain organism. However, several reports indicate that variants of a certain type or more than one type can occur in one and the same organism:

1. Two genes, containing the consensus sequence for TPQ-containing amine oxidases, occur in *Arthrobacter* P1, the amino acid sequences derived from the gene sequences showing that only a few amino acids are different. However, sofar under the growth conditions applied, only one gene came to expression [36].

2. *Klebsiella aerogenes* W70 contains a gene encoding TPQ-containing amine oxidase, which is not expressed but has been cloned and brought to expression, as well as a gene (*Tyn A*) which comes to expression when the organism is grown on tyramine (not on tyrosine). Tyn A is a membrane-integrated oxidase with unidentified cofactor which oxidizes tyramine but not 2-phenylethylamine [37].

3. Depending on the type of amine used as inducer, two, most probably TPQ-containing, amine oxidases, indicated as methylamine oxidase and benzylamine oxidase, with different substrate specificity can be induced in the yeasts *Hansenula polymorpha* [38,39], *Candida boidinii* [40], and *Pichia pastoris* [41]. However, exceptions occur to the rule as only one enzyme was induced in a few cases [42,43]. Also two quinoprotein (TPQ-containing) amine oxidases can be induced in the bacterium *Arthrobacter globiformis* (2-phenylethylamine oxidase [44] and histamine oxidase [45,46]). Whether the existence of two enzyme species having the same cofactor in a certain organism is due to just adaptation of one ancestral enzyme to the requirements posed by the different substrates by slight modification of the active site or to acquirement of two enzymes originating from different sources, remains to be elucidated. Support for the latter possibility can be derived from differences among quinoprotein amine oxidases with respect to enantioselectivity [47], intermediates occurring in the oxidative part of the catalytic cycle [7], topologies of the Cu^{2+}/TPQ cofactors [48], and low similarity of the amino acid sequences [6]. Thus the possibility that an organism acquired two different ancestral genes for a quinoprotein amine oxidase during its evolutionary history, is not unlikely, in line with e.g. the low amino acid sequence similarity of the two quinoprotein amine oxidases in *A. globiformis* [44,45,46]. However, diversity is not only dictated by the substrate to be converted: e.g. *Aspergillus niger* when grown on n-butylamine as N-source is not only able to produce the quinoprotein benzylamine oxidase but also a flavoprotein amine oxidase [17]. Since the former enzyme is probably localized in the cell wall [6] and the gene for the latter enzyme contains the targeting signal for peroxisomal enzymes [18], it is tempting to speculate that the reason for the use of two different types of amine oxidases is due to different roles requiring different localizations. However, in this explanation it is difficult to understand why the peroxisome-located quinoprotein methylamine oxidase of this organism, being able to convert n-butylamine, does not take over the role of the flavoprotein amine oxidase, which is the actual n-butylamine converting enzyme.

2.5. INDUCTION

In general, induction of an amine oxidoreductase occurs with the amine which is the physiological substrate. However, when the amine is used as a C-source, catabolite repression has been found to occur in several cases. On the other hand, when the amine is used as an N-source, the induction of the amine oxidoreductase is frequently repressed by the presence of ammonia. A gene encoding a regulator protein for quinoprotein arylalkylamine oxidase induction has been found in the case of *E. coli* [37,49] but the role of 2-phenylethylamine as inducer is under debate. An example of the surprising consequences of regulation of amine utilization is shown by yeasts. Although e.g. methylamine and ethylamine are utilized by many yeasts as N-source, they do not serve as C-source [50], although the capability to utilize methanol and ethanol as C-source is widespread among these organisms (the alcohols yielding formaldehyde and acetaldehyde, respectively, just as the amines). However, exceptions occur to the rule, some organisms being able to utilize the amines as C-source [42,43]. Whether the sole amine oxidase induced in these exceptions is a new type of amine oxidase, as suggested by its exceptionally small subunits, is presently unknown.

Finally, it is still unclear whether the aldehyde produced by periplasmic amine oxidation in Gram-negative bacteria freely diffuses to the cytoplasm or is transported by a specific carrier so that it can be assimilated in the cytoplasm. In the case of *Paracoccus denitrificans*, indications for the existence of a transport system have been reported [51]. In cases where the amine is metabolized in the cytoplasm, an uptake system for the amine may exist, as was found for *Arthrobacter* P1 [52].

3. Structural and Mechanistic Aspects

3.1. QUINOPROTEIN (TPQ-CONTAINING) AMINE OXIDASES

Depending on the source from which they originate, these homodimeric enzymes have subunits ranging from 80 to 100 kDa in size, and contain 2 Cu^{2+} ions and 1 to 2 TPQ (Fig. 1) molecules per enzyme molecule. Determination of TPQ is usually performed by spectrophotometric titration of the enzyme with phenylhydrazine or its derivatives [53]. The observation that only one TPQ becomes derivatized with hydrazine has been ascribed to negative cooperativity, assuming that when one of the two subunits in the enzyme molecule is inactivated by the reaction of TPQ with the hydrazine, the TPQ in the other subunit will be practically untitratable [54]. In this view, each subunit contains 1 TPQ and full derivatization requires long incubation with an excess of the hydrazine. However, another explanation for the low value can be given. On inspecting the absorption spectra of quinoprotein amine oxidases, it appears that the ratio of the absorbance at 470 nm to that at 340 nm, the former one thought to be representative for the amount of TPQ in the enzyme, varies considerably among the enzymes, the ratio being linearly related to the amount of TPQ determined by titration [55]. This suggests that the processing of the specific tyrosyl residue in the precursor enzyme to TPQ, yielding the mature enzyme, is suboptimal in certain enzymes. To explain that a long incubation time with excess hydrazine gave a higher value, it could be imagined that no further derivatization but further TPQ generation takes place during this incubation. However, since no systematic investigations of the enzymes have been performed, at present it cannot be decided which explanation is correct.

Amino acid sequence similarity for the quinoprotein amine oxidases is low between enzymes having different substrate specificities or between enzymes having similar substrate specificities but who originate from unrelated organisms [6]. However, all have the consensus sequence NYD/E [1], Y being the precursor of TPQ. Also 3 specific histidyl residues are conserved, functioning as ligands for the Cu^{2+}. Although only one gene has been found for it, two different quinoprotein amine oxidases were isolated from *Aspergillus niger* [56]. Most probably, the occurrence of the minor enzyme is caused by inadequate or still unfinished processing of the pro-enzyme. This is another indication that processing of the specific tyrosyl residue to TPQ, occurring in a spontaneous way in several steps [57,58,59], does not always give the full complement of TPQ.

In our attempts to investigate the topology of the Cu/TPQ couple, *E. coli* amine oxidase was derivatized with fluoridated phenylhydrazine probes and ^{19}F-NMR spectroscopy was used to determine the distance between the two cofactors [48]. Recently, the topology was also determined for the active site of amine oxidase crystals from *E. coli* [47] and pea seedling [60]. No direct interaction was observed between the two cofactors, the apical H_2O liganded to Cu^{2+} being situated between the metal ion and the 2-O position of the TPQ ring. Since the distances are much shorter than those determined by us, this casts doubt on the suitability of the "F-NMR method" to determine the distance between a F-atom and a paramagnetic centre.

Investigations on the mechanism of *E. coli* amine oxidase revealed [7] the absorption spectrum of a species in the catalytic cycle which appears to be novel since such a spectrum has not been observed so far. Although a proposition has been made, the precise identity and role of this species in the present view on the mechanism [61-66] remain to be elucidated.

3.2. QUINOPROTEIN (TTQ-CONTAINING) AMINE DEHYDROGENASES

Methylamine dehydrogenase (MADH) consists of 2 small subunits (15 kDa), 2 large subunits (47 kDa), and 2 TTQ cofactor molecules. TTQ (Fig. 1) forms part of the protein chain of the small subunit. The structures of the genes involved in methylamine oxidation and organized in the so-called *mau* cluster have been nearly completely elucidated for several organisms [67-72]. It appears that this cluster comprises the structural genes for MADH, the genes for biosynthesis of TTQ, as well as the structural genes for some of the components in the respiratory chain to which MADH is connected. With respect to the genes in the *mau* cluster of *P. denitrificans*, organized in the transcription order *mauRFBEDACJGMN*, the following functions have been assigned, as indicated below.

MauA and *B* are the structural genes for the small and large subunit of MADH, respectively. *MauC* encodes for amicyanin, the small blue copper protein acting as primary electron acceptor for MADH in the respiratory chain. *MauD* and *E* are involved in transport and processing of the small subunit of MADH [73]. The functions of *mauN, M,* and *J* remain obscure because mutants lacking these genes did not show complete blocking of methylamine oxidation. *MauR* is transcribed in the opposite direction and codes a LysR-type activator of transcription [71]. *MauG* is clearly involved in biosynthesis of TTQ and thus in maturation of the small subunit [72]. The amino acid sequence of MauG, deduced from the gene sequence, shows some similarity with that of cytochrome *c* peroxidase isolated from *Pseudomonas aeruginosa*. A strain in which *mauG* was mutated was unable to grow on methylamine, did not produce active MADH but produced all other components of the MADH respiratory chain. Moreover, a protein was produced showing the same chromatographic behaviour and antibody reactivity as MADH. Purification and characterization of this immature MADH [C.J.N.M. van der Palen, H. Dekker, R.J.M. van Spanning, A.O. Muijsers, S. de Vries and J.A. Duine, unpublished results] showed that it had not the characteristic absorption spectrum of MADH above 300 nm, and did not contain the quinone moiety, as deduced from the negative result in the assay for a quinone cofactor. Mass spectroscopy of the small subunit from mutant as well as wild type MADH revealed a difference of 13 to 14 units, suggesting that the precursor of TTQ in the mutant MADH was still in the phenolic form (the difference between the quinone and the phenol being due to insertion of one oxygen and release of two hydrogens). Conversion of immature MADH into enzymatically active MADH, giving a positive result in the quinone assay, with the typical absorption spectrum of MADH, and a mass value for the small subunit equivalent to that belonging to the wild type one, could be achieved by incubation with chemical oxidants like $KMnO_4$, $K_3Fe(CN)_6$, or Cu^{2+} (in decreasing order of effectiveness). Whether the cross-link between the 2 tryptophyl residues is already present in immature MADH or is formed during oxidation of the phenolic OH group of the TTQ precursor in mutant MADH is presently unknown. However, the fact that the conversion into TTQ was successful for immature MADH but not for its isolated small subunit, probably indicating that the quaternary structure in the former case induces the apropriate conformation in the small subunit for cross-link formation (it is known [74] that the conformation of the small subunit in the enzyme is different from that of the isolated form), suggests that the latter hypothesis is more likely. Model studies in which coupling of the indole rings as well as quinone formation was achieved by chemical oxidation of skatole in the presence of 6- or 7-monohydroxyskatole with Fremy's salt, indicates that such a reaction is feasible [75].

MADH is localized in the periplasm of the bacteria in which it occurs. Two of the products formed from the oxidation of methylamine in that compartment, protons and ammonium ions, have a tremendous effect on the catalytic performance of the enzyme, as revealed by detailed kinetic investigations [76-79]. The underlying reasons for this are now emerging [80,81]. Stopped-flow kinetics, and experiments with inactive substrate analogs and deuterated methylamine revealed that the substrate reacts in its ammonium form with MADH [79].

All components of the MADH respiratory chain of *P. versutus* are known and the sequence of reactions is as follows:
MADH -> amicyanin -> cytochrome c_{550} -> cytochrome *c* oxidase (*aa₃* type) -> O_2.
In vitro kinetic studies have also been performed for amicyanin/cytochrome c_{550} [82]

and cytochrome c_{550}/cytochrome c oxidase/O_2 [P. van Bastelaere, J. van Beumen, S. de Vries and J.A. Duine, unpublished results]. Since a cytochrome c_1-like component has been found which enhanced the *in vitro* reaction between cytochrome c_{550} and cytochrome c oxidase, and other cytochromes c and oxidases can substitute [83], the chain may be more complex than indicated here and alternative chains may exist suited for other growth conditions. Although the structure of a co-crystal of MADH/amicyanin/cytochrome c_{551i} from *P. denitrificans* has been elucidated [84], the relevance of it for the *in vivo* mechanism of action is questionable [80].

4. Conclusions

1. Microbial amine oxidation is catalyzed by oxidases as well as (NAD(P)-independent) dehydrogenases. Since a) aldehydes (and also H_2O_2 in the case of oxidases) are toxic for the cell, b) aldehydes have to be assimilated in the cytoplasm (in case the amine serves as a C-source), c) extracellular H_2O_2 production is benificial for certain other processes to occur (in fungi), and d) electron transfer occurs from the amine dehydrogenases to the respiratory chain, the localization of the amine oxidoreductases is of utmost importance. For the dehydrogenases investigated (TTQ-containing), the localization appears to be in the periplasm of the Gram-negative bacteria in which they occur, which is benificial for energy generation. For the oxidases investigated (TPQ-containing), the localization is in the peroxisome, cell wall, periplasm or cytoplasm. At present, the aforementioned physiological consequences of this are not clear. In the case of methylamine, the localization of the amine oxidoreductase seems unimportant since the formaldehyde produced is scavenged by a nucleophile in the cytoplasm (GSH, mycothiol, THF, depending on the type of organism) and the harmless adduct is subsequently oxidized by a specific NAD(P)-dependent dehydrogenase to the corresponding formate ester. Undoubtedly, regulatory aspects of enzyme induction and repression for amine oxidation and aldehyde conversion have also to do with the physiological consequences mentioned above.

2. The survey of amine oxidoreductases given here indicates that an enormous diversity exists with respect to the type of cofactor used by these enzymes. Whether a correlation exists between the type of cofactor and the cellular function of the particular enzyme using it, is presently unknown. Studies on the physiology of organisms having several types of amine oxidoreductases and extensive knowledge on the mechanisms of these enzymes, may eventually give the answer.

3. Quinoprotein amine oxidoreductases have a "transaminase-like" mechanism of amine oxidation in common. Another common feature is that their cofactor forms part of the protein chain and is formed by oxidation of a specific aromatic amino acid in the precursor enzyme. In the case of TPQ, it is now well established that formation of this cofactor is an autocatalytic process, the bound copper ion which eventually functions as cofactor in the catalytic cycle also being responsible for catalysis of the conversion of the tyrosyl residue into TPQ by molecular oxygen in several steps. Indications exist

that this process is not always optimal, perhaps explaining the controversial TPQ content of certain quinoprotein amine oxidases. On the other hand, in the case of TTQ the final biosynthesis step is catalyzed by an enzyme (mauG) which converts the monophenolic TTQ precursor into the o-quinone form, probably with concomitant dimerization of the 2 tryptophyl residues. How the tryptophyl residue eventually bearing the o-quinone moiety is oxidized to the monophenolic form and how mauG is able to convert the buried tryptophyl residues into TTQ, is presently unknown. However, the aspect of enzymatic specificity in the latter step seems to be rather unimportant since chemical oxidants could convert the precursor into TTQ, *in vitro* as well as *in vivo*.

Quinoprotein amine oxidoreductases are interesting enzymes to study inter- (MADH) and intra-molecular (amine oxidases) long range electron transfer. All the components of the MADH respiratory chain are known now, the 3-D structure of co-crystals of MADH and amicyanin has been elucidated, and kinetic studies on mixtures of 2-components systems have been performed. Therefore, the *in vitro* mechanism of action of a respiratory chain can be studied now using these components. Since the topology of the Cu^{2+}/TPQ cofactor set in quinoprotein amine oxidase has recently been elucidated and spectroscopic studies have revealed the existence of several species of the reduced enzyme form ($TPQH_2/Cu^{2+}$, $TPQH^-/Cu^{1+}$, a Zwitter-ionic form of the latter), it is tempting to speculate that the mechanism of intramolecular electron transfer and the role of Cu^{2+} in reoxidation of the reduced quinone cofactor and in O_2 activation will be elucidated in the near future.

5. References

1. Janes, S.M., Mu, D., Wemmer, D., Smith, A.J., Kaur, S., Maltby, D., Burlingame, A.L. and Klinman, J.P. (1990) A new cofactor in eukaryotic enzymes: 6-hydroxydopa at the active site of bovine amine oxidase, *Science* 248, 981-987.
2. Wang, S.X., Mure, M., Medzihradszky, K.F., Burlingame, A.L., Brown, D.E., Dooley, D.M., Smith, A.J., Kagan, H.M. and Klinman, J.P. (1996) A cross-linked cofactor in lysyl oxidase: redox function for amino acid side chains, *Science* 273, 1078-1084.
3. Zwart, K., Veenhuis, M., van Dijken, J.P. and Harder, W. (1980) Development of amine oxidase-containing peroxisomes in yeasts during growth on glucose in the presence of methylamine as the sole source of nitrogen, *Arch. Microbiol.* 126, 117-126.
4. Van Dijken, J.P. and Veenhuis, M. (1980) Cytochemical localization of glucose oxidase in peroxisomes of *Aspergillus niger*, *Eur. J. Appl. Microbiol. Biotechnol.* 9, 275-283.
5. Faber, K.N., Keizer-Gunnink, I, Pluim, D., Harder, W. and Veenhuis, M. (1995) The N-terminus of amine oxidase from *Hansenula polymorpha* contains a peroxisomal targeting signal, *FEBS Lett.* 357, 115-120.
6. Frébort I, (1997) Structure of the active site and cellular location of the copper/topaquinone-containing amine oxidases from the fungus *Aspergillus niger* AKU 3302, Thesis Yamaguchi University, Japan.
7. Steinebach, V., de Vries, S. and Duine, J.A. (1996) Intermediates in the catalytic cycle of copper-quinoprotein amine oxidase from *Escherichia coli*, *J. Biol. Chem.* 271, 5580-5588.
8. Van Iersel, J., van der Meer, R.A. and Duine, J.A. (1986) Methylamine oxidase from *Arthrobacter* P1: a bacterial copper-quinoprotein amine oxidase, *Eur. J. Biochem.* 161, 415-419.
9. Hacisalihoglu, A., Jongejan, J.A. and Duine, J.A. (1997) Distribution of amine oxidases and amine dehydrogenases in bacteria grown on primary amines and characterization of the amine oxidase from *Klebsiella oxytoca*, *Microbiology* 143, 505-512.
10. Hanlon, S.P. and Cooper, R.A. (1995) Cellular location influences copper-dependent topaquinone formation for phenylethylamine oxidase expressed in *Escherichia coli* K-12, *FEMS Microbiol. Lett.* 133, 271-275.

11. Hanlon, S.P., Hill, T.K., Flavell, M.A., Stringfellow, J.M. and Cooper, R.A. (1997) 2-Phenylethylamine catabolism by *Escherichia coli* K-12: gene organization and expression. *Microbiology* **143**, 513-518.

12. Shimizu, E., Ichise, H., Odawara, T. and Yorifuji, T. (1993) NADP-dependent phenylacetaldehyde dehydrogenase for degradation of phenylethylamine in *Arthrobacter globiformis, Biosci. Biotechnol. Biochem.* **57**, 852-853.

13. Misset-Smits, M., van Ophem, P.W., Sakuda, S. and Duine, J.A. (1997) Mycothiol, 1-O-(2'[*N*-L-cysteinyl]amido-2'-deoxy-a-D-glucopyranosyl)-D-*myo*-inositol, is the Factor for NAD/Factor-dependent formaldehyde dehydrogenase, *FEBS Lett.* **409**, 221-222.

14. Chistoserdova, L. (1996) Metabolism of formaldehyde in *Methylobacterium extorquens* AM1, *Microbial Growth on C1-compounds*, Kluwer Academic Publishers, Dordrecht.

15. Kim, S.W., Luykx, D.M.A.M., de Vries, S and Duine, J.A. (1996) A second molybdoprotein aldehyde dehydrogenase from *Amycolatopsis methanolica* NCIB 11946, *Arch. Biochem. Biophys.* **325**, 1-7.

16. Ras, J., van Ophem, P.W., Reijnders, W.N.M., van Spanning, R.J.M., Duine, J.A., Stouthamer, A.H. and Harms, N. (1995) Isolation, sequencing and mutagenesis of the gene encoding NAD- and glutathione-dependent formaldehyde dehydrogenase (GD-FAlDH) from *Paracoccus denitrificans* in which GD-FAlDH is essential for methylotrophic growth, *J. Bacteriol.* **177**, 247-251.

17. Schilling, B. and Lerch, K. (1995) Amine oxidases from *Aspergillus niger*: identification of a novel flavin-dependent enzyme, *Biochim. Biophys. Acta* **1243**, 529-537.

18. Schilling, B. and Lerch, K. (1995) Cloning, sequencing and heterologous expression of the monoamine oxidase gene from *Aspergillus niger, Mol. Gen. Genet.* **247**, 430-438.

19. Yamada, H., Kumagai, H. and Uwajima, T. (1971) Tyramine oxidase (*Sarcina lutea*), *Meth. Enzymol.* **17B**, 722-726.

20. Isobe, K., Tani, Y. and Yamada, H. (1980) Crystallization and characterization of polyamine oxidase from *Penicillium chrysogenum* and *Aspergillus terreus, Agric. Biol. Chem.* **44**, 2651-2658 and 2749-2751.

21. Adachi, O., Yamada, H. and Ogata, K. (1966) Purification and properties of putrecine oxidase from *Micrococcus rubens, Agric. Biol. Chem.* **30**, 1202-1210.

22. DeSa, R.J. (1972) Putrescine oxidase from *Micrococcus rubens, J. Biol. Chem.* **247**, 5527-5534.

23. Tokieda, T., Niimura, T., Takamura, F. and Yamaha, T. (1977) Purification and some properties of cyclohexylamine oxidase from a *Pseudomonas* sp., *J. Biochem.* **81**, 851-858].

24. Tabor, C.W. and Kellogg, P.D. (1971) Spermidine dehydrogenase (*Serratia marcescens*) *Meth. Enzymol.* **17B**, 746-753.

25. Nozaki, M. (1987) Aromatic amine dehydrogenase from *Alcaligenes faecalis, Meth. Enzymol.* **142**, 650-655.

26. Durham, D.R. and Perry, J.J. (1978) Purification and characterization of a heme-containing amine dehydrogenase from *Pseudomonas putida, J. Bacteriol.* **134**, 837-843; **135**, 981-986.

27. Shinagawa, E., Matsushita, K., Nakashima, K., Adachi, O. and Ameyama, M. (1988) Crystallization and properties of amine dehydrogenase from *Pseudomonas* sp., *Agric. Biol. Chem.* **52**, 2255-2263.

28. Hisano, T., Murata, K., Kimura, A., Matsushita, K., Toyama, H. and Adachi, O. (1992) Characterization of membrane-bound spermidine dehydrogenase of *Citrobacter freundii, Biosci. Biotechnol. Biochem.* **56**, 1916-1920.

29. Hisano, T., Abe, S., Wakashiro, M., Kimura, A. and Murata, K. (1990) Microbial spermidine dehydrogenases: Purification and properties of the enzyme in *Pseudomonas aeruginosa* and *Citrobacter freundii, J. Ferm. Bioeng.* **69**, 335-340.

30. Klinman, J.P. (1996) New quinocofactors in eukaryotes, *J. Biol. Chem.* **271**, 27189-27192.

31. Tabor, C.W., Tabor, H. and Bachrach, U. (1964) Identification of the aminoaldehydes produced by the oxidation of spermine and spermidine with purified plasma amine oxidase, *J. Biol. Chem.* **239**, 2194-2203.

32. Lee, Y., Huang, H. and Sayre, L.M. (1996) Model studies support pyrrolation of the topaquinone cofactor to explain inactivation of bovine plasma amine oxidase by 3-pyrrolines. Unusual processing of a secondary amine, *J. Am. Chem. Soc.* **118**, 7241-7242.

33. Cerniglia, C.E. and Perry, J.J. (1975) Metabolism of n-propylamine, isopropylamine, and 1,3-diaminopropane by *Mycobacterium convolutum, J. Bacteriol.* **124**, 285-289.

34. Niimura, Y., Omori, T. and Minoda, Y. (1986) Purification and properties of an amine dehydrogenase from *Pseudomonas* K95 grown on 1,12-diaminododecane, *Agric. Biol. Chem.* **50**, 1445-1451.

35. Cuskey, S.M., Peccoraro, V. and Olsen, R.H. (1987) Initial catabolism of aromatic biogenic amines by *Pseudomonas aeruginosa* PAO: pathway description, mapping of mutations, and cloning of essential genes, *J. Bacteriol.* **169**, 2398-2404.

36. Zhang, X., Fuller, J.H. and McIntire, W.S. (1993) Cloning, sequencing, expression, and regulation of the structural gene for the copper/topaquinone-containing methylamine oxidase from *Arthrobacter* strain P1, a

162

Gram-positive facultative methylotroph, *J. Bacteriol.* **175**, 5617-5627.

37. Murooka, Y., Akazami, H. and Yamashita, M. (1996) The monoamine regulon including syntheses of arylsulfatase and monoamine oxidase in bacteria, *Biosci. Biotechnol. Biochem.* **60**, 935-941.

38. Bruinenberg, P.G., Evers, M., Waterham, H.R., Kuipers, J., Arnberg, A.C. and Ab, G. (1989) Cloning and sequencing of the peroxisomal amine oxidase gene from *Hansenula polymorpha*, *Biochim. Biophys. Acta* **1008**, 157-167.

39. Mu, D., Janes, S.M., Smith, A.J., Brown, D.E., Dooley, D.M. and Klinman J.P. (1992) Tyrosine codon corresponds to topaquinone at the active site of copper amine oxidases, *J. Biol. Chem.* **267**, 7979-7982.

40. Haywood, G.W. and Large, P.J. (1981) Microbial oxidations of amines. Distribution, purification and properties of two primary amine oxidases from the yeast *Candida boidinii* grown on amines as sole nitrogen source, *Biochem. J.* **199**, 187-201.

41. Green, J., Haywood, G.W. and Large, P.J. (1983) Serological differences between the multiple amine oxidases of yeasts and comparison of the specificities of the purified enzymes from *Candida utilis* and *Pichia pastoris*, *Biochem. J.* **211**, 481-493.

42. Large, P.J. and Sherlock, L.A. (1987) Characterization of the amine oxidase involved in the growth of *Trichosporon cutaneum* X4 on ethylamine as source of carbon, nitrogen and energy, *Arch. Microbiol.* **147**, 64-67.

43. Sherlock, L.A., Large, P.J. and Whitaker, R.G. (1986) A new type of methylamine oxidase: the sole oxidase produced during growth of *Sporobolomyces albo-rubescens* on primary alkylamines, *Yeast* **2**, 87-92.

44. Tanizawa, K., Matsuzaki, R., Shimizu, E., Yorifuji, T. and Fukui, T. (1994) Cloning and sequencing of phenylethylamine oxidase from *Arthrobacter globiformis* and implication of Tyr-382 as the precursor to its covalently bound quinone cofactor, *Biochem. Biophys. Res. Commun.* **199**, 1096-1102.

45. Shimizu, E., Odawara, T., Tanizawa K. and Yorifuji, T. (1994) Histamine oxidase, a Cu^{2+}-quinoprotein enzyme of *Arthrobacter globiformis*, *Biosci. Biotechnol. Biochem.* **58**, 2118-2120.

46. Choi, Y-H., Matsuzaki, R., Fukui, T., Shimizu, E., Yorifuji, T., Sato, H., Ozaki, Y. and Tanizawa, K. (1995) Copper/topaquinone-containing histamine oxidase from *Arthrobacter globiformis*. Molecular cloning and sequencing, overproduction of precursor enzyme, and generation of topaquinone cofactor, *Biol. Chem.* **270**, 4712-4720.

47. Wilmot, C.M., Murray, J.M., Alton, G., Parsons, M.R., Convery, M.A., Blakeley, V, Corner, A.S., Palcic, M.M., Knowles, P.F., McPherson, M.J. and Phillips, S.E.V. (1997) Catalytic mechanism of the quinoenzyme amine oxidase from *Escherichia coli*: exploring the reductive half-reaction, *Biochemistry* **36**, 1608-1620.

48. Steinebach, V., de Jong, G.A.H., Wijmenga, S.S., de Vries, S. and Duine, J.A. (1996) The copper-topaquinone-phenylhydrazine-adduct geometry in *Escherichia coli* amine oxidase derivatized with phenylhydrazines substituted with trifluoromethyl groups, as determined with ^{19}F-NMR relaxation measurements, *Eur. J. Biochem.* **238**, 683-689.

49. Cooper, R.A. (1997) On the amine oxidases of *Klebsiella aerogenes* strain W70, *FEMS Microbiol. Lett.* **146**, 85-89.

50. van Dijken, J.P. and Bos, P. (1981) Utilization of amines by yeasts, *Arch. Microbiol.* **128**, 320-324.

51. Köstler, M. and Kleiner, D. (1989) Assimilation of methylamine by *Paracoccus denitrificans* involves formaldehyde transport by a special carrier, *FEMS Microbiol. Lett.* **65**, 1-4.

52. Dijkhuizen, L, de Boer, L., Boers, R.H., Harder, W. and Konings, W.N. (1982) Uptake of methylamine via an inducible, energy-dependent transport system in the facultative methylotroph *Arthrobacter* P1, *Arch. Microbiol.* **133**, 261-266.

53. Steinebach, V., Groen, B.W., Wijmenga, S.S., Niessen, W.M.A., Jongejan, J.A. and Duine, J.A. (1995) Identification of topaquinone, as illustrated for pig kidney diamine oxidase and *Escherichia coli* amine oxidase, *Anal. Biochem.* **230**, 159-166.

54. De Biase, D., Agostinelli, E., De Matteis, G., Mondovi, B. and Morpurgo, L. (1996) Half-of-the-sites reactivity of bovine serum amine oxidase. Reactivity and chemical identity of the second site, *Eur. J. Biochem.* **237**, 93-99.

55. Steinebach, V., Benen, J.A.E., Bader, R., Postma, P.W., de Vries, S. and Duine, J.A. (1996) Cloning of the *maoA* gene that encodes aromatic amine oxidase of *Escherichia coli* W3350 and characterization of the overexpressed enzyme, *Eur. J. Biochem.* **237**, 584-591.

56. Frebort, I., Pec, P., Luhova, L., Toyama, H., Matsushita, K., Hirota, S., Kitagawa, T., Ueno, T., Asano, Y., Kata, Y. and Adachi, O. (1996) Two amine oxidases from *Aspergillus niger* AKU 3302 contain topaquinone as the cofactor: unusual cofactor link to the glutamyl residue occurs at only one of the enzymes, *Biochim. Biophys. Acta* **1295**, 59-72.

57. Ruggiero, C.E., Smith, J.A., Tanizawa, K. and Dooley, D.M. (1997) Mechanistic studies of topa quinone

biosynthesis in phenylethylamine oxidase, *Biochemistry* **36**, 1953-1959.

58. Nakamura, N., Matsuzaki, R., Choi, Y.-H., Tanaizawa, K. and Sanders-Loehr, J. (1996) Biosynthesis of topaquinone cofactor in bacterial amine oxidases. Solvent origin of C-2 oxygen determined by Raman spectroscopy, *J. Biol. Chem.* **271**, 4718-4724.

59. Matsuzaki, R., Suzuki, S., Yamaguchi, K., Fukui, T. and Tanizawa, K. (1995) Spectroscopic studies on the mechanism of the topaquinone generation in bacterial monoamine oxidase, *Biochemistry* **34**, 4524-4530.

60. Kumar, M., Dooley, D.M., Freeman, H.C., Guss, J.M., Harvey, I., McGuirl, M.A., Wilce, M.C.J. and Zubak, V.M. (1996) Crystal structure of a eukaryotic (pea seedling) copper-containing amine oxidase at 2.2 Å resolution, *Structure* **4**, 943-955.

61. Pedersen, J.Z., El-Sherbini, S., Finazzi-Agro, A. and Rotilio, G. (1992) A substrate-cofactor free radical intermediate in the reaction mechanism of copper amine oxidase, *Biochemistry* **31**, 8-12.

62. Dooley, D.M., McGuirl, M.A., Brown, D.E., Turowski, P.N., McIntire, W.S. and Knowles, P.F. (1991) A Cu(I)-semiquinone state in substrate-reduced amine oxidases, *Nature* **349**, 262-264.

63. Turowski, P.N., McGuirl, M.A. and Dooley, D.M. (1993) Intramolecular electron transfer rate between active-site copper and topaquinone in pea seedling amine oxidase, *J. Biol. Chem.* **268**, 17680-17682.

64. McCracken, J., Peisach, J., Cote, C.E., McGuirl, M.A. and Dooley, D.M. (1992) Pulsed EPR studies of the semiquinone state of copper-containing amine oxidases, *J. Am. Chem. Soc.* **114**, 3715-3720.

65. Warncke, K., Babcock, G.T., Dooley, D.M., McGuirl, M.A. and McCracken, J. (1994) Structure of the topa-semiquinone catalytic intermediate of amine oxidase as revealed by magnetic interactions with exchangeable H-2 and H-1 nuclei, *J. Am. Chem. Soc.* **116**, 4028-4037.

66. Medda, R., Padiglia, A., Pedersen, J.Z., Rotilio, G., Finazzi-Agro, A. and Floris, G. (1995) The reaction mechanism of copper amine oxidase: detection of intermediates by the use of substrates and inhibitors, *Biochemistry* **34**, 16375-16381.

67. Chistoserdov, A.Y., Chistoserdova, L.V., McIntire, W.S. and Lidstrom, M.E. (1994) Genetic organization of the *mau* gene cluster in *Methylobacterium extorquens* AM1: complete nucleotide sequence and generation and characteristics of *mau* mutants, *J. Bacteriol.* **176**, 4052-4062.

68. Chistoserdov, A.Y., McIntire, W.S., Mathews, F.S. and Lidstrom, M.E. (1994) Organization of the methylamine utilization (*mau*) genes in *Methylophilus methylotrophus* W3A1-NS, *J. Bacteriol.* **176**, 4073-4080.

69. Ubbink, M., van Kleef, M.A.G., Kleinjan, D.J., Hoitink, C.W., Huitema, F., Beintema, J.J., Duine, J.A. and Canters, G.W. (1991) Cloning, sequencing and expression studies of the genes encoding amicyanin and the b-subunit of methylamine dehydrogenase from *Thiobacillus versutus*, *Eur. J. Biochem.* **202**, 1003-1012.

70. Huitema, F., van Beeumen, J. van Driessche, G., Duine, J.A. and Canters, G.W. (1993) Cloning and sequencing of the gene coding for the large subunit of methylamine dehydrogenase from *Thiobacillus versutus*, *J. Bacteriol.* **175**, 6254-6259.

71. Van Spanning, R.J.M., van der Palen, C.J.N.M., Slotboom, D.J., Reijnders, W.N., Stouthamer, A.H. and Duine, J.A. (1994) Expression of the *mau* genes involved in methylamine metabolism in *Paracoccus denitrificans* is under control of a LysR-type transcriptional activator, *Eur. J. Biochem.* **226**, 201-210.

72. van der Palen, C.J.N.M., Slotboom,D-J., Jongejan, L., Reijnders, W.N.M., Harms, N., Duine, J.A. and van Spanning, R.J.M. (1995) Mutational analysis of the *mau* genes involved in methylamine metabolism in *Paracoccus denitrificans*, *Eur. J. Biochem.* **230**, 860-871.

73. Van der Palen, C.J.N.M., Reijnders, W.N.M., de Vries, S., van Spanning, R.J.M. and Duine, J.A. (1997) MauE and MauD are essential in methylamine metabolism of *Paracoccus denitrificans*, *Antonie van Leeuwenhoek*, **72**, 219-228.

74. Huitema (1994) The primary structure of methylamine dehydrogenase from *Thiobacillus versutus*, Dissertation Delft University of Technology.

75. Itoh, S., Ogino, M., Haranou, S., Terasaka, T., Ando, T., Komatsu, M., Ohshiro, Y., Fukuzumi, S., Kano, K., Takagi, K. and Ikeda, T. (1995) A model compound of novel cofactor tryptophan tryptophylquinone of bacterial methylamine dehydrogenase. Synthesis and physicochemical properties, *J. Am. Chem. Soc.* **117**, 1485-1493.

76. Gorren, A.C.F and Duine, J.A. (1994) The effects of pH and cations on the spectral and kinetic properties of methylamine dehydrogenase from *Thiobacillus versutus*, *Biochemistry* **33**, 12202-12209.

77. Gorren, A.C.F., de Vries, S. and Duine, J.A. (1995) Binding of monovalent cations to methylamine dehydrogenase in the semiquinone state and its effect on electron transfer, *Biochemistry* **34**, 9748-9754.

78. Moenne-Loccoz, P., Nakamura, N., Itoh, S., Fukuzumi, S., Gorren, A.C.F., Duine, J.A. and Sanders-Loehr, J. (1996) Electrostatic environment of the tryptophylquinone cofactor in methylamine dehydrogenase: evidence from resonance Raman spectroscopy of model compounds, *Biochemistry* **35**, 4713-4720.

79. Gorren, A.C.F., Moenne-Loccoz, P., Backes, G., de Vries, S., Sanders-Loehr, J. and Duine, J.A. (1995)

Evidence for a methylammonium-binding site on methylamine dehydrogenase of *Thiobacillus versutus*, *Biochemistry* **34**, 12926-12931.

80. Gorren, A.C.F., van der Palen, C.J.N.M., van Spanning R.J.M. and Duine, J.A. (1996) Enzymology of methylamine dehydrogenase, *Microbial Growth on C1-*, Kluwer Academic Publishers, Dordrecht,.

81. Bishop, G.R. and Davidson, V.L. (1995) Intermolecular electron transfer from substrate-reduced methylamine dehydrogenase to amicyanin is linked to proton transfer, *Biochemistry* **34**, 12082-12086.

82. Ubbink, M., Hunt, N.I., Hill, H.A.O. and Canters, G.W. (1994) Kinetics of the reduction of wild type and mutant cytochrome c_{550} by methylamine dehydrogenase and amicyanin from *Thiobacillus versutus*, *Eur. J. Biochem.* **222**, 561-571.

83. De Gier, J-W.L., van Spanning, R.J.M., Oltman, L.F. and Stouthamer, A.H. (1992) Oxidation of methylamine by a *Paracoccus denitrificans* mutant impaired in the synthesis of the bc_1 complex and the aa_3-type oxidase, *FEBS Lett.* **306**, 23-26.

84. Chen, L., Durley, R.C.E., Mathews, F.S. and Davidson, V.L. (1994) Structure of an electron transfer complex: methylamine dehydrogenase, amicyanin, and cytochrome c_{551i}, *Science* **264**, 86-90.

FLAVOCYTOCHROMES: NATURE'S ELECTRICAL TRANSFORMERS

S.K. CHAPMAN[1], G.A. REID[2] and A.W. MUNRO[1]
[1] Department of Chemistry and
[2] Institute of Cell & Molecular Biology
University of Edinburgh, West Mains Road,
Edinburgh EH9 3JJ, UK.

1. Introduction

Flavocytochromes, as the name suggests, contain both flavin and heme groups as cofactors. The flavin can be flavin adenine dinucleotide (FAD) or flavin mononucleotide (FMN) and the heme is usually heme-*b* or heme-*c*. Typical cofactors are illustrated in Figure 1.

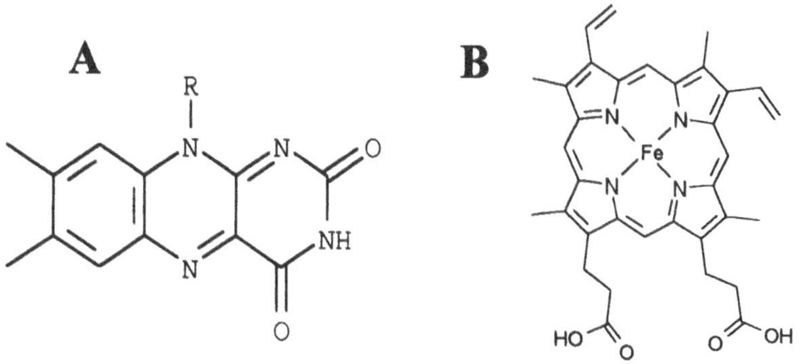

Figure 1. Typical cofactors found in flavocytochromes:
A = flavin (the sidechain R differs between FMN and FAD); **B** = protoheme IX

Flavins have the ability to act as either 1- or 2- electron redox centers since they have three accessible oxidation states (oxidized, semiquinone and reduced). Heme groups on the other hand are usually facile 1-electron transfer centers (cycling between Fe^{2+} and Fe^{3+}). In addition to their electron transfer abilities, both flavin and heme cofactors can activate and reduce molecular oxygen. Hence, enzymes containing flavin and heme exhibit a huge diversity of function and may act in different cases as

G.W. Canters and E. Vijgenboom (eds.),
Biological Electron Transfer Chains: Genetics, Composition and Mode of Operation, 165-184.
© 1998 Kluwer Academic Publishers.

dehydrogenases, electron transferases, oxidases or monooxygenases. It is, therefore, appropriate to think of flavocytochromes as natural electrical transformers, allowing two-electron processes, such as the oxidation or reduction of organic substrates, to be coupled to one-electron transfers to, or from, a variety of biologically important redox partners.

To illustrate the versatility in function of flavocytochromes three examples will be discussed. These are: Flavocytochrome b_2; flavocytochrome c_3; and flavocytochrome P-450. Details of these three enzymes are listed in Table 1.

TABLE 1. Examples of three different flavocytochromes

ENZYME	COFACTORS	FUNCTION	SOURCE
Flavocytochrome b_2	1 FMN 1 heme-b	L-Lactate dehydrogenase	Saccharomyces cerevisiae Hansenula anomala
		L-Mandelate dehydrogenase	Rhodotorula graminis
Flavocytochrome c_3	1 FAD 4 heme-c	Fumarate reductase	Shewanella putrefaciens
Flavocytochrome P-450	1 FAD, 1 FMN 1 heme-b	Fatty acid monooxygenase	Bacillus megaterium

2. Flavocytochrome b_2

Flavocytochrome b_2 from the yeast Saccharomyces cerevisiae couples L-Lactate dehydrogenation to cytochrome c reduction in the mitochondrial intermembrane space, as shown in Figure 2.

The DNA encoding S. cerevisiae flavocytochrome b_2 has been cloned and expressed at a high-level in E. coli [1]. The enzyme is a homotetramer with subunit M_r of 57.5 kDa. The X-ray crystal structures have been determined for the native enzyme from S. cerevisiae [2], and for the recombinant enzyme from E. coli [3]. These are isostructural and clearly demonstrate that each subunit consists of two distinct domains. The first 100 residues fold to form a b-type cytochrome domain which is connected via a short hinge region to a larger (400 residue) flavodehydrogenase domain. The domain structure of a single flavocytochrome b_2 subunit is illustrated in Figure 3.

The electron flow through flavocytochrome b_2 can be described in terms of five consecutive electron transfer events [4]. L-Lactate dehydrogenation results in the 2-electron reduction of flavin. These two electrons are individually passed to b_2-heme (intramolecular electron transfer) and finally from b_2-heme to cytochrome c (intermolecular electron transfer). Thus, this enzyme provides an ideal model system to study both intra- and inter- molecular electron transfer.

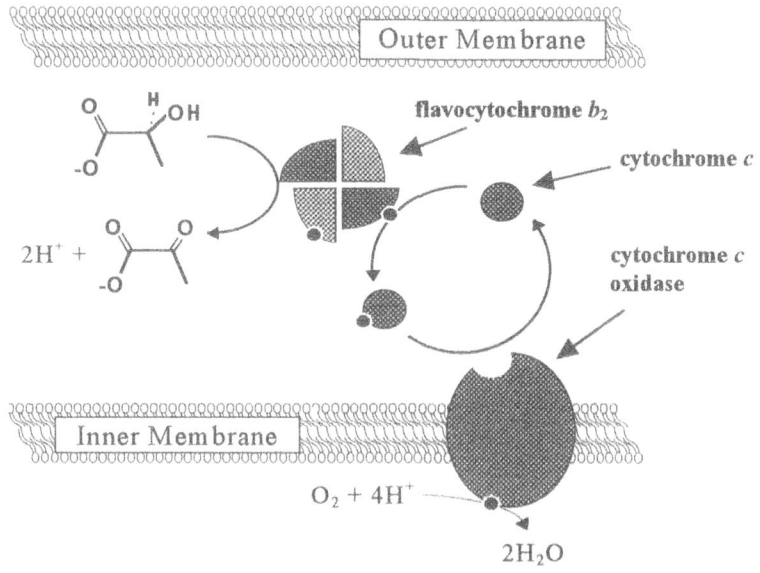

Figure 2. Schematic representation of the physiological role of *S. cerevisiae* flavocytochrome b_2. L-Lactate dehydrogenation is coupled to cytochrome c reduction by the enzyme. The transport of electrons (represented as small dots) to cytochrome c oxidase completes the short respiratory chain

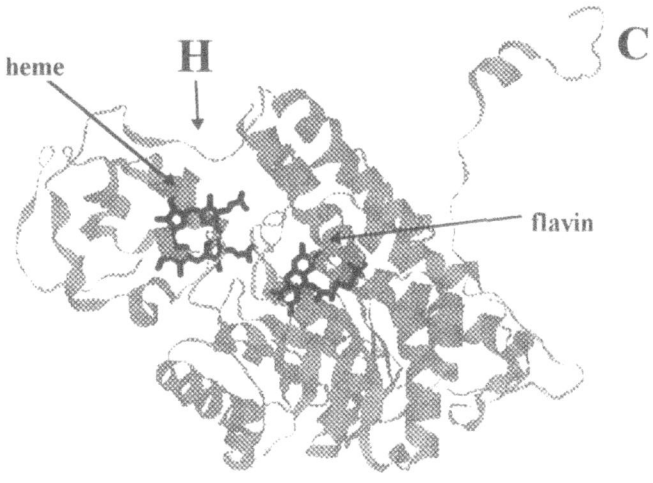

Figure 3. A ribbon diagram of a single subunit of flavocytochrome b_2 from *S. cerevisiae*. The heme (b-type) and flavin (FMN) groups are arrowed. **H**, hinge region; **C**, extended *C*-terminal tail.

2.1 INTRAMOLECULAR ELECTRON TRANSFER

The catalytic cycle for flavocytochrome b_2 involves two separate intramolecular electron transfer events. The first is from fully reduced flavin to b_2-heme which generates reduced b_2-heme and flavin semiquinone. The second is from flavin semiquinone to b_2-heme (which occurs only after the b_2-heme has been reoxidized by cytochrome c). Although the distance between the cofactors (9.7 Å) and the intervening medium is the same for both of these electron transfer steps, they occur at very different rates.

The electron transfer from reduced flavin to b_2-heme is reversible with the position of the equilibrium lying 85% in favor of reduced b_2-heme. Detailed pre-steady-state analysis of this step has determined the value of the forward rate constant to be approximately 1500 s^{-1} (at 25°C, pH 7.5, $I = 0.10$) [5]. However, the second intramolecular electron transfer step, from flavin semiquinone to b_2-heme, has been shown to be much slower, with a value of 120 s^{-1} [4]. This slow electron transfer from semiquinone to b_2-heme is in fact the major rate-limiting step in the catalytic cycle. Indeed, quenched-flow EPR has demonstrated that during steady-state turnover of the enzyme approximately 75% of the flavin exists in the semiquinone state [4]. The availability of these kinetic data makes it appropriate to consider which particular structural elements and amino acid residues in the enzyme are important in controlling these intramolecular electron transfers.

The interface region between the flavin and heme groups is illustrated in Figure 4. The roles of the residues highlighted in Figure 4 have been examined by detailed kinetic analyses of various mutant forms of the enzyme with the following results:

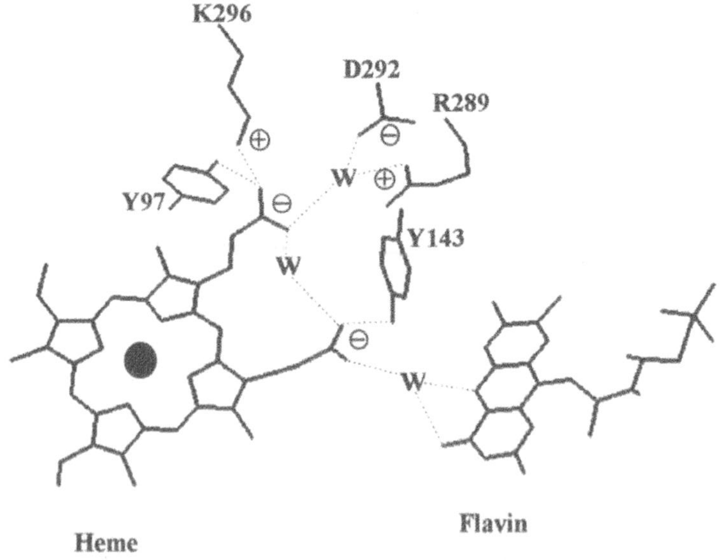

Figure 4. The interface region between the flavin- and heme- containing domains of flavocytochrome b_2. Selected sidechains are labeled. Water molecules are indicated as W.

2.1.1. *Tyrosine-143*

The importance of this residue for intramolecular electron transfer has been examined by analysis of a mutant enzyme in which tyrosine-143 has been replaced by phenylalanine [6]. The 3-dimensional structure of flavocytochrome b_2 gave the first indication that tyrosine-143 might have a pivotal role to play in catalysis. Although flavocytochrome b_2 is a homotetramer, crystallization of the enzyme resulted in two crystallographically distinguishable subunits in the asymmetric unit [2]. In subunit 1, the electron density clearly shows both cytochrome and flavodehydrogenase domains with tyrosine-143 hydrogen bonding to a heme propionate as in Figure 4. However, in subunit 2 electron density for the cytochrome domain is not seen (presumably due to positional disorder) with glycine-100 being the first defined residue. In this case tyrosine-143 hydrogen bonds to the carboxylate group of pyruvate which is found at the active site. Thus it seems possible that tyrosine-143 might be an active site "switch" facilitating either reduction of the flavin by lactate or electron transfer from flavin to heme. Studies on the Y143F-b_2 enzyme confirm that this is in fact the case [6,7]. The rate constant for electron transfer from fully reduced flavin to b_2-heme falls from 1500 s^{-1} in the wild-type enzyme down to 20 s^{-1} in Y143F-b_2. This effect of nearly two orders of magnitude clearly demonstrates that tyrosine-143 is a crucial residue for flavin to heme electron transfer. The X-ray crystal structure of the Y143F-b_2 confirms that this large effect on rate is not due to gross structural changes in the enzyme [8]. The most likely explanation for the effect of the mutation on electron transfer rate relates to the fact that the cytochrome domain is believed to be mobile with respect to the flavodehydrogenase domain [9,10]. Thus, the removal of a key interface hydrogen bond between tyrosine-143 and the heme propionate (Figure 4) curtails the number of productive encounters between the two domains and hence lowers the efficiency of interdomain electron transfer [5].

2.1.2. *Lysine-296*

The electrostatic interaction between lysine-296 and a heme propionate (Figure 4) represents the only salt bridge at the interface of the two domains. To examine the importance of this interaction on the rate of electron transfer from flavin to heme, a mutant enzyme was generated in which the lysine was changed to methionine. This substitution removes the charge whilst retaining a similar steric bulk. Kinetic analysis of the K296M-b_2 enzyme indicated that the interdomain electron transfer rate was essentially the same as that seen for the wild-type enzyme [11]. This clearly indicates that lysine-296 and the salt bridge it forms are not essential for flavin to heme electron transfer.

2.1.3. *Arginine-289*

Arginine-289 forms several hydrogen bonds; one of which is to a water molecule which, in turn, hydrogen bonds to a heme propionate (Figure 4). Hence, this residue might have a role to play in interdomain interaction and electron transfer. In addition, arginine-289 has a stacking interaction with arginine-376, which is known to bind and orientate the substrate, L-lactate. Thus, arginine-289 could be important for both lactate dehydrogenation and flavin to heme electron transfer. To test these ideas arginine-289 was changed to lysine, generating the R289K-b_2 enzyme. Kinetic analysis of R289K-b_2 shows dramatic effects on the lactate dehydrogenase function of the enzyme, with a 10-fold decrease in the rate of lactate turnover using ferricyanide or cytochrome c as electron acceptors [12]. However, an estimation of the rate of electron transfer from flavin to heme indicates that the mutation has little effect. The obvious conclusion is that arginine-289 is crucial for substrate dehydrogenation but is not important for interdomain electron transfer.

2.1.4. *The interdomain hinge*

The idea that the cytochrome domain is mobile with respect to the flavodehydrogenase domain has already been discussed briefly in Section 2.1.1. This mobility is made possible due to an interdomain hinge region which links the two domains together (Figure 3). This hinge region has been the subject of a detailed examination involving hinge swaps [9], hinge deletions [13,14] and hinge insertions [15]. All of the studies on the hinge region are consistent with the idea that the structural integrity of the hinge is essential for efficient electron transfer from flavin to heme.

2.2. INTERMOLECULAR ELECTRON TRANSFER

To understand the intermolecular electron transfer from flavocytochrome b_2 to cytochrome c it is essential to first characterize the complex formed between these two proteins. There have been many attempts to do this but contradictory results from various studies have made interpretation difficult (see [16] and references therein). One thing that can be deduced with confidence is that electrostatics play an important part in complex formation. In 1993 Tegoni *et al.* [17] reported a computer generated model of what the flavocytochrome b_2:cytochrome c complex might look like. This model predicted a number of flavocytochrome b_2 residues to be important for complex formation and interprotein electron transfer [17]. These included glutamate-92 (predicted to salt-bridge with arginine-13 on cytochrome c) and phenylalanine-52 (predicted to form part of a tunneling pathway for electron transfer between b_2- and c- hemes). In order to test these and other predictions from the model an extensive re-examination of the flavocytochrome b_2:cytochrome c reaction was performed [18]. A number of flavocytochrome b_2 mutations were generated including changing glutamate-92 to both glutamine and lysine and changing phenylalanine-52 to alanine. If the hypothetical complex was correct these mutations would have been expected to have significant effects on the flavocytochrome b_2:cytochrome c reaction. In fact results have indicated that these residues are not important for interprotein electron transfer [10,18]. Data from studies on other mutant flavocytochromes b_2 (for example mutations in the C-terminal tail and hinge regions) were also inconsistent with the model. Thus it would appear that the hypothetical complex reported by Tegoni *et al.* [17] is not a realistic representation of a catalytically competent interaction between these two proteins and should be disregarded.

Recently a new model for the flavocytochrome b_2:cytochrome c complex has been proposed [19]. This model has the advantage that it has already been kinetically tested and is consistent with the results from solution studies. In this new model two residues on flavocytochrome b_2; glutamate-63 and aspartate-72, are predicted to be important for complex formation. This prediction is supported by results from studies on mutant enzymes in which glutamate-63 and aspartate-72 have been changed to lysines. In both cases significant effects are seen on the rate of electron transfer from flavocytochrome b_2 to cytochrome c [19]. A representation of this proposed complex is shown in Figure 5.

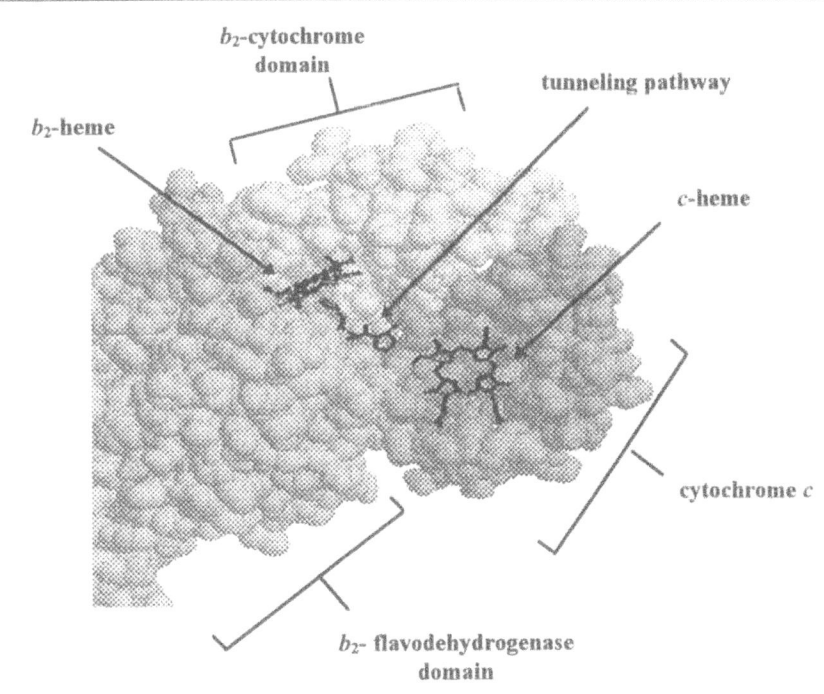

In the model, one cytochrome c binds to each subunit of flavocytochrome b_2 . This binding site is positioned at the interdomain border of flavocytochrome b_2 (Figure 5). The model predicts that aspartate-72, glutamate-63 and glutamate-237 of flavocytochrome b_2 form electrostatic interactions with arginine-13 (lysine in cytochrome c from most species), lysine-27 and lysine-19 of yeast cytochrome c [19]. Thus, these residues form a triangular recognition site and through the middle of this triangle runs a possible σ-tunneling pathway (arrowed in Figure 5). This pathway involves the imidazole ring of histidine-66 (a ligand to the iron), the backbone of alanine-67 and the ring of proline-68. There is then a through-space jump of only 3 Å to the cytochrome *c* heme. For the direct through-space route, the shortest edge-to-edge distance is 13.1 Å from the imidazole ring of histidine-66 to the heme of cytochrome *c* [19]. The fact that kinetic studies are completely consistent with the above model (which allows for three possible orientations for cytochrome *c*) suggests that it does indeed represent a catalytically competent complex. Thus this new model [19] completely supersedes the previous one [17].

3. Flavocytochrome c_3

Flavocytochrome c_3 from the bacterium *Shewanella putrefaciens* is a fumarate reductase that is required for anaerobic respiration with fumarate as the terminal electron acceptor. Fumarate respiration is common among facultative anaerobes but the fumarate reductases from most of these organisms (*S. putrefaciens* is the only known exception to date) are membrane-bound enzymes that consist of either three or four subunits and accept electrons directly from quinols in the cytoplasmic membrane [20]. The membrane-bound enzymes are closely related to succinate dehydrogenases in structure and are catalytically freely reversible. The fumarate reductase from *Escherichia coli* has been extensively characterized and its subunit structure and topology are shown digrammatically in Figure 6. The quinol binding site is associated with the small, membrane anchor subunits (FrdC, FrdD), though these contain no redox cofactors. The FrdB subunit contains 3 iron-sulfur clusters and the FrdA subunit contains covalently bound FAD at the active site for fumarate reduction.

Inner membrane

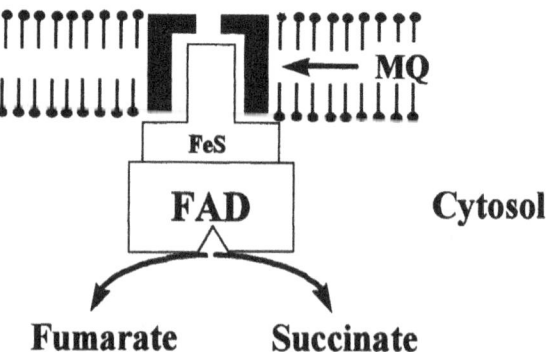

Figure 6. Schematic representation of the *E. coli* fumarate reductase. The membrane anchor subunits are shown in black. The iron/sulfur containing subunit is labeled FeS and the flavin containing subunit is labeled FAD. MQ is menaquinone.

In contrast, flavocytochrome c_3 is a structurally much simpler enzyme, consisting of a single subunit. It is a soluble, periplasmic protein of 64 kDa [21,22] that is abundant in anaerobically grown *S. putrefaciens*. The isolated protein is an efficient fumarate reductase (using methyl viologen as reductant) but this reaction is effectively unidirectional since the enzyme has a high K_m for succinate and a very low k_{cat} for succinate dehydrogenation. The *fcc* gene encoding flavocytochrome c_3 has been cloned and sequenced [22]. The protein is apparently composed of a small, N-terminal tetraheme cytochrome domain that bears some, though very limited, resemblance to cytochrome c_3 from sulfate reducing bacteria and a larger flavoprotein domain that is clearly related to the FrdA subunit of the membrane-bound fumarate reductases. The subunit structure and location of *S. putrefaciens* flavocytochrome c_3 is shown diagrammatically in Figure 7.

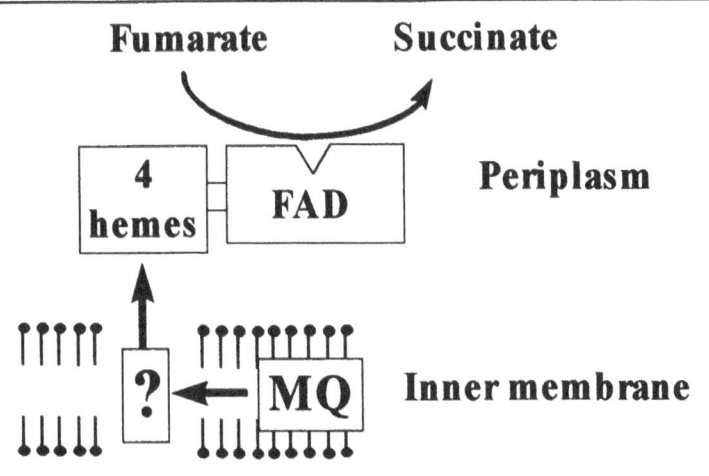

Figure 7. Schematic representation of the *S. putrefaciens* fumarate reductase. The FAD and heme containing domains are labeled. The (as yet) unknown physiological electron donor is indicated by the question mark. MQ is menaquinone.

The flavocytochrome c_3 flavoprotein domain shares 26% sequence identity with *E. coli* FrdA. This degree of sequence conservation indicates a closely related structure and presumably a similar mechanism for fumarate reduction. This conjecture is supported by the observation that several amino acid residues that have previously been implicated in catalytic activity in the membrane-bound enzymes, either by chemical modification [23,24] or by site-directed mutagenesis [25], are conserved in flavocytochrome c_3.

S. putrefaciens is remarkable for the diversity of terminal oxidants that can be used during anaerobic respiration, including fumarate, nitrate, TMAO, thiosulfate and, more unusually, Mn(IV) and Fe(III). Relatively little is known about the electron transfer pathways in *Shewanella*, though several cytochromes have been shown to be produced during anaerobic growth [26]. The location of flavocytochrome c_3 in the periplasm (Figure 7) raises questions as to how reducing equivalents are transferred from membrane-bound components at earlier steps in the pathway.

3.1 FUMARATE RESPIRATION

Quinones have been implicated in fumarate respiration in *S. putrefaciens* [27] since a mutant strain that was deficient in quinones was also unable to grow anaerobically with fumarate. It is unclear whether flavocytochrome c_3 can oxidize quinols directly or whether other electron carriers are involved. However, since flavocytochrome c_3 is soluble and because the mid-point potential for the fumarate/succinate couple is only +30 mV, it seems unlikely that proton translocation can be associated with the fumarate reductase reaction. The reduction potentials of the heme groups in flavocytochrome c_3 are extremely low for a *c*-type cytochrome, titrating between -320 and -220 mV. This poses further bioenergetic problems since, if we assume that electrons are transferred to the flavocytochrome c_3 flavin group directly from one or more of the hemes, then the reductant for the hemes must have a sufficiently low reduction potential to drive this reaction. It is not obvious how this can be reconciled (a) with energy conservation at the dehydrogenase end of the respiratory chain since the available redox potential gap is quite small and (b) with the involvement of menaquinone in the respiratory chain to fumarate. Fumarate respiration can indeed support growth in *S. putrefaciens*, demonstrating that this pathway is coupled to energy conservation. Furthermore, it has been demonstrated that proton translocation occurs in response to a fumarate pulse with lactate as reductant in whole *S. putrefaciens* cells [28].

On the basis of EPR. spectroscopy with whole cells, it has been proposed that *S. putrefaciens* actually produces a membrane-bound fumarate reductase in addition to the soluble flavocytochrome c_3 [29]. If this was the case, then it could be argued that flavocytochrome c_3 is not itself involved in an energy conserving pathway. However, our genetic experiments demonstrate that flavocytochrome c_3 is the only respiratory fumarate reductase in *S. putrefcaiens*. We constructed a null mutant in which the *fcc* gene has been disrupted by insertion of a kanamycin resistance marker. This strain, EG301, produces no immunochemically detectable flavocytochrome c_3, has no detectable fumarate reductase activity and is unable to grow anaerobically with fumarate [30]. Growth with other acceptors is unaffected. Thus flavocytochrome c_3 is required as a respiratory fumarate reductase and, furthermore, is the only fumarate reductase in *S. putrefaciens*.

3.2 SPECTROSCOPIC ANALYSIS

The sequence of flavocytochrome c_3 indicates the presence of four typical *c*-type heme attachment sites (CxxCH motifs) and extinction coefficients for the purified protein indicate the presence of four hemes and one FAD molecule per flavocytochrome c_3 monomer [21]. The presence of four hemes in a relatively small cytochrome domain is quite remarkable, but a good precedent is found in cytochromes c_3 from

Desulphovibvrio species. These soluble, low potential periplasmic proteins apparently mediate electron transfer in the pathway from H_2 to SO_4^{2-}. The crystal structure of cytochrome c_3 shows that each of the four heme irons is ligated by two histidine side chains. EPR and magnetic circular dichroism spectroscopy of flavocytochrome c_3 from *S. putrefaciens* demonstrate that the hemes here are also bis-histidine ligated [30]. The ligation pattern, the very low redox potentials and the periplasmic location of each of these proteins suggests an underlying relatedness. However, apart from the heme attachment motifs there is little sequence conservation between them. Determination of the three dimensional structure of flavocytochrome c_3 (underway in our laboratory) will allow a more informed comparison to be made.

4. Flavocytochrome P-450 BM3

The cytochromes P-450 are a superfamily of hemoproteins, catalyzing an array of oxdative reactions with a plethora of organic substrates [31]. They are found in organisms as diverse as bacteria in man, they are an excellent example of how nature has tampered with a similar enzyme structure and mechanism to perform numerous different functions. Aside from their roles in eukaryote physiology, which include the generation of bioactive fatty acid derivatives and synthesis and interconversions of steroid molecules, the P-450s are also vital in the detoxification of exogenous drugs and toxins. In mammals, the vast majority of this function is performed by microsomal P-450s in the liver, with the oxygenations designed to make the organic products more soluble than the original substrates to facilitate their excretion in the aqueous solutions of urine or bile. Indeed, the synthesis of many P-450s is controlled at the genetic level by exposure to the very substrates that they oxidize [32]. However, it is apparent that, while the P-450s play a vital role in phase 1 metabolism of xenobiotics, the evoutionary rate has been outstripped by man's production of novel toxins and organic drugs and his pollution of the environment with the same. Thus, certain P-450s (many of which have quite broad and overlapping specificity) have been recognized to catalyze the oxidation of a number of xenobiotics into reactive and/or mutagenic products [33]. Thus the P-450s are a "double edged sword" - essential for cellular function but prone to inflict self-damage if provided with the wrong substrates.

The critical function of the P-450s is the activation of molecular oxygen with (in most cases) the monooxygenation of a bound substrate. This is achieved by means of two successive one-electron transfers to the heme center from the redox partner. The first transfer (reducing ferric resting heme to the ferrous state) occurs after substrate is bound to the P-450 active site. The binding of substrate usually induces a low-spin (S = 1/2) to high-spin (S = 5/2) shift in the heme iron spin-state equilibrium, probably through removal of a water ligand from the 6th (axial) coordination position of the heme iron. Following this reduction, dioxygen binds to the Fe^{2+} heme. Reduction by a 2nd electron from the redox partner results in the completion of the P-450 catalytic cycle, although steps subsequent to the 2nd electron transfer remain to be clarified - since the subsequent intermediates are so short-lived that they have been near-impossible to study by spectroscopic methods. The 2nd electron transfer leads to

176

the transient formation of an unstable oxyferryl iron intermediate which is the species ultimately responsible for P-450-mediated catalysis [34] (Figure 8).

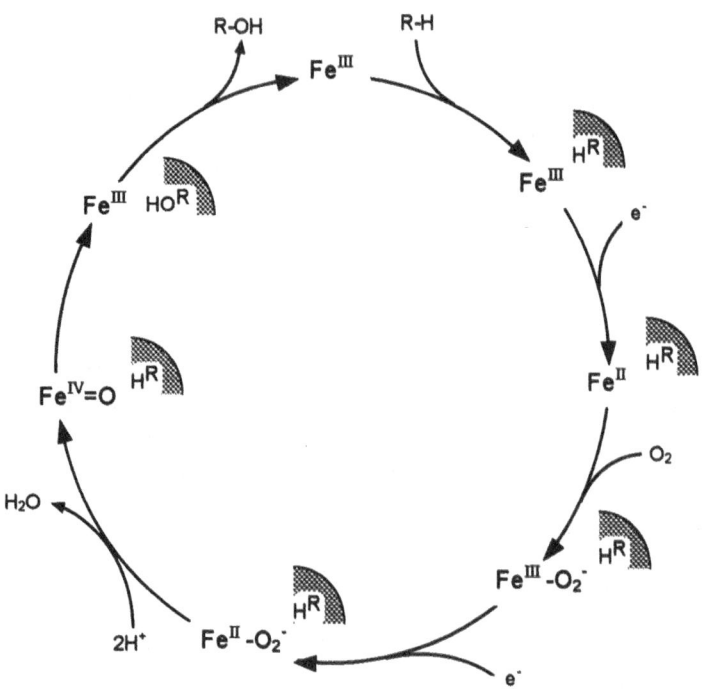

Figure 8. The postulated catalytic cycle of cytochromes P-450. The Fe (III) heme is not reduced until substrate is bound at the active site. Oxygen does not bind to the heme until the iron is reduced [Fe (II)]. Thereafter, a second electron transfer to the heme center initiates the rapid sequence events which lead to the monooxygenation of substrate and reduction of the second oxygen to water.
RH = substrate; ROH = product (= monooxygenated substrate).

From both biotechnological and enzymological points of view, there are many interesting bacterial P-450s. In many cases, bacteria have evolved the versatile P-450 systems to oxygenate environmental pollutants and other unusual organic chemicals, permitting the bacteria to make use of these materials as carbon sources for growth. Examples are the linalool 8-methyl hydroxylase from *Pseudomonas incognita* and the camphor hydroxylase from *Pseudomonas putida* [35]. The bacterial P-450s are soluble enzymes, unlike their eukaryotic counterparts. This fact means that bacterial forms are more simply purified and can be maintained in solution at higher concentration than eukaryotic P-450s. The bacterial forms, therefore, are more suitable for studies with a variety of spectroscopic techniques. Indeed, the four available P-450 atomic structures are all derived from crystals obtained from soluble bacterial P-450s [36-39]. The intron-free bacterial genes are also more simply overexpressed in an *E. coli* system, meaning that cloned bacterial P-450s can generally be isolated in large quantities for biophysical analysis. Thus, the use of soluble bacterial P-450s as models for their eukaryotic homologues is highly attractive. A major problem, however, is that all bar one bacterial P-450 uses a different electron transfer apparatus to that employed by the mammalian microsomal enzymes. For example, the best characterized bacterial P-450 system, the P-450cam camphor hydroxylase from *Pseudomonas putida*, is reduced by single electron transfers from a small iron sulfur protein (putidaredoxin) which, in turn, is reduced by an NADH-dependent FAD-containing enzyme (putidaredoxin reductase). This so-called class I system is analogous to that used by eukaryotic mitochondrial steroid hydroxylases (perhaps reflecting the bacterial origin of this organelle). By contrast the microsomal enzymes are reduced by a single enzyme - an NAD(P)H-dependent FAD- and FMN-containing cytochrome P-450 reductase [35]. Obviously, a bacterial P-450 with an analogous reductase system would represent an extremely attractive model system. Such a P-450 system is flavocytochrome P-450 BM3.

4.1. PROPERTIES OF FLAVOCYTOCHROME P-450 BM3

Flavocytochrome P-450 BM3 is the third P-450 isolated from the soil bacterium *Bacillus megaterium*. The enzyme is a fatty acid hydroxylase, catalyzing the sub-terminal oxidation of a wide range of fatty acids (carbon chain length approx. C12 - C20). The specific activity is extremely high [40], on the order of 100-1000 fold higher than that of most mammalian forms. Underlying this extremely high activity is the fact that P-450 BM3 has its redox partner fused to the hemoprotein in a single 119kDa polypeptide. Moreover, this redox partner was shown to contain both FAD and FMN and to catalyze the reduction of cytochrome *c* [41]. Isolation and sequencing of the gene (*cyp*102) confirmed that the redox partner showed extensive similarity to eukaryotic P-450 reductases and that the enzyme was therefore not only the first characterized bacterial class II P-450 but also that the reductase was fused to the P-450 (Figure 9) in a manner analogous to the nitric oxide synthases [42].

PCR-generated sub-genes encoding the heme (P-450) and diflavin (reductase) domain have been overexpressed in *E. coli* [43], as have, more recently, the sub-genes encoding the FAD (ferredoxin reductase-like) and FMN (flavodoxin-like) domains of the reductase. The independently expressed heme and diflavin domains retain spectroscopic and structural integrity. However, perhaps not surprisingly, they retain little affinity for one another after scission. In addition, visible region circular dichroism studies indicate that there may be alterations in heme environment caused by separation of the domains [44]. Studies of the electron transfer between the isolated heme and diflavin domains (following the rate of formation of the CO adduct step in fatty acid hydroxylation catalyzed by the isolated domains is clearly flavin to which is formed by rapid CO binding to Fe^{2+} heme) indicate that the rate-limiting step in fatty acid hydroxylation catalyzed by the isolated domains is clearly flavin to heme transfer. The maximal rate of monooxygenation achievable with the isolated domains is less than 10/min which, although >100 fold less than that feasible with the holoenzyme, is in the same range as that found for the majority of mammalian class II P-450 systems.

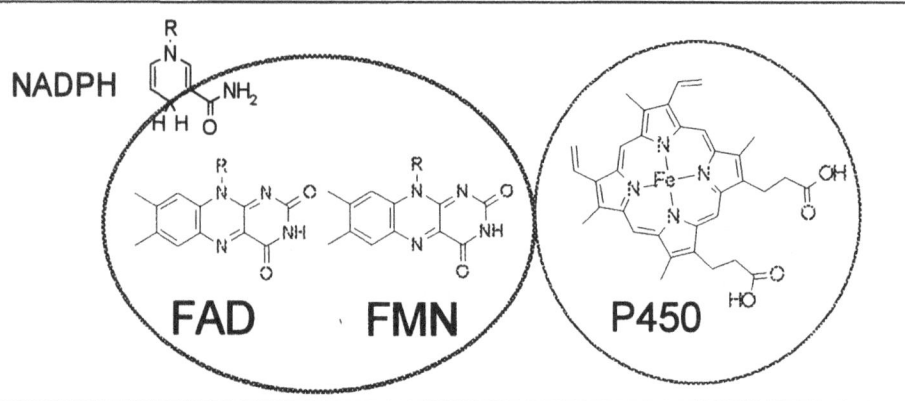

Figure 9. Schematic representation of flavocytochrome P-450 BM3

4.1.1. *Electron transfer in flavocytochrome P-450 BM3*

Recently, potentiometry has been used to measure the reduction potentials of the flavins and heme in flavocytochrome P-450 BM3 and its component domains [45]. It is clear that dissection of the domains does not significantly affect the reduction potentials of any of the bound cofactors. These studies have allowed a clearer understanding of the route of electron transfer through the enzyme and of the mechanism by which the enzyme controls its activity. Electron transfer occurs from NADPH to the FAD to form the hydroquinone. However, the potentials of both the FMN oxidized/semiquinone and semiquinone/hydroquinone couples are considerably more positive than those of the FAD, so electrons are rapidly transferred to the FMN. While the midpoint potential for the two-electron reduction of FAD is of the same order as that of the NADPH/NADP⁺ couple, the reduction potential for the FAD semiquinone/hydroquinone couple is considerably more negative than that of the FAD oxidized/semiquinone couple. In the diflavin domain (and in the fatty acid-free flavocytochrome) the two flavins can only be reduced as far as a 3 electron form; with the FMN stable in the hydroquinone form and the FAD in a blue (neutral) single electron-reduced semiquinone state [46]. EPR measurements confirm that there is a single semiquinone seen in the NADPH-reduced substrate-free enzymes. The reduction potential of the heme is rather lower than that of either of the reduced forms of the FMN. However, on binding of fatty acid substrates (e.g. arachidonic acid), the reduction potential of the heme is considerably increased (by more than 130mV), and electron transfer from FMN to heme becomes feasible. The binding of fatty acid is thought to cause the removal of the 6th heme ligand (a water molecule) and it is this phenomenon which likely underlies the reduction potential alteration.

This elegant control mechanism prevents electron transfer to the heme iron in the absence of substrate for monooxygenation. This prevents futile cycling of the enzyme, which would lead to the production of oxygen radicals and hydrogen peroxide. The different catalytic rates seen with different substrates appear to be related to the efficiency with which these substrates bind to P-450 BM3 to induce a heme iron spin-state shift. For instance, palmitic acid has a much lower K_d (approx. $2\mu M$) than does lauric acid, and the V_{max} for its hydroxylation is 4-fold higher. Interestingly, it has been shown that lauric acid binds to the active site initially at too great a distance for oxygenation to occur at the iron center. Only after the first electron transfer to the iron does the fatty acid move to a position adjacent to the heme [47].

The kinetics observed with P-450 BM3 are highly interesting. Electron transfers through the flavins are very fast and are likely to be too rapid to play a significant role in the limitation of fatty acid hydroxylation activity [46]. On stopped-flow analysis of NADPH reduction of the diflavin domain, the reaction kinetic are seen to be at least biphasic - with an extremely fast absorbance decrease at 459nm (reflecting electron transfer to FMN and formation of the FMN hydroquinone) and a slower phase in which the blue semiquinone forms.

It has been shown that cytochrome c reduction (mediated by electron transfer from the FMN) can occur at full rate simultaneously with fatty acid hydroxylation and without decreasing the rate of the latter [48]. Not only does this demonstrate that the rate limiting step does not involve electron transfers within the flavin domain, but also that the sites of interaction between the linked reductase and heme domain and the reductase and cytochrome c are probably different. Reducing P-450 BM3 fully with NADPH prior to addition of fatty acid substrate leads to a marked reduction in the rate of fatty acid hydroxylation. The reason for this appears to be that the hydroquinone FMN is not an efficient electron transferase to the P-450 heme. Whether this is due to a slow conformational change that occurs when the reductase of P-450 BM3 becomes fully reduced in the absence of fatty acids is not clear, but it is known that the enzyme can be "rescued" by the addition of cytochrome c or other artificial electron acceptors which are still reduced efficiently by the FMN hydroquinone [49]. It may be the case that during active turnover of the flavocytochrome, electron transfer from FMN semiquinone to heme takes place more quickly than that from FAD to FMN semiquinone. Electron transfer to the heme can be monitored in carbon monoxide-saturated solutions using stopped-flow spectrophotometry, following the absorbance increase at 450nm [46]. This method indicates that there are substrate-dependent differences in the rates of electron transfer, with those substrates with lower K_d values/higher monooxygenation rates giving faster flavin-to-heme electron transfer rates. For example, stopped-flow mixing of fatty acid-saturated solutions of P-450 BM3 with solutions of NADPH give rates (at $25°C$) of $130s^{-1}$ for lauric acid and $230s^{-1}$ for myristic acid [46]. A clear rate-limiting step has not been identified for the catalytic cycle of P-450 BM3, but it may be that product dissociation is a major factor. Redox potentiometry and stopped-flow kinetics have allowed us to build up a clear picture of the pathway of electron transfer through the enzyme and this is illustrated in Figure 10.

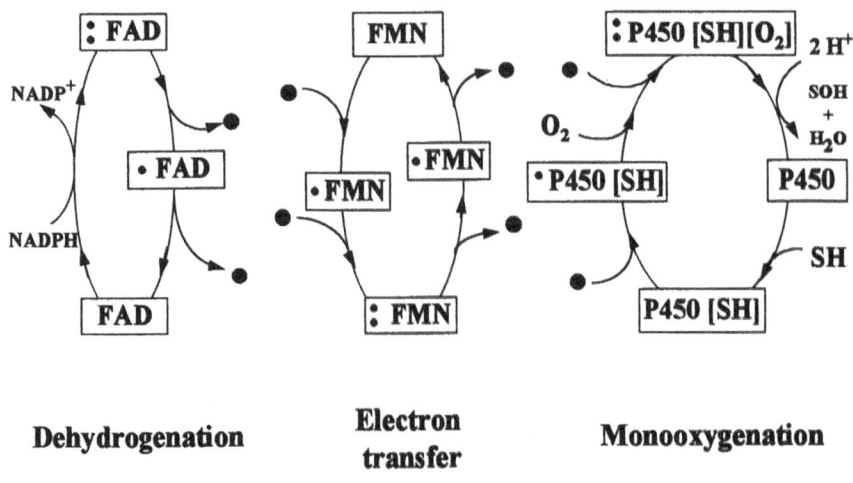

Dehydrogenation   **Electron transfer**   **Monooxygenation**

Figure 10. A schematic representation of the catalytic cycle of P-450 BM3. Electrons are represented as black circles. Hydride transfer from NADPH results in transient formation of FAD hydroquinone. This is unstable and electrons are transferred to the FMN. Electron transfer from FMN to heme cannot occur until substrate (SH) is bound.

4.2. MUTAGENESIS STUDIES

Studies on P-450 BM3 are at an interesting stage, with the generation of atomic structures for both the substrate-bound and substrate-free forms highlighting many potentially important residues for catalysis (Figure 11) Already, site-directed mutagenesis has been used to investigate the roles of potential key residues.

Mutagenesis has been used to examine the role of the phylogenetically conserved residue W96 (see Figure 11) in the binding of heme and maintenance of heme iron in the low-spin state [50]. This residue was proposed to act as part of a conserved electron transfer route to the heme iron in the P-450s [51]. However, stopped-flow data for W96A, Y and F mutants of P-450 BM3 indicate that flavin to heme electron transfer is nearly as fast in the mutants as it is in wild-type P-450 BM3 (Munro *et al.*, unpublished data). Alteration of residue F87 shows that this residue plays a vital role in substrate recognition and orientation; with mutant F87G showing a marked increase in oxygenation of aromatic compounds. The F87A mutation favors hydroxylation of lauric acid at the omega position (rather than the omega-2 favored in the wild-type) whereas the F87V mutation converts the enzyme into a regio- and stereo-selective (14S, 15R) arachidonic acid epoxygenase [52,53]. It is interesting to note that the phenylalanine side chain at position 87 undergoes significant reorientation on substrate binding (Figure 11) [54]. Residue R47 is important for docking with the carboxylate of fatty acid substrates and alteration of this residue leads to significant uncoupling of fatty acid hydroxylation from NADPH oxidation. Electron transfer is also uncoupled from substrate hydroxylation in the mutant T268A and residue T268 has been implicated in oxygen binding and activation [55]. In the diflavin domain of P-450 BM3, mutants have been created in a number of residues which destabilize or altogether abolish binding of the FMN coenzyme. Mutations to residues G570 and W574 have particularly severe effects on FMN binding. In addition, electron transfer rates to both the P-450 domain and cytochrome c are orders of magnitude lower than in the wild-type enzyme [56]. Various alterations of the hydrophilic linker region between the heme and diflavin domains in P-450 BM3 have indicated that the length of this peptide is critical to efficient FMN-to-heme electron transfer and have suggested that the linker may not adopt any regular secondary structure [44,57].

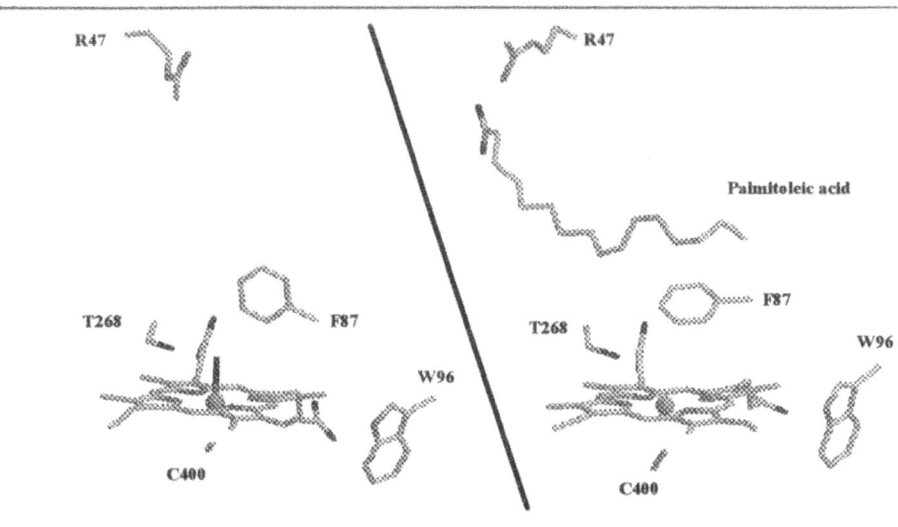

Figure 11. Critical residues in the active site of P-450 BM3. Substrate-free on the left, substrate-bound on the right.

5. Conclusions

It should be clear from this article that flavocytochromes represent an ideal class of enzymes for the study of electron transfer. In addition, the ability of flavocytochromes to function as electrical transformers, linking the two-electron oxidation/reduction of organic compounds to single electron transfers, opens up many possibilities in biosensing and biotransformations.

6. References

1.   Black, M.T., White, S.A., Reid, G.A. and Chapman, S.K. (1989) High Level expression of fully active yeast flavocytochrome b_2 in *Escherichia coli, Biochem.J.* **258**, 255-259.
2.   Xia, Z.-X. and Mathews, F.S. (1990) Molecular structure of flavocytochrome b_2 at 2.4 Å resolution, *J.Mol.Biol.* **212**, 837-863.
3.   Tegoni, M. and Cambillau, C. (1994) The 2.6 Å refined structure of the *Escherichia coli* recombinant *Saccharomyces cerevisiae* flavocytochrome b_2-sulfite complex, *Protein.Sci.* **3**, 303-313.

4. Daff, S., Ingledew, W.J., Reid, G.A. and Chapman S.K. (1996) New insights into the catalytic cycle of flavocytochrome b_2, *Biochemistry* **35**, 6345-6450

5. Chapman, S.K., Reid, G.A., Daff, S., Sharp, R.E., White, P., Manson F.D.C. and Lederer, F. (1994) Flavin to heme electron transfer in flavocytochrome b_2, *Biochem.Soc.Trans.* **22**, 713-718.

6. Miles, C.S., Rouvière-Fourmy, N., Lederer, F., Mathews, F.S., Reid, G.A., Black, M.T. and Chapman, S.K. (1992) Tyr-143 facilitates interdomain electron transfer in flavocytochrome b_2. *Biochem.J.* **285**, 187-192.

7. Rouvière-Fourmy, N, Capeillère-Blandin, C. and Lederer, F. (1994) Role of tyrosine 143 in lactate dehydrogenation by flavocytochrome b_2. Primary kinetic isotope effect studies with a phenylalanine mutant. *Biochemistry* **33**, 798-806.

8. Tegoni, M., Begotti, S. and Cambillau, C. (1995) X-ray structure of two complexes of the Y143F flavocytochrome b_2 mutant crystallized in the presence of lactate or phenyl lactate. *Biochemistry* **34**, 9840-9850.

9. White, P., Manson, F.D.C., Brunt, C.E., Chapman, S.K. and Reid, G.A. (1993) The importance of the interdomain hinge in intramolecular electron transfer in flavocytochrome b_2. *Biochem.J.* **291**, 89-94.

10. Chapman, S.K., Reid, G.A., Bell, C., Short, D. and Daff, S. (1996) Flavocytochrome b_2: An ideal model for studying protein-mediated electron transfer. *Biochem.Soc.Trans.* **24**, 73-77.

11. Pike, A.D., Manson, F.D.C., Chapman, S.K. and Reid G.A. (1995) Investigating the importance of a salt-bridge in flavocytochrome b_2 interdomain electron transfer. *J.Inorg.Biochem.* **59**, 270.

12. Pike, A.D., Chapman, S.K., Manson, F.D.C., Reid, G.A., Gondry, M. and Lederer, F. (1997) Investigating the importance of an interface residue in interdomain electron transfer. in K.J. Stevenson, V. Massey and C.H. Williams (eds.), *Flavins and Flavoproteins*, University of Calgary Press, Calgary, in press.

13. Sharp, R.E., White, P., Chapman, S.K. and Reid, G.A. (1994) Role of the interdomain hinge of flavocytochrome b_2 in intra- and inter- protein electron transfer. *Biochemistry* **33**, 5115-5120.

14. Sharp, R.E., Chapman, S.K. and Reid, G.A. (1996) Deletions in the interdomain hinge region of flavocytochrome b_2 : Effects on intraprotein electron transfer. *Biochemistry* **35**, 891-899.

15. Sharp, R.E., Chapman, S.K. and Reid, G.A. (1996) Modulation of flavocytochrome b_2 intraprotein electron transfer in an interdomain hinge region. *Biochem.J.* **316**, 507-513.

16. Capeillère-Blandin, C. (1995) Flavocytochrome b_2-cytochrome c interactions: The electron transfer reaction revisited. *Biochimie* **77**, 516-530.

17. Tegoni, M., White, S.A., Roussel, A., Mathews, F.S. and Cambillau, C. (1993) A hypothetical complex between crystalline flavocytochrome b_2 and cytochrome c. *Proteins.Struct.Function.Genet.* **16**, 408-422.

18. Daff, S., Sharp, R.E., Short, D.M., Bell, C., White, P., Manson, F.D.C., Reid, G.A. and Chapman, S.K. (1996) Interaction of cytochrome c with flavocytochrome b_2. *Biochemistry* **35**, 6351-6357.

19. Short, D.M., Walkinshaw, M., Taylor, P., Reid, G.A. and Chapman S.K. (1997) Discovery of the kinetically relevant binding site on flavocytochrome b_2 for cytochrome c. in K.J. Stevenson, V. Massey and C.H. Williams (eds.), *Flavins and Flavoproteins*, University of Calgary Press, Calgary, in press.

20. Ackrell, B.A.C., Johnson, M.K., Gunsalus, R. P. and Cecchini, G. (1992) Structure and function of succinate dehydrogenase and fumarate reductase. in F. Müller (ed.) *Chemistry and Biochemistry of Flavoenzymes*) CRC Press, Boca Raton, Florida.

21. Morris, C.J., Black, A.C., Pealing, S.L., Manson, F.D.C., Chapman, S.K., Reid, G.A., Gibson, D.M. and Ward, F.B. (1994). Purification and properties of a novel cytochrome: flavocytochrome c from *Shewanella putrefaciens*. *Biochem. J.* **302**, 587-593.

22. Pealing, S.L., Black, A.C., Manson, F.D.C., Ward, F.B., Chapman, S.K. and Reid, G.A. (1992). Sequence of the gene encoding flavocytochrome c from *Shewanella putrefaciens*: a tetraheme flavoenzyme that is a fumarate reductase related to the membrane-bound enzymes from other bacteria. *Biochemistry* **31**, 12132-12140.

23. Kotlyar, A.B. & Vinogradov, A.D. (1984) Evidence for an essential arginine residue in the substrate-binding site of the mammalian succinate dehydrogenase. *Biochem. Int.* **8**, 545-552.

24. Vinogradov, A.D. (1986) Succinate-ubiquinone reductase segment of respiratory chain *Biochemistry USSR* **51**, 1663-1668.

25. Schröder, I., Gunsalus, R.P., Ackrell, B.A.C., Cochran, B. and Cecchini, G. (1991) Identification of active site residues of *Escherichia coli* fumarate reductase by site-directed mutagenesis. *J. Biol. Chem.* **266**, 13572-13579.

26. Morris, C.J., Gibson, D.M. and Ward, F.B. (1990) Influence of respiratory substrate on the cytochrome content of *Shewanella putrefaciens*. *FEMS Microbiol. Letts.* **69**, 259-262.

27. Myers, C.R. and Myers, J.M. (1993). Role of menaquinone in the reduction of fumarate, nitrate, iron(III) and manganese(IV) by *Shewanella putrefaciens* MR-1. *FEMS Microbiol. Lett.* **114**, 215-222.

28. Myers, C.R. and Nealson, K.H. (1990) Respiration-linked proton translocation coupled to anaerobic reduction of manganese(IV) and iron(III) in *Shewanella putrefaciens* MR-1. *J. Bacteriol.* **172**, 6232-6238.

29. Tsapin, A.I., Burbaev, D.S., Nealson, K.H. and Keppen, O.I. (1995) Investigations of succinate dehydrogenase and fumarate reductase in whole cells of *Shewanella putrefaciens* (strains MR-1 and MR-7) using electron spin resonance spectroscopy. *Appl. Mag. Res.* **9**, 509-516.

30. Pealing, S.L., Cheesman, M.R., Reid, G.A., Thomson, A.J., Ward, F.B. and Chapman, S.K. (1995). Spectroscopic and kinetic-studies of the tetraheme flavocytochrome *c* from *Shewanella putrefaciens* NCIMB400. *Biochemistry* **34**, 6153-6158.

31. Nelson, D.W., Kamataki, T., Waxman, D.J., Guengerich, F.P., Estabrook, R.W.,Feyereisen, R., Gonzalez, F.J., Coon, M.J., Gunsalus, I.C., Gotoh, O., Okuda, K. and Nebert, D.W. (1993). The P-450 superfamily - update on new sequences, gene mapping, accession numbers, early trivial names of enzymes and nomenclature. *DNA Cell. Biol.* **12**, 1-51.

32. Gonzalez, F.J. (1989). The molecular biology of cytochrome P-450s. *Pharmacol. Rev.* **40**, 243-288.

33. Guengerich, F.P. (1988). Roles of cytochrome P-450 enzymes in chemical carcinogenesis and cancer chemotherapy. *Cancer Res.* **48**, 2946-2954.

34. Mueller, J., Loida, P.L. and Sligar, S.G. (1995) in P.R.O. Ortiz de Montellano (ed.) *Cytochrome P450; Structure, Mechanism and Biochemistry* 2nd edition, Plenum Press, New York, pp. 83-124.

35. Munro, A.W. and Lindsay, J.G. (1996). Bacterial cytochromes P-450. *Molec. Microbiol.* **20**, 1115-1125.

36. Poulos, T.L., Finzel, B.C. and Howard, A.J. (1987). High resolution crystal structure of cytochrome P-450cam. *J. Mol. Biol.* **195**, 687-700.

37. Ravichandran, K.G., Boddupalli, S.S., Hasemann, C.A., Peterson, J.A. and Deisenhofer, J. (1993). Crystal structure of hemoprotein domain of P-450 BM3, a prototype for microsomal P-450s. *Science* **261**, 170-176.

38. Hasemann, C.A., Ravichandran, K.G., Peterson, J.A. and Deisenhofer, J. (1994). Crystal structure and refinement of cytochrome P-450terp at 2.3 Ångstrom resolution. *J. Mol. Biol.* **236**, 1169-1185.

39. Cupp-Vickery, J.R. and Poulos, T.L. (1995). Structure of cytochrome P-450eryF involved in erythromycin biosynthesis. *Nature Struct. Biol.* **2**, 144-153.

40. Narhi, L.O. and Fulco, A.J. (1986). Characterization of a catalytically self-sufficient 119,000 Dalton cytochrome P-450 monooxygenase induced by barbiturates in *Bacillus megaterium. J. Biol. Chem.* **261**, 7160-7169.

41. Narhi, L.O. and Fulco, A.J. (1987). Identification and characterization of two functional domains in cytochrome P-450 BM-3, a catalytically self-sufficient monooxygenase induced by barbiturates in *Bacillus megaterium. J. Biol. Chem.* **262**, 6683-6690.

42. Bredt, D.S., Hwang, P.M., Glatt, C.E., Lowenstein, C., Reed, R.R. and Snyder, S.H. (1991). Cloned and expressed nitric oxide synthase structurall resembles cytochrome P-450 reductase. *Nature, London,* **351**, 714-718.

43. Miles, J.S., Munro, A.W., Rospendowski, B.N., Smith, W.E., McKnight, J. and Thomson, A.J. (1992). Domains of the catalytically self-sufficient cytochrome P-450 BM3 - genetic construction, overexpression, purification and spectroscopic characterization. *Biochem. J.* **288**, 503-509.

44. Munro, A.W., Lindsay, J.G., Coggins, J.R., Kelly, S.M. and Price, N.C. (1994). Structural and enzymological analysis of the interaction of isolated domains of cytochrome P-450 BM3. *FEBS Lett.* **343**, 70-74.

45. Daff, S.N., Chapman, S.K., Turner, K.L., Holt, R., Govindaraj. S., Poulos, T.L. and Munro, A.W. (1997). Redox control of the catalytic cycle of P-450 BM3. *Biochemistry* (submitted).

46. Munro, A.W., Daff, S., Coggins, J.R., Lindsay, J.G. and Chapman, S.K. (1996). Probing electron transfer in flavocytochrome P-450 BM3 and its component domains. *Eur. J. Biochem.* **239**, 403-409.

47. Modi, S., Sutcliffe, M.J., Primrose, W.U., Lian, L.-Y. and Roberts, G.C.K. (1996). The catalytic mechanism of cytochrome P-450 BM3 involves a 6 Ångstrom movement of the bound substrate on reduction. *Nature Struct. Biol.* **3**, 414-417.

48. Klein, M.L. and Fulco, A.J. (1994). The interaction of cytochrome c and the heme domain of cytochrome P-450 (BM-3) with the reductase domain of cytochrome P-450 (BM-3). *Biochim. Biophys. Acta* **1201**, 245-250.

49. Murataliev, M.B. and Feyereisen, R. (1996). Functional interactions in cytochrome P-450 BM3 - fatty acid substrate-binding alters electron transfer properties of the flavoprotein domain. *Biochemistry* **35**, 15029-15037.

50. Munro, A.W., Malarkey, K., McKnight, J., Thomson, A.J., Kelly, S.M., Price, N.C., Lindsay, J.G., Coggins, J.R and Miles, J.S. (1994) The effect of replacement of tryptophan 96 of cytochrome P-450 BM3 from *Bacillus megaterium* on catalytic function. *Biochem. J.* **303**, 423-428.

51. Baldwin, J.E., Morris, G.M. and Richards W.G. (1991). Electron transport in cytochromes P450 by covalent switching. *Proc.Roy.Soc.Lond.Bio.Sci.* **245**, 43-51.

52. Oliver, C.F., Modi, S., Sutcliffe, M.J., Primrose, W.U., Lian, L.Y. and Roberts, G.C.K. (1997). A single mutation in cytochrome P-450 BM3 changes substrate orientation in a catalytic intermediate and the regiospecificity of hydroxylation. *Biochemistry* **36**, 1567-1572.

53. Graham Lorence, S., Truan, G., Peterson, J.A., Falck, J.R., Wei, S.Z., Helvig, C. and Capdevila, J.H. (1997). An active site substitution, F87V, converts cytochrome P-450 BM-3 into a regio- and stereoselective (14S, 15R)-arachidonic acid epoxygenase. *J. Biol. Chem.* **272**, 1127-1135.

54. Li, H.Y. and Poulos, T.L. (1997). The structure of the cytochrome P-450 BM3 haem domain complexed with the fatty acid substrate palmitoleic acid. *Nature Struct. Biol.* **4**, 140-146.

55. Yeom, H., Sligar, S.G., Li H.Y. and Fulco A.J. (1995). The role of Thr268 in oxygen activation of cytochrome P450 (BM-3). *Biochemistry* **34**, 14733-14740

56. Klein, M.L. and Fulco, A.J. (1993). Critical residues involved in FMN binding and catalytic activity in cytochrome P-450 (BM-3). *J. Biol. Chem.* **268**, 7553-7661.

57. Govindaraj, S. and Poulos, T.L. (1996). Probing the structure of the linker connecting the reductase and heme domains of cytochrome P-450 BM-3 using site-directed mutagenesis. *Protein Sci.* **5**, 1389-1393.

THE CHEMISTRY OF BIOLOGICAL DENITRIFICATION

Spectroscopic Studies Provide Insights into the Mechanism of Dissimilatory Heme cd_1 and Copper-containing Nitrite Reductases

B. A. AVERILL*, Y. WANG†, J. O. KA‡, I. N.
ROUBLEVSKAIA‡, and J. M. TIEDJE‡
*E. C. Slater Institute, University of Amsterdam, Plantage
Muidergracht 12 ,1018 TV Amsterdam, The Netherlands;
†Department of Chemistry, University of Virginia,
Charlottesville, VA 22901 USA; ‡Center for Microbial
Ecology and Department of Crop and Soil Science,
Michigan State University, East Lansing, MI 48824 USA

1. Abstract

Results of spectroscopic studies of the heme cd_1 nitrite reductase from *Pseudomonas stutzeri* JM 300 and the copper-containing nitrite reductase from *Achromobacter cycloclastes* and their complexes with the substrate (NO_2^-) and inhibitors (CO, CN^-) are described. The heme d_1 in the former and the Type 2 Cu in the latter are found to form complexes with exogenous ligands, as expected since these groups are presumed to constitute the active sites of the enzymes, at which nitrite is reduced to NO. In addition, however, the heme c group of the heme cd_1 nitrite reductase exhibits unexpectedly high reactivity toward exogenous ligands. Preliminary results of site-directed mutagenesis studies of the copper-containing nitrite reductase from *Pseudomonas* G-⁻⁻⁻ ⸗ ⸗⸗ also ⸗⸗ ⸗. The M189A mutant, in which the Met ligand to the Type 1 Cu center is transformed to Ala, unexpectedly appears to retain nearly wild-type levels of activity both *in vivo* and *in vitro*.

2. Introduction

The pathway by which NO_2^- is reduced to N_2O in denitrifying bacteria has long been controversial, but evidence from several laboratories has now established that NO is an obligatory intermediate in most, if not all, bacteria [1, 2]. Among the most important unresolved questions in this field are the following. How is NO_2^- reduced to NO (or N_2O) by these enzymes? Why are two redox centers needed in both cases to effect the one-electron transformation of NO_2^- to NO [3]? The results of traditional spectroscopic and mechanistic studies of the native enzymes, combined with studies of selected mutant

185

G.W. Canters and E. Vijgenboom (eds.),
Biological Electron Transfer Chains: Genetics, Composition and Mode of Operation, 185-196.
© 1998 *Kluwer Academic Publishers.*

enzymes and the recent reports of the X-ray structures of examples of both copper [4, 5] [6] and heme cd_1 [7] nitrite reductases, should allow us to gain new insights into these problems.

3. Materials and Methods

The heme cd_1 nitrite reductase (heme cd_1 NiR) from *Pseudomonas stutzeri* JM300 [8] and the native and Type 2-depleted (T2D) copper-containing nitrite reductase (Cu NiR) from *Achromobacter cycloclastes* [9] were prepared as described elsewhere. Optical spectra were measured on either a Cary 219 or an HP-8452 diode array spectrophotometer at room temperature. EPR spectra were obtained on a Bruker ESP 300 spectrometer operating at X-band, using either a liquid nitrogen-cooled glass Dewar insert for 77 K operation or an Oxford ESR-900 continuous flow cryostat with an Oxford ITC4 temperature controller for lower temperatures. FTIR spectra were measured with a Mattson Cygnus 100 spectrophotometer operating at room temperature, using CaF_2 transmission cells. Data were accumulated on a PC using Mattson's FIRST software for collection and analysis; background spectra of a D_2O or H_2O blank were subtracted.

A 1.9 kb EcoRI/BamHI DNA fragment from *Pseudomonas* G179 containing the nitrite reductase structural gene [10] was cloned in a pUC119 vector. To generate the specific mutants M189A and M189C, synthetic single-stranded oligonucleotides were prepared and used as primers in the polymerase chain reaction (PCR). Three resulting plasmids, designated pJMT11, pJMT12, and pJMT13, contained the wild type NiR sequence, and the M189A and M189C mutants, respectively, and were subcloned into the broad host-range vector pBS329 to give plasmids pJMT31, pJMT32, and pJM33, respectively. These last three plasmids were used for transformation of the NiR-deficient strain PG179NiR⁻, generated by Tn5 transposon mutagenesis of the wild type [11].

4. Results and Discussion

4.1 LIGAND BINDING TO THE HEME cd_1 NITRITE REDUCTASE

We have previously used FTIR spectroscopy to probe the binding of the product NO to the oxidized heme cd_1 nitrite reductase, and obtained evidence for formation of a ferric heme nitrosyl (Fe^{3+}-NO) complex that is presumably the same species formed by dehydration of coordinated nitrite during the catalytic process [12]. Although these results demonstrated that the enzyme could form such a species, they did little to answer the question of how and why the enzyme utilizes the unique heme d_1 chromophore at the catalytic site, largely because very few vibrational data on ferric heme nitrosyl complexes are available for comparison. We have therefore examined complexes of the enzyme with other ligands as probes of the electronic environment of the heme iron in order to address such questions.

Figure 1 shows the FTIR spectrum of the reduced heme cd_1 NiR in the presence of ^{12}CO. The most obvious feature is a sharp ($\Delta v_{1/2} = 8$ cm^{-1}) band at 1970 cm^{-1}, which is attributable to a heme Fe^{2+}–CO complex. That the species responsible for the absorption was a metal-carbonyl complex was confirmed by use of ^{13}CO, which gave the expected shift of 31 cm^{-1} to lower energy. Attribution to the heme d_1 site is based on the parallels to the behavior of the enzyme from *Ps. aeruginosa*, which has been shown to contain a high-spin heme d_1 and a low-spin heme c in the reduced form [13], and on the optical spectra of the *Ps. stutzeri* enzyme, which clearly show no change in the optical spectrum of the heme c upon reaction with CO (data not shown).

Available data for CO complexes of reduced heme proteins are summarized in Table 1 below. It is clear that the heme d_1 chromophore results in the highest frequency yet reported for a ferrous heme-CO protein complex, indicating that the presence of the two carbonyl groups on the periphery of the porphyrin results in a significant decrease in electron density at the Fe center. As a result, there is decreased donation of π electron density from the ferrous iron to the π^* orbitals of the CO, resulting in a higher C-O stretching frequency. Thus, the heme d_1 produces a relatively electron-deficient heme, in comparison to the protoporphyrin IX present in myoglobin and hemoglobin, for example. This electronic effect may well be important during the catalytic process, possibly by polarizing the coordinated nitrite ion and facilitating protonation and dehydration to the ferric-NO complex.

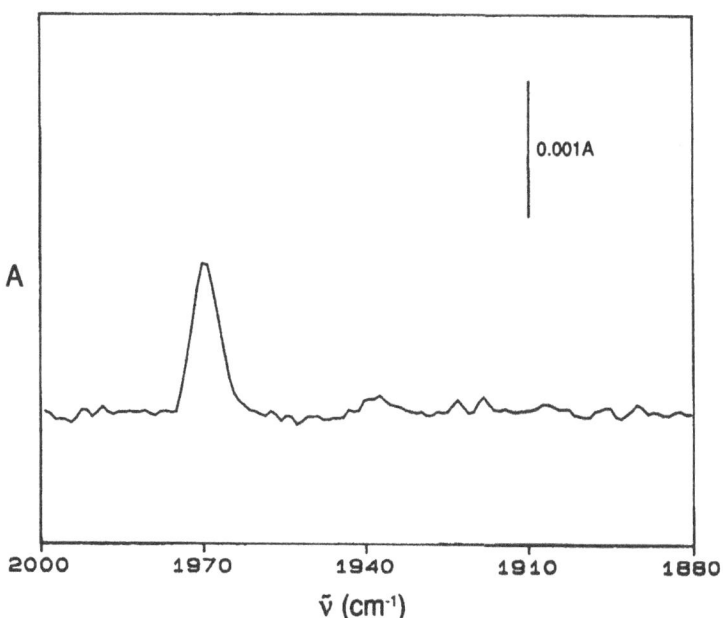

Figure 1. FTIR spectrum of the carbon monoxide complex of the reduced heme cd_1 nitrite reductase from *Ps. stutzeri* JM300. The sample contained 2 mM protein in 0.2 M phosphate buffer at pH 7.3. The scan time was about 30 min. at room temperature with a resolution of 2 cm^{-1}. An IR cell with CaF$_2$ windows and a Teflon spacer with a path length was 0.025 mm was used.

TABLE 1. C–O Stretching frequencies in reduced heme–CO complexes

Species	Frequency (cm^{-1}) (^{12}C^{16}O)	Half-width (cm^{-1})	ε x 10^{-3} (M^{-1} cm^{-1})
cd_1 NiR·CO	1970	6	0.42
HRP·CO[a]	1933	12	NR[b]
P450$_{cam}$·CO	1940	14	NR
Mb·CO	1944	12	2.0
Hb·CO[a]	1951	8	3.7
Cyt c ox.·CO[a]	1963.5	5.5	4.9
P420$_{cam}$·CO	1965	25	NR
CO	2143		

[a] Ref. [14] [b] Not reported.

Similar results are observed for the cyanide complex of the heme cd_1 nitrite reductase, although the situation is complicated by several factors: (i) cyanide forms complexes with both the oxidized and reduced forms of the enzyme; (ii) cyanide appears to coordinate to both the heme d_1 and the heme c groups in the enzyme in both oxidation states; and (iii) the vibrational frequency of coordinated cyanide is not as sensitive to the nature of the metal center, making it a poorer probe of the electronic structure of a metal center than is carbon monoxide. Points (i) and (ii) are clearly illustrated in Figures 2 and 3, which show the FTIR spectra of both the oxidized and reduced forms of the enzyme in the presence of ^{12}C^{14}N$^-$. That both peaks in both spectra are due to metal-cyanide species was confirmed by measuring the spectra of the corresponding ^{13}C^{14}N$^-$ and ^{12}C^{15}N$^-$, which showed the shifts expected based on a simple diatomic oscillator model. These bands are considerably weaker than those observed for CO, and required a much longer pathlength to obtain data of comparable quality in a reasonable time.

For the oxidized protein with ^{12}C^{14}N$^-$ the observed frequencies are 2128 and 2103 cm^{-1}, while for the reduced enzyme the corresponding figures are 2070 and 2058 cm^{-1}. Based on the results obtained with a sample of the half-apo enzyme from which the heme d_1 has been extracted [8], for which a C-N stretch of 2099 cm^{-1} was observed, the higher frequency band in the spectra of the oxidized enzyme is attributed to the heme d_1-CN$^-$ complex, while the lower frequency band is attributed to the heme c-CN$^-$ complex. In contrast, both peaks in the spectrum of the reduced NiR appear to be due to heme d_1-CN$^-$ species, inasmuch as optical spectra of the reduced enzyme in the presence of cyanide clearly show that the heme c does not react with cyanide. We have previously observed evidence for heterogeneity in heme d_1 content/environment in the Ps. stutzeri NiR (Wang & Averill, unpublished observations), and it may well be that the spectra of the reduced enzyme-CN$^-$ complex reflect this. Available C-N vibrational data for heme-CN$^-$ complexes are summarized in Table 2 below. Once again, the frequencies for the heme d_1-CN$^-$ complex in both the oxidized and reduced enzyme are the highest yet observed for protein heme-cyanide complexes [15, 16]. This finding is consistent with the CO data described above, and constitutes further evidence for the

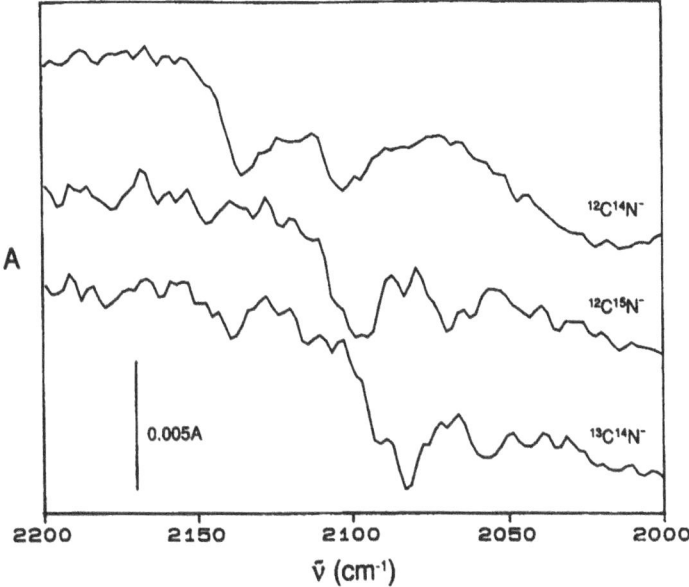

Figure 2. FTIR spectra of the cyanide complex of the oxidized heme cd_1 nitrite reductase from *Ps. stutzeri* JM300. Samples contained ca. 2 mM protein 0.2 M phosphate buffer, pH 7.3. The scan time was about 35 min. at room temperature with a resolution of 4 cm^{-1}. FT-IR measurements were carried out in an IR cell having CaF$_2$ windows with a Teflon spacer and a path length of 0.066 mm.

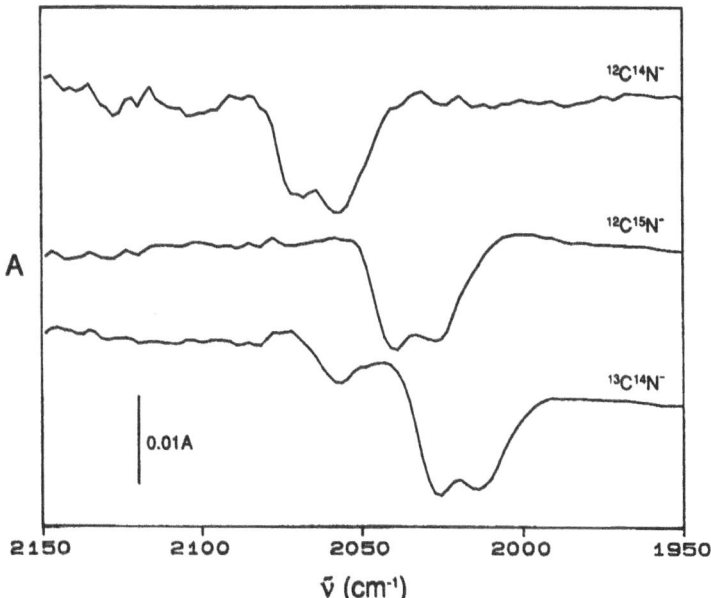

Figure 3. FTIR spectra of the cyanide complex of the reduced heme cd_1 nitrite reductase from *Ps. stutzeri* JM300. Experimental conditions were as in Figure 2.

TABLE 2. C–N Stretching frequencies in heme-CN⁻ complexes[a]

Species	Frequency ($^{12}C^{14}N^-$)	Frequency shift (^{15}N) obs.	calc.	Frequency shift (^{13}C) obs.	calc.
ox. cd_1	2128 (d_1)	29	33.1	47	44.2
NiR·CN⁻	2103 (c)	33	33.0	47	44.0
MetHb·CN^{-b}	2122	30	32.9	44.4	44.5
MetMb·CN^{-b}	2126	31	33.0	46	44.5
ox. HRP·CN^{-b}	2131	33	33.1	47	44.6
red. cd_1	2070 (d_1)	31	32.1	45	43.3
NiR·CN^{-b}	2058 (c)	32	31.9	42	43.1
red HRP·CN^{-b}	2024	29	31.4	41	42.4
CN^{-b}	2080	31	32.2	42	43.5
HCN^b	2092	33	32.5	33	43.8

[a] All frequencies and frequency shifts are reported in cm⁻¹. [b] Ref. [15]. HRP is horseradish peroxidase.

hypothesis that a heme d_1 environment results in an electron-deficient heme Fe.

The fact that both hemes in the oxidized enzyme react with an exogenous ligand such as cyanide is not particularly surprising. Evidence for reaction of the heme c in the *Ps. aeruginosa* enzyme with imidazole and cyanide has been presented previously [17, 18]. Further, we have found that the EPR spectra of both the heme c and heme d_1 are perturbed by ligands such as imidazole, cyanide, azide, and nitrite (Figure 4). While a detailed analysis of the spectra presented in Figure 4 is beyond the scope of the present paper, it is clear that both the high-spin heme (d_1) at g 6 and the low-spin species observed at g 3-4 react with virtually all the ligands examined. Thus, the anomalous reactivity observed with cyanide and the oxidized NiR is not unique to cyanide, but reflects an unusual and intrinsically high reactivity of the heme c in this enzyme.

4.2 LIGAND BINDING TO THE COPPER NITRITE REDUCTASE

Since direct vibrational spectroscopy has proven to be a very useful tool for probing the interaction of the heme cd_1 NiR with small molecules and anions, the same approach was applied to the copper-containing nitrite reductase. Samples of both T2D Cu NiR and the Type 2 enriched enzyme were reduced and treated with CO separately in order to provide unequivocal proof of the identity of the CO binding site. The FTIR spectrum of the reduced Type 2 enriched Cu NiR in the presence of ^{12}CO is shown in Fig. 5. A very intense band is observed at 2050.2 cm⁻¹ with a line width of 14 cm⁻¹, neither the position nor the band width vary significantly in D_2O. The 2050 cm⁻¹ band was barely visible in the spectrum of the reduced T2D enzyme under similar conditions (data not shown). The identification of the band as a C-O stretch was confirmed by examining identical samples in the presence of ^{13}CO, where a corresponding band at 2003 cm⁻¹ was observed (not shown). The ^{13}CO band was somewhat less intense than that observed with ^{12}CO band, presumably due to the lower pressure of CO used in the

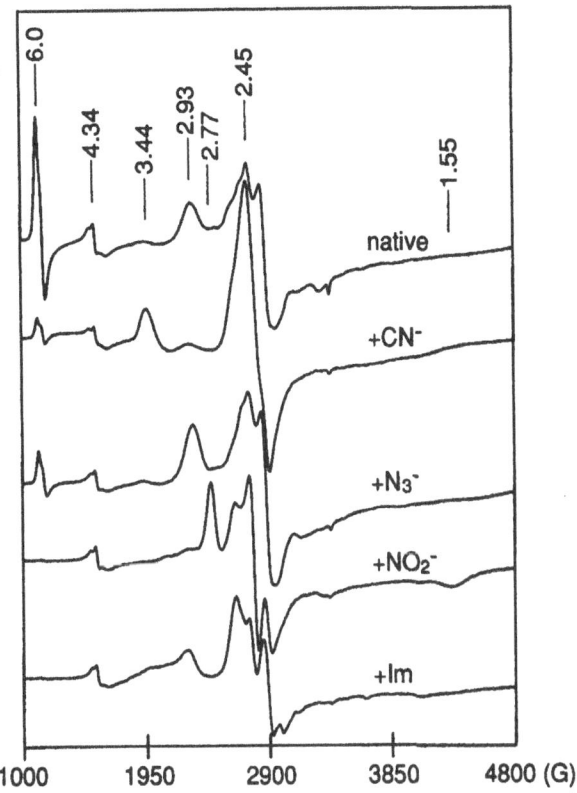

Figure 4. EPR spectra of the native form of the oxidized heme cd_1 nitrite reductase from *Ps. stutzeri* JM300 (top scan) and in the presence of cyanide (150 mM), azide (300 mM), nitrite (150 mM), and imidazole (1 M) (from top to bottom). All samples contained ca. 0.050 mM enzyme in 100 mM phosphate buffer, pH 8.0, for the cyanide derivative and pH 7.3 for the rest. Experimental conditions: microwave frequency, 9.45 GHz; microwave power, 0.2 milliwatts; temperature, 10K; modulation frequency, 100 kHz; modulation amplitude, 10 Gauss; receiver gain, 1.0×10^4.

[13]CO experiment. The observed isotope shift is 47.2 cm^{-1}, which compares reasonably well with the calculated value of 45.8 cm-1 based on a simple diatomic model.

Since the only difference between the two enzyme samples studied was the content of Type 2 copper, the band at 2050.2 cm-1 must be due to a reduced Type 2 copper–CO complex, Cu$^+$–CO. This result constitutes the clearest spectroscopic evidence yet obtained for this enzyme that an exogenous ligand is selectively bound to a particular Cu site. Since CO is a strong inhibitor of this enzyme, our results provide further evidence that the Type 2 copper is indeed the substrate binding site of the enzyme.

There have been several reports of FTIR studies of CO bound copper-containing protein, including hemocyanin [19], cytochrome *c* oxidase [20], and dopamine ß-hydroxylase [21]; these data are summarized in Table 3. The position of the 2050.2 cm^{-1} band observed for the Cu NiR-CO complex compares very well with that of CO bound hemocyanins, which have a C-O stretch ranging from 2043 to 2063 cm^{-1}

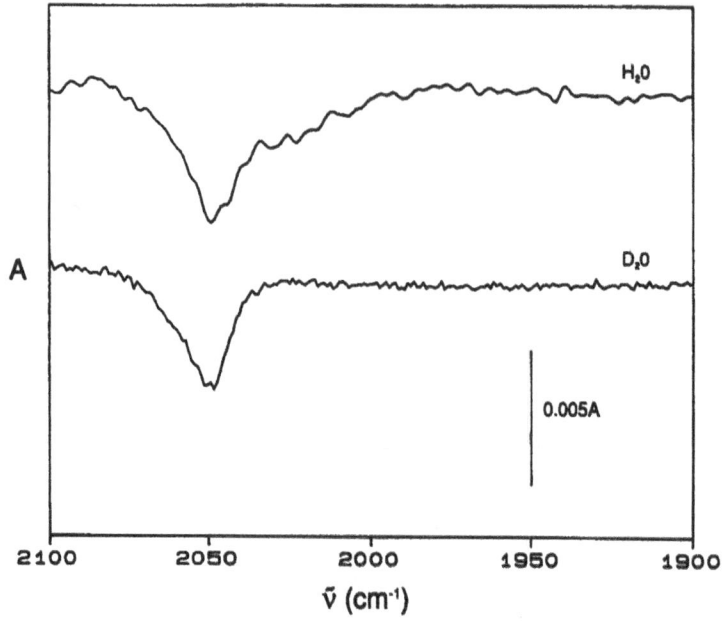

Figure 5. FTIR spectra of the CO complex of the reduced copper-containing nitrite reductase from *A. cycloclastes* in H_2O (top) and D_2O (bottom) solution. The samples were ca. 1.5-2.0 mM protein in 0.1 M TrisHCl buffer, pH 7.5. The scan time was about 20 min. at room temperature with a resolution of 2 cm^{-1}; an IR cell having CaF_2 windows with a Teflon spacer and a path length of 0.025 mm was used.

depending on the origin of the enzyme [19]. Care must be taken, however, in trying to extract structural information based on the similarity of CO vibrational frequencies. Both hemocyanin and cytochrome *c* oxidase contain dinuclear centers at which CO is bound, even though the primary mode of interaction of CO appears to be with a single Cu atom. Such is clearly not the case in this copper-containing nitrite reductase, where the two copper centers are separated by ca. 12 Å [5]. It seems to be reasonable to compare the Cu NiR results with those obtained for the CO complex dopamine ß-hydroxylase, which is reported to be mononuclear even though the enzyme has two

TABLE 3. A comparison of C–O stretching frequencies observed for Cu^+–CO complexes in proteins

Species	C–O frequency (cm^{-1})	half-width (cm^{-1})	10^3 ε (M^{-1} cm^{-1})
A. c. Cu NiR· CO	2050	15	0.96
Limulus Hc· CO[a]	2054.3	11	1.30
Cancer Hc· CO[a]	2043.4	10	—
Dopamine ß-hydroxylase·CO[b]	2089	18	0.27
CO	2143		

[a] Ref. [20]; [b] Ref. [21]

coppers per active site [21]. The reduced enzyme–CO complex exhibited a CO stretch at 2089 cm^{-1}, with a line width of 18 cm^{-1}. Inspection of the CO stretching frequencies reveals that, unlike the case with the heme cd_1 NiR, there is nothing unusual about the electronic environment of the Type 2 Cu center of the Cu NiR; its CO stretching frequency falls squarely in the middle of the normal range observed for Cu^{+}–CO complexes in proteins.

4.3 SITE-DIRECTED MUTANTS OF THE COPPER NITRITE REDUCTASE

In order to confirm that the site of NO$_2^-$ reduction in the Cu NiR's is indeed the Type 2 Cu centers and that the Type 1 centers are simply spectroscopically unusual variants [22] of a standard blue copper electron transfer site, we have prepared a variety of mutants in which the ligands to the Type 1 center are altered. The basic idea behind this approach is to alter the redox potentials of the Type 1 Cu site sufficiently so that electron transfer from the Type 1 to the Type 2 center will become very slow or impossible. In principle, this could be accomplished either by lowering the redox potential of the Type 1 site, so that it cannot be reduced readily by the physiological electron donor(s) or by raising the redox potential such that the Type 1 site is too poor a reductant to reduce the Type 2 center. In either case, the resulting enzyme would be expected to be inactive both *in vivo* and *in vitro* with the physiological electron donor(s), even though a catalytically competent center is still present. In contrast, it should be possible to "short-circuit" the internal electron transfer chain by use of appropriate artificial electron donors *in vitro*.

As outlined under Materials and Methods, a 1.9kb EcoRI/BamHI fragment of *Ps*. G-179 DNA was cloned into a pUC119 vector and used for site-directed mutagenesis. Subsequent to mutagenesis, the wild-type and mutant NiR's were cloned into an *nir⁻* strain of *Ps*. G-179 (prepared via Tn5 transposon mutagenesis) for expression using the broad host-range vector pBS329. Table 4 lists the mutants that have been prepared to date and gives some of their properties, as well as those of some of the controls utilized. It is clear form the data in Table 4 that it is possible to use the Tn5-generated *nir⁻* strain of *Ps*. G-179 as a host for expression of both the wild-type Cu NiR and the mutants. Neither the *nir⁻* strain nor that containing the pBS329 vector alone gave any cross-reacting material or activity *in vivo* or *in vitro*.

Of the mutants listed in Table 4, five contain mutations of ligands to the Type 1 Cu site, and three contain mutations of ligands to the Type 2 Cu site. All of the latter produced a stable protein that showed no activity, presumably a Type 2-depleted form of the enzyme. Disruption of the tri-histidine coordination at the Type 2 site is expected to affect the metal-binding properties dramatically. These results are comparable to those obtained with *Alcaligenes faecalis* S-6 [23], and will not be discussed further. Of more interest are the mutants containing altered Type 1 Cu sites. Three of the mutants prepared contained amino acid residues in place of the Cys residue. Since it is well-documented that the electronic properties of blue Cu centers are dominated by the highly covalent CysS$^-$–Cu interaction [24], it is not particularly surprising that all three of these mutants exhibited no enzymatic activity, although (somewhat surprisingly) stable cross-reacting protein was observed in all cases. Whether these proteins contain

TABLE 4. Nitrite reductase activity *in vivo* and *in vitro* and presence of cross-reacting material for *Pseudomonas* G-179 mutants prepared in this work

Strain	Bubbles[a]	Enzyme activity[b]	Protein on Western blot[c]
179+ (WT)	+	+	+ (2)
179- (Tn5)	-	-	-
179- + pBS329	-	-	-
179- + 1.8kb + pBS329	+	+	+ (2)
C175A (T1)	-	-	+ (2)
C175S (T1)	-	-	+ (2)
C175D (T1)	-	ND[d]	+ (2)
M189A (T1)	+ ($\approx$2/3)	+ ($\approx$2/3)	+ (2)
M189C (T1)	-	-	+ (2)
H139L (T2)	-	ND[d]	+ (1)
H174L (T2)	-	ND[d]	+ (1)
H345L (T2)	-	-	+ (1)

[a] Bubbles are indicative of NiR activity *in vivo* due to reduction of NO_2^- to gaseous products by growing cells. [b] Enzyme activity was measured in crude extracts of the mutants grown aerobically and then induced by addition of nitrate. [c] The presence of cross-reacting material was determined by SDS-PAGE and Western blotting using polyclonal rabbit antibodies generated to the Cu NiR from *A. cycloclastes*. A (2) indicates that two bands, apparently due to partial proteolysis, were observed; a (1) indicates that a single band was observed. [d] ND = not determined.

Zn^{2+} in place of Cu, as has been observed for a similar mutant of the *Al. faecalis* NiR [6], will require isolation and characterization of the mutant proteins.

In addition, two mutants were prepared in which the Met ligand to the Type 1 Cu center was removed (M189A) or replaced by a potential strong ligand (M189C). The current model of the spectroscopic properties of the perturbed Type 1 Cu site in Cu NiR's [22] suggests that constraints on the Cu-S(Met) bond distance are important determinants of the electronic properties of the Cu site, and these two mutants are expected to affect this interaction strongly, albeit in opposite directions. The M189C mutant is similar to the C175 mutants, in that stable protein is synthesized but no activity is observed. Most surprisingly, however, the M189A mutant appears to be essentially fully functional both *in vivo* (as evidenced by the appearance of bubbles of N_2 when grown on NO_3^-) and *in vitro* as measured by direct assay of enzyme activity in crude extracts. The apparent level of enzyme activity is about 2/3 of that of the wild-type enzyme, when the observed activity is normalized normalized to relative intensity of the bands observed on Western blots. Whether this reflects a real difference in enzymatic properties or a difference in expression levels will require isolation and characterization of both the wild-type and M189A forms of the enzyme. In any case, these results constitute the first evidence that a functionally important Cu center in an enzyme can be altered by site-directed mutagenesis *with retention of enzymatic activity*. It is clear that the system requires and will repay more detailed study.

5. Acknowledgements

This research was supported by the U. S. National Science Foundation (DMB-8917427 and MCB-9307501) and the U. S. Department of Agriculture (USDA-NRIGP 91-37305-6663).

6. References

1. Berks, B. C., Ferguson, S. J., Moir, J. W. B. and Richardson, D. J. (1995) Enzymes and associated electron transport systems that catalyze the respiratory reduction of nitrogen oxides and oxyanions, *Biochimica et Biophysica Acta* **1232**, 97-173.
2. Ye, R. W., Averill, B. A. and Tiedje, J. M. (1994) Denitrification: production and consumption of nitric oxide, *Applied and Environmental Microbiology* **60**, 1053-58.
3. Averill, B. A. (1996) Dissimilatory nitrite and nitric oxide reductases, *Chemical Reviews* **96**, 2951-2964.
4. Adman, E. T., Godden, J. W. and Turley, S. (1995) The structure of copper-nitrite reductase from *Achromobacter cycloclastes* at five pH values, with NO_2^- bound and with type II copper depleted, *Journal of Biological Chemistry* **270**, 27458-27474.
5. Godden, J. W., Turley, S., Teller, D. C., Adman, E. T., Liu, M. Y., Payne, W. J. and LeGall, J. (1991) The 2.3 Angstrom X-ray structure of nitrite reductase from *Achromobacter cycloclastes*, *Science* **253**, 438-442.
6. Murphy, M. E. P., Turley, S., Kukimoto, M., Nishiyama, M., Horinouchi, S., Sasaki, H., Tanokura, M. and Adman, E.T. (1995) Structure of *Alcaligenes faecalis* nitrite reductase and a copper site mutant, M150E, that contains zinc, *Biochemistry* **34**,12107-12117.
7. Fülöp, V., Moir, J. W. B., Ferguson, S. J. and Hajdu, J. (1995) The anatomy of a bifunctional enzyme: structural basis for reduction of oxygen to water and synthesis of nitric oxide by cytochrome cd_1, *Cell* **81**, 360-377.
8. Weeg-Aerssens, E., Wu, W., Ye, R. W., Tiedje, J. M. and Chang, C. K. (1991) Purification of cytochrome cd_1 nitrite reductase from *Pseudomonas stutzeri* JM300 and reconstitution with native and synthetic heme d_1, *Journal of Biological Chemistry* **266**, 7496-7502.
9. Libby, E. and Averill, B. A. (1992) Evidence that the type 2 copper centres are the site of nitrite reduction by *Achromobacter cycloclastes* nitrite reductase, *Biochemical and Biophysical Research Communications* **187**, 1529-1535.
10. Ye, R. W., Fries, M. R., Bezborodnikov, S. G., Averill, B. A. and Tiedje, J. M. (1993) Characterization of the structural gene encoding a copper-containing nitrite reductase and homology of this gene to DNA of other denitrifiers. *Applied and Environmental Microbiology* **59**, 250-254.
11. Ye, R. W., Averill, B. A. and Tiedje, J. M. (1992) Characterization of Tn5 mutants deficient in dissimilatory nitrite reduction in *Pseudomonas* sp. strain G-179, which contains a copper nitrite reductase, *Journal of Bacteriology* **174**, 6653-6658.
12. Wang, Y. and Averill, B. A. (1996) Direct observation by FTIR spectroscopy of the ferrous heme NO+ intermediate in reduction of nitrite by a dissimilatory heme cd_1 nitrite reductase, *Journal of the American Chemical Society* **118**, 3972-3973.
13. Sutherland, J., Greenwood, C., Peterson, J. and Thomson, A. J. (1986) An investigation of the ligand-binding properties of *Pseudomonas aeruginosa* nitrite reductase, *Biochemical Journal* **233**, 893-898.
14. Maxwell, J. C. and Caughey, W. S. (1978) Infraredspectroscopy of ligands, gases, and other groups in aqueous solutions and tissue, *Methods in Enzymology* **54**, 302-323.
15. Yoshikawa, S., O'Keeffe, D. H.and Caughey, W. S. (1985) Investigations of cyanide as an infrared probe of hemeprotein ligand binding sites,*Journal of Biological Chemistry* **260**, 3518-3528.
16. Tsubaki, M. and S. Yoshikawa (1985) Fourier-transform infrared study of cyanide binding to the Fe_{a3}–Cu_B binuclear site of bovine heart cytochrome *c* oxidase: implication of the redox-linked conformational change at the binuclear site, *Biochemistry* **32**, 164-173.
17. Walsh, T. A., Johnson, M. K., Thomson, A. J., Barber, D. and Greenwood, C. (1981) The characterization and magnetic properties of the azide and imidazole derivatives of *Pseudomonas* nitrite reductase. *Journal of Inorganic Biochemistry* **14**,1-14.
18. Barber, D., Parr, S. R.and Greenwood, C. (1978) The reactions of *Pseudomonas* cytochrome c_{551} oxidase with potassium cyanide, *Biochemical Journal* **175**, 239-249.

19. Alben, J. O., Moh, P. P., Fiamingo, F. G. and Altschuld, R. A. (1981) Cytochrome oxidase (a_3) heme and copper observed by low-temperature Fourier transform infrared spectroscopy of the CO complex, *Proceedings of the National Academy of Sciences of the U.S.A* **78**, 234-237.
20. Fager, L. Y. and Alben, J. O. (1972) Structure of the carbon monoxide binding site of hemocyanins studied by Fourier transform infrared spectroscopy, *Biochemistry* **11**, 4786-4782.
21. Pettingill, T. M., Strange, R. W. and Blackburn, N. J. (1991) Carbonmonoxy dopamine β-hydroxylase. Structural characterization by Fourier transform infrared, fluorescence, and X-ray absorption spectroscopy *Journal of Biological Chemistry* **266**, 16996-17003.
22. LaCroix, L. B., Shadle, S. E., Wang, Y., Averill, B. A., Hedman, B., Hodgson, K. O. and Solomon, E. I. (1996) Electronic structure of the perturbed blue copper site in nitrite reductase: spectroscopic properties, bonding, and implications for the entatic/rack state, *Journal of the American Chemical Society* **118**, 7755-7768.
23. Kukimoto, M., Nishiyama, M., Murphy, M. E. P., Turley, S., Adman, E. T., Horinouchi, S. and Beppu, T. (1994) X-ray structure and site-directed mutagenesis of a nitrite reductase from *Alcaligenes faecalis* S-6: Roles of two copper atoms in nitrite reduction, *Biochemistry* **33**, 5246-5252.
24. Solomon, E. I., Penfield, K. W., Gewirth, A. A., Lowery, M. D., Shadle, S. E., Guckert, J. A. and LaCroix, L. B. (1996) Electronic structure of the oxidized and reduced blue copper sites: contributions to the electron transfer pathway, reduction potential, and geometry, *Inorganica Chimica Acta* **243**, 67-78.

CYTOCHROME *C* NITRITE REDUCTASE FROM *SULFUROSPIRILLUM DELEYIANUM* AND *WOLINELLA SUCCINOGENES*

Molecular and spectroscopic properties of the multihaem enzyme

O. EINSLE , W. SCHUMACHER, E. KURUN, U. NATH, P.M.H. KRONECK

Fakultät für Biologie, Universität Konstanz, D-78464 Konstanz, Germany

1. Respiratory Transformation of Nitrogen Oxides and Nitrogen Oxyanions

Within the biogeochemical nitrogen cycle N_xO_y compounds function as electron acceptors of anaerobic respiratory chains. Prokaryotes carry out their energy conservation during reduction of these nitrogen compounds, and use complex transition metal enzymes for their transformation [1,2]. Dissimilatory nitrate reduction proceeds via two different pathways, i.e. denitrification, and nitrate- ammonification (Scheme 1). In denitrification, the nitrite reductase is either a cytochrome cd_1, or a copper protein [3,4]. During nitrate-ammonification nitrate is reduced to nitrite as the only liberated intermediate that is subsequently converted to ammonia in a six-electron step by a cytochrome *c* nitrite reductase [4]. H_2 and formate are the predominant electron donors that are oxidized by nitrate-ammonifying bacteria [5-7]. Eisenmann *et al.* [8] reported on the use of sulfide as electron donor to support bacterial growth by nitrate-ammonification, thus connecting the biogeochemical cycles of nitrogen and sulfur.

Scheme 1

$$^{+5}NO_3^- \longrightarrow {}^{+3}NO_2^- \longrightarrow {}^{+2}NO \longrightarrow {}^{+1}N_2O \longrightarrow {}^{0}N_2 \qquad (1)$$

$$^{+5}NO_3^- \longrightarrow {}^{+3}NO_2^- \longrightarrow {}^{-3}NH_4^+ \qquad (2)$$

In this contribution some aspects of the bioenergetics and respiratory chains in nitrate-ammonifying bacteria will be discussed. Special emphasis will be directed towards the biochemical and spectroscopic properties of the multihaem protein cytochrome *c* nitrite reductase (NiR) from *Sulfurospirillum deleyianum* and *Wolinella succinogenes*. This enzyme appears to have novel haem centers which catalyze an important step within the nitrogen cycle.

G.W. Canters and E. Vijgenboom (eds.),
Biological Electron Transfer Chains: Genetics, Composition and Mode of Operation, 197-208.
© *1998 Kluwer Academic Publishers.*

2. Respiratory Chains and Bioenergetics of Nitrate-Ammonifying Bacteria

The organization of the enzymes participating in respiratory nitrate-ammonification has been reviewed most recently [9]. The reduction of nitrate to nitrite or ammonia was described for bacteria with a fermentative rather than a respiratory metabolism [10,11]. However, growth of various nitrate-ammonifying bacteria by oxidation of non-fermentable substrates like H_2 or formate linked to the reduction of nitrate and nitrite to ammonia demonstrated that nitrate-ammonification may also function as respiratory energy conserving process [6, 8, 12-18]. With *S. deleyianum* nitrate was quantitatively reduced to nitrite; reduction of nitrite started only after nitrate was exhausted [8,18].

In *E. coli* the reduction of nitrate to nitrite with formate (or H_2) is linked to the generation of a proton electrochemical potential [12,13]. The formate dehydrogenase of *E. coli* appears to have its hydrophilic active site subunit exposed to the periplasm [19-21]. A low potential cytochrome *b* was coupled to the formate dehydrogenase of *E. coli* [5], *W. succinogenes* [22,23] and *S. deleyianum* [24]. The quinone involved in the oxidation of formate (or H_2) generated a transmembrane proton electrochemical potential through a redox loop mechanism. Uptake of two H $^+$ from the cytoplasm during the reduction of quinone and release of two H^+ into the periplasm during reoxidation of the quinol at the periplasmic phase of the cytoplasmic membrane was shown [12,13]. Reduction of nitrate resulted in the consumption of two H^+ from the cytoplasm and the formation of nitrite and water. The H^+/e^- ratio of one for quinol-nitrate oxidoreduction and of two for H_2-nitrate and formate-nitrate oxidoreduction are fully consistent with this model [12,13].

In nitrate-ammonifying bacteria, especially in enterobacteria, there may exist two independently regulated dissimilatory ways of nitrite reduction to ammonia with two different physiological functions [11,25]. A membrane-bound formate:nitrite oxidoreductase complex with a respiratory function was first described for *E. coli* [6,7]. A cytochrome *c* NiR similar to the enzyme of *E. coli* was isolated from *D. desulfuricans* and its involvement as the terminal reductase of a formate (or H_2)-dependent nitrite reduction to ammonia was shown [26,27]. Proton translocation coupled to the ammonification of nitrite was found for *D. desulfuricans* [27], *C. sputorum* biovar *bubulus* [28] and *E. coli* [29]. Electron transport from H_2 to nitrite and the synthesis of ATP are chemiosmotically coupled by the generation of a proton electrochemical potential as demonstrated for the first time for *D. gigas* [30] and *W. succinogenes* [16].

The model for the organization of the respiratory chain, and for the direction of proton exchange associated with the electron transport from formate to nitrite in nitrite-ammonifying bacteria, represents the combined results obtained from various bacteria [9]. A periplasmically oriented membrane-bound formate dehydrogenase or a hydrogenase are involved. Most likely these primary dehydrogenases functioning in the reduction of nitrate→nitrite and nitrite→ammonia are identical entities [16, 19-21, 23, 27, 28, 31-34]. A cytochrome *c* NiR is involved as membrane-associated terminal reductase facing the periplasm as shown for *E. coli* [35, 36], *D. desulfuricans* [27], *C. sputorum* biovar *bubulus* [28], and *S. deleyianum* [18]. There is good evidence that a pool of quinone (probably menaquinone) is mediating the electron flow between the formate dehydrogenase, or the hydrogenase, and the cytochrome *c* NiR. Menaquinone was also the major quinone after cultivation of *S. deleyianum* with formate as electron donor [24,37].

The *E. coli* cytochrome *c* NiR was the first among these enzymes to be completely sequenced [36,38]. It is a heterooligomeric complex encoded by the *nrfABCDEFG* operon. *NrfA* is encoding for the periplasmic nitrite-reactive cytochrome *c* (M_r 51 kDa), *nrfB* represents a small hydrophilic cytochrome *c* (M_r 18 kDa) and *nrfC* is encoding for an iron-sulfur protein carrying 16 cysteine residues (M_r 24.6 kDa). *NrfD* was suggested to be involved in electron transfer to NiR as the quinol oxidase [38]. Because of sequence homology, the membrane-integrated hydrophobic *nrfE* (M_r 60.8 kDa) and the periplasmic hydrophilic *nrfF* (M_r 14.4 kDa) were proposed to be involved in haem transport across the cytoplasmic membrane and in periplasmic cytochrome *c* assembly, respectively [11,38].

From present biochemical and genetic evidence, the following electron transfer sequence is suggested for electrogenic nitrite-ammonification as discussed by Schumacher *et al.* [9]. The oxidation of formate (or H_2) releases two H^+ to the periplasm per substrate oxidized while two e^- are transferred to a quinone via an iron-sulfur protein and a cytochrome *b* [21]. The quinol is reoxidized at the site of the nitrite reductase complex and the electrons are transferred to the cytochrome *c* NiR via an iron-sulfur protein and a *c*-type cytochrome. Reduction of nitrite to ammonia results in the consumption of protons from the periplasm and the formation of ammonia and two water molecules. A H^+/e^- ratio of one was determined for the electron transport of formate→nitrite in *C. sputorum* biovar *bubulus* [28], and for H_2→nitrite in *D. desulfuricans* [27] and *D. gigas* [30]. However, the two periplasmic reactions would neither produce a net gain of protons ($3H_2 \rightarrow 6H^+ + 6e^-$; $NO_2^- + 8H^+ + 6e^- \rightarrow NH_4^+ + 2H_2O$), nor would they produce a net gain of charges translocated across the cytoplasmic membrane. Consequently, an additional process has to be considered that leads to a net transfer of negative charge to drive the observed electron transport phosphorylation. It was shown for *W. succinogenes* that the protons required for quinone reduction through the formate dehydrogenase and hydrogenase are consumed from the cytoplasm [39]. The involvement of a quinone in proton translocation during nitrate reduction by a redox loop mechanism in *E. coli* was shown [12,13]. Most likely, the two protons required for protonation of the reduced quinone are taken up from the cytoplasm, and the reoxidation of quinol at the periplasmic face is releasing two protons into the periplasm during nitrite-ammonification as suggested by de Vries *et al.* [28].

3. The Multihaem Cytochrome *c* Nitrite Reductase (NiR)

3.1. MOLECULAR PARAMETERS

NiR converts nitrite to ammonia in a six-electron transfer reaction [4,9]:

$$NO_2^- + 8H^+ + 6e^- \rightarrow NH_4^+ + 2H_2O \qquad E_o' (NO_2^-/NH_4^+) = +0.34 \text{ V}$$

TABLE 1. Molecular properties of dissimilatory nitrite reductases

	Fraction	M_r[a] [kDa]	heme/M_r[b]	Fe/M_r[c]	ε_{red}(553 nm) [mM^{-1} cm^{-1}]	Sp. activity[d] [U mg^{-1}]	Reference
Sulfurospirillum deleyianum	soluble	55±2	3.5±0.1	2.9±0.4	85[e]	1070	[49]
Sulfurospirillum deleyianum	membrane	55±2	3.9±0.2	3.5±0.3	98	1050	[49]
Heterooligomer	membrane	245±15	17.8±1.3	16.5±2.1	397	970	[49]
Wolinella succinogenes	membrane	55±2	3.4±0.2	3.2±0.4	89	850	[49]
Heterooligomer	membrane	245±15	13.6±2.1	13.6±2.4	361	810	[49]
Wolinella succinogenes	soluble	63	5.44	5.2			[45]
Wolinella succinogenes	membrane	63	3.5	8.3			[34]
Wolinella succinogenes	membrane	63	>5	n.d.		708	[46]
Escherichia coli	soluble	69	6.33	6.11		518	[41]
Escherichia coli	periplasmic	53.05[f]	4xCXYCH	4			[36]
Desulfovibrio desulfuricans	membrane	66	5.96	5.49		345	[44]
Vibrio fischeri	soluble	80-95	1.5-1.8	≈2			[42]
Vibrio fischeri	soluble	57	5.45	5.62		682	[43]

[a] According to SDS-PAGE and gelfiltration in 0.7% OcGlc; [b] Alkaline pyridine Fe(II)-hemochromogen extraction; [c] Fe-AAS; [d] 1 U=1 μmol NO$_2^-$ min^{-1} at pH 7.0 (optimum) at 37°C, k_{cat} = 6000 e$^-$ s^{-1}; [e] Tetraheme cytochrome c_3: ε_{red}(552) = 128 mM^{-1} cm^{-1}; monoheme horse heart cytochrome c : ε_{red}(550) = 27.6 mM^{-1} cm^{-1}; [f] Processed apoprotein (50.58 kDa), processed holoprotein (53.05 kDa) according to the DNA sequence

So far, respiratory ammonia-forming NiRs have been isolated from *E. coli* [40,41], *Vibrio fischeri* [42,43], *D. desulfuricans* [44], *W. succinogenes* [34, 45,46], *Vibrio alginolyticus* [47], and *S. deleyianum* [48,49]. A comparison of the molecular properties (Table 1) suggests that the respiratory ammonia-forming nitrite reductases constitute a group of homologous enzymes. Most NiRs were described as monomeric proteins (M_r 50-70 kDa) with a tendency to aggregate *in vitro* [34,44]. However, Blackmore *et al.* [46] isolated a low (M_r 63 kDa) and a high molecular mass enzyme complex (M_r 360 kDa) of NiR from the membrane fraction of *W. succinogenes*. For clarification of the quaternary arrangement, in a comparative study NiR was isolated from the membrane fraction of both *S. deleyianum* and *W. succinogenes* as a monomer (M_r 55±2 kDa), and as a hetero-oligomeric complex [49]. The heterooligomeric nature of NiR was confirmed by cross-linking experiments which gave 5 distinct bands on SDS-PAGE. The complex was built from four 55 kDa units and contained additionally a small *c*-type cytochrome (M_r 22 kDa) as revealed by SDS-PAGE and haem-staining. Hereby, the possibility was excluded that the 22 kDa *c*-type cytochrome is a contaminant protein [49]. SDS helped to dissociate the complex into its components. Reductants, such as dithiothreitol, had no influence on the dissociation of the complex indicating that disulfide bridges were not involved which would also explain that the 22 kDa cytochrome *c* might be partially lost during the purification of the membranous enzyme. Schröder *et al.* [34] had already noticed a 22 kDa *c*-type cytochrome enriched from solubilized membranes of *W. succinogenes* but did not assign this protein as part of the cytochrome *c* nitrite reductase complex. A 22 kDa *c*-type cytochrome component was not detected in the high M_r preparations of *W. succinogenes* [46] while the suggested 120 kDa putative NiR anchor protein was never observed in our preparations. For the purification of nitrite reductase from *E. coli* a strain was used that over produced cytochrome *c* nitrite reductase in addition to a smaller c-type cytochrome with an estimated molecular mass of 30 kDa [41]. The seven gene operon encoding the formate-dependent nitrite reductase of *E. coli* [38] contained downstream of the cytochrome *c* NiR structural gene the sequence of a 18 kDa c-type cytochrome that was suggested to be identical to the 30 kDa cytochrome revealed by Kajie and Anraku [41]. Although the 18 kDa cytochrome *c* that was proposed to be the electron donor of *E. coli* NiR is carrying five Cys-X-Y-Cys-His haem binding motifs it must not necessarily be a pentahaem because of a low haem staining activity [38]. It was only recently, that the cytochrome *c* NiR was isolated from the membrane fraction of *D. desulfuricans* as a heterooligomeric complex consisting of a 18.8 kDa cytochrome *c* in addition to the NiR (M_r 62 kDa) [50].

3.2. NUMBER OF HAEMS AND HAEM BINDING MOTIFS

Cytochrome *c* nitrite reductases were considered to contain six covalently bound haem prosthetic groups. Actually, the reported haem iron content varied between two and six (Table 1). Darwin *et al.* [36] reported on the sequence of the structural gene of the *E. coli* NiR, carrying four Cys-X-Y-Cys-His haem binding motifs, which were also reported for NiR from *Haemophilus influenzae* [51], and which led to the proposal of a tetrahaem structure for the *E. coli*. Most recently, the sequences of the enzymes from *S. deleyianum* and *W. succinogenes* have been completed, again revealing the presence of the four characteristic Cys-X-X-Cys-His motifs (O. Einsle, A. Messerschmidt, P.M.H. Kroneck, unpublished data).

In the pyridine Fe(II)-hemochrome assay, NiR of *S. deleyianum* and *W. succinogenes* gave absorption maxima at 415, 520, and 550 nm [49]. From difference spectra (enzyme, dithionite-reduced minus ferricyanide-oxidized) a value of ≈ 4 haem/M_r 55 kDa was obtained for the monomer, and ≈ 18 haem/M_r 245 kDa for the complex which is in excellent agreement with the Fe content obtained by the atomic absorption technique (Table 1). These analytical values data fit very nicely the sequence data showing four haem-binding consensus sequences. Within that sequence, there is also a conserved Cys-X-Y-Cys-Lys motif which was not assigned to a haem-binding site in the case of the *E. coli* enzyme [36]. Schröder *et al.* [34] reported 3.5 haem, 4.8 non-haem Fe and 2.1 acid-labile sulfide/M_r 63 kDa for their preparation of the membranous enzyme from *W. succinogenes*; a single peptide chain carrying four haem *c* groups, and probably an iron-sulfur cluster, was proposed. In the purest and most active samples of cytochrome *c* NiR from *S. deleyianum* and *W. succinogenes*, Fe-S centers were not present.

Both the monomeric and the complex form of the enzymes from *S. deleyianum* and *W. succinogenes* exhibited a high specific activity, with up to 1050 μmol NO_2^- min^{-1} mg^{-1} which led to the development of an electrochemical nitrite sensor [52,53].

3.3. SPECTROSCOPIC PROPERTIES

Cytochrome *c* nitrite reductase is a multihaem enzyme with rather unique spectroscopic properties which makes it an interesting molecule for the application of various physical techniques, such as multifrequency electron paramagnetic resonance (EPR) [9,48,49,54-56], Mössbauer spectroscopy [55,56] and magnetic circular dichroism (MCD) [57]. NiR from both *S. deleyianum* and *W. succinogenes* (as isolated, phosphate buffer pH 7.5) gave absorption maxima at 280, 409, and 534 nm, a shoulder at 615 nm and maxima at 420.5, 523.5, and 553.3 nm in the reduced state (dithionite). The molar absorption coefficient ε_{553} of the monomer was $\approx 4x$ the coefficient ε_{550} of reduced horse heart cytochrome *c* [58], but was slightly smaller than ε_{552} of the tetrahaem cytochrome c_3 from *D. desulfuricans* [59].

The micro-aerophilic sulfur-reducing *S. deleyianum*, when cultivated with elemental sulfur, nitrate or nitrite as terminal electron acceptor, expressed the cytochrome *c*-dependent NiR [8,24]. As isolated NiR showed a characteristic intense resonance at g 3.8 in the EPR spectrum at ≈ 9.30 GHz, 10 K, perpendicular mode (Figure 1). This resonance was even detectable in whole cells and membranes of *S. deleyianum* when grown on elemental sulfur [24], and led to the purification of the enzyme [48,49]. In addition to the strong signal at g 3.85 and a broad line at g 9.8, the EPR spectrum showed low-spin Fe(III) resonances at g 2.94, 2.29 and ≈ 1.50 (Figure 1). Depending on the the sample there was a more or less minor signal at g 6.1 from high-spin Fe(III) haem [48] which became more intense at pH>10, or in the SDS-treated sample [57]. The characteristic features at g 9.8 and 3.85 were also reported for

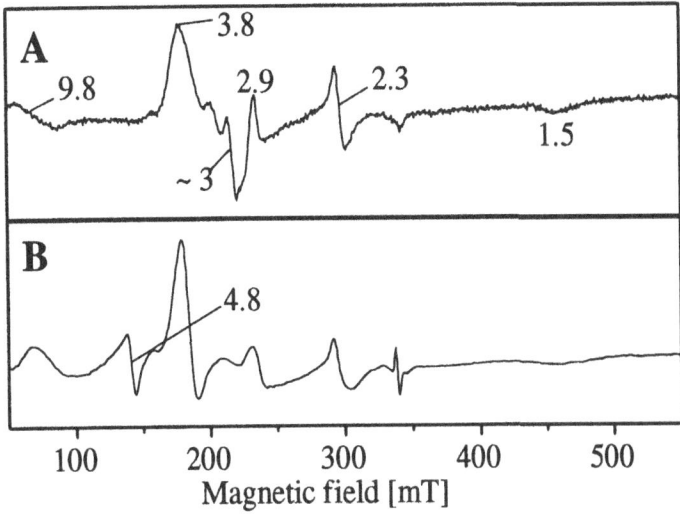

Figure 1. First derivative EPR spectrum spectra of soluble (A) and membranous (B) cytochrome *c* nitrite reductase from *S. deleyianum.* Enzyme 150 μM in 60 mM phosphate buffer, pH 7.2. Perpendicular mode, 9.5 GHz microwave frequency, 0.15 mW power, temperature 5 K.

the tetrahaem cytochrome c_{554} from *Nitrosomonas europaea,* and assigned to an exchange-coupled pair of low and high-spin Fe(III) haem centers [54,60,61]. Most important, EPR spectra of the NiR complex of *S. deleyianum* and *W. succinogenes* displayed a significant resonance around g 4.83 at X-band (perpendicular mode; baseline crossing) which was absent in the spectra of the monomeric enzyme. This resonance was assigned to the 22 kDa cytochrome *c* component of the nitrite reductase complex [49]. In agreement with this assignment, the soluble monomeric enzyme from *E. coli* as well as from *W. succinogenes* prepared by Liu *et al.* [45] did not show the EPR signal at g 4.83 [62]. Blackmore *et al.* [57] also described the unusual signal at g 4.8 in the X-band EPR spectrum of NiR from the membrane fraction of *W. succinogenes.*

The resonances observed in the X-band EPR spectrum (≈ 9.30 GHz) at g 9.8, 4.83 and 3.85 were not detected in the S-band spectrum (≈ 3.5 GHz, loop-gap resonator), 10 K (W.E. Antholine, P.M.H. Kroneck, unpublished results). On the other hand, a strong signal at g 9.8 (baseline crossing) was detected in the parallel mode at X-band (dual mode cavity, ≈ 9.1 GHz), 5 K (Figure 2). From spin quantitation and computer simulations (CuSO$_4$ standard [63]) of the signals at g 2.94, 2.29 and 1.50, 1.80 ± 0.2 low-spin Fe(III) haem centers were determined as reported for the enzyme from *W. succinogenes* (Fe(III) myoglobin cyanide standard [57]). Assuming a tetrahaem configuration for NiR from *S. deleyianum* this would leave two haems in a spin-coupled system, most likely built of a S =5/2 high-spin and a S = 1/2 low-spin

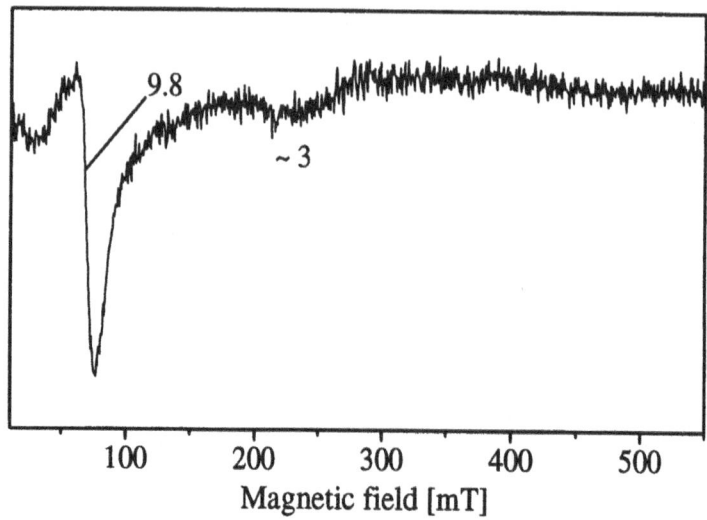

Figure 2. First derivative EPR spectrum of soluble cytochrome *c* nitrite reductase from *S. deleyianum.* Enzyme as in Figure 1. Parallel mode, 9.1 GHz microwave frequency, 2.0 mW power, temperature 5 K.

Fe(III). The two Fe (III) centers are exchange-coupled in a dihaem cluster which is the origin for the integer-spin EPR signal observed at g 9.8 (Figure 2). A similar situation has been reported recently for the hydroxylamine oxidoreductase from *Nitrosomonas* where the integer-spin signal can be associated with the reactive P_{460} haem site. In this case the interacting spins are assumed to be both S =1/2 [60]. Note that hydroxylamine oxidoreductase contains seven *c*-type cytochromes and catalyzes the four-electron oxidation of hydroxylamine to nitrite.

4. Acknowledgements

We wish to acknowledge the valuable contributions of several former co-workers: M. Gandbhir, U. Hole, D.H.W. Kastrau, R. Kümmerle and M. Meininghaus. Discussions with W.E. Antholine, H. Beinert, O. Farver, B. Gründig, M. Kroder, I. Pecht and B. Strehlitz have been very helpful. This work was supported by Deutsche Forschungsgemeinschaft, Fonds der Chemischen Industrie, Volkswagenstiftung, and the EU network MASIMO within Human Capital and Mobility.

5. References

1. Zumft, W.G. (1992) The denitrifying prokaryotes. In A.. Balows , H.G. Trüper, M. Dworkin, W. Harder, and K.-H. Schleifer (eds.) *The prokaryotes. A Handbook on the biology of bacteria: ecophysiology, isolation, identification, applications*, Springer Verlag ,New York, Berlin, Heidelberg, London, Paris, Tokyo, Hong-Kong, Barcelona, Budapest,Vol. 1, pp. 554-582.

2. Berks, B.C., Ferguson, S.J., Moir, J.W.B., and Richardson, D.J. (1995) Enzymes and associated electron transport systems that catalyse the respiratory reduction of nitrogen oxides and oxyanions. *Biochim. Biophys. Acta*, **1232**, 97-173.

3. Godden, J.W., Turley, S., Teller, D.C., Adman, E.T., Liu, M.-Y., Payne, W.J., and LeGall, J. (1991) The 2.3 Ångstrøm X-ray structure of nitrite reductase from *Achromobacter cycloclastes*, *Science* **253**, 438-44.

4. Brittain, T., Blackmore, R., Greenwood, C., and Thomson, A.J. (1992) Bacterial nitrite-reducing enzymes, *Eur. J. Biochem.* **209**, 793-802.

5. Enoch, H.G. and Lester, R.L. (1974) The role of a novel cytochrome *b*-containing nitrate reductase and quinone in the *in vitro* reconstitution of formate-nitrate reductase activity in *E. coli. Biochem. Biophys. Res. Commun.*, 61, 1234-1241.

6. Abou-Jaoudé, A., Chippaux, M., and Pascal, M.-C. (1979) Formate-nitrite reduction in *Escherichia coli* K12. 1. Physiological study of the system, *Eur. J. Biochem.* **95**, 309-314.

7. Abou-Jaoudé, A., Pascal, M.-C., and Chippaux, M. (1979) Formate-nitrite reduction in *Escherichia coli* K12. 2. Identification of components involved in the electron transfer, *Eur. J. Biochem.* **95**, 315-321.

8. Eisenmann, E., Beuerle, J., Sulger, K., Kroneck, P.M.H., and Schumacher, W. (1995) Lithotrophic growth of *Sulfurospirillum deleyianum* with sulfide as electron donor coupled to respiratory reduction of nitrate to ammonia, *Arch. Microbiol.* **164**, 180-185.

9. Schumacher, W., Neese, F., Hole, U., and Kroneck, P.M.H. (1997) Cytochrome *c* nitrite reductase and nitrous oxide reductase: two metallo enzymes of the nitrogen cycle with novel metal sites. In G. Winkelmann and C.J. Carrano (eds.), *Transition Metals in Microbial Metabolism*, Harwood Academic Publishers, London, in press.

10. Cole, J.A. (1988) Assimilatory and dissimilatory reduction of nitrate to ammonia. In J.A. Cole and S.J. Ferguson (eds.), *The nitrogen and sulphur cycles*, Society for General Microbiology, University Press, Cambridge, pp. 281-329.

11. Cole, J.A. (1996) Nitrate reduction to ammonia by enteric bacteria: redundancy, or a strategy for survival during oxygen starvation ? *FEMS Microbiol. Lett.*, 136, 1-11.

12. Jones, R.W. (1980) The role of the membrane-bound hydrogenase in the energy-conserving oxidation of molecular hydrogen by *Escherichia coli*, *Biochem. J.* **188**, 345-350.

13. Jones, R.W., Lamont, A., and Garland, P.B. (1980) The mechanism of proton translocation driven by the respiratory nitrate reductase complex of *Escherichia coli*, *Biochem. J.* **190**, 79-94.

14. de Vries, W., Niekus, H.G.D., Boellaard, M., and Stouthamer, A.H. (1980) Growth yields and energy generation by *Campylobacter sputorum* subspecies *bubulus* during growth in continuous culture with different hydrogen acceptors, *Arch. Microbiol.* **124**, 221-227.

15. Yoshinari, T. (1980) N_2O reduction by *Wolinella succinogenes*, *Appl. Environm. Microbiol.* **39**, 81-84.

16. Bokranz, M.J., Katz, J., Schröder, I., Roberton, A.M., and Kröger, A. (1983) Energy metabolism and biosynthesis of *Vibrio succinogenes* growing with nitrate or nitrite as terminal electron acceptor, *Arch. Microbiol.* **135**, 36-41.

17. Seitz, H.-J. and Cypionka, H. (1986) Chemolithotrophic growth of *Desulfovibrio desulfuricans* with hydrogen coupled to ammonification of nitrate or nitrite, *Arch. Microbiol.* **146**, 63-67.

18. Schumacher, W. and Kroneck, P.M.H. (1992) Anaerobic energy metabolism of the sulfur-reducing bacterium "Spirillum" 5175 during dissimilatory nitrate reduction to ammonia, *Arch. Microbiol.* **157**, 464-470.

19. Hooper, A. B. and DiSpirito, A.A. (1985) In bacteria which grow on simple reductants: generation of a proton gradient involves extracytoplasmic oxidation of substrate, *Microbiol. Rev.* **49**, 140-157.

20. Berg, B.L., Li J., Heider, J., and Stewart, V. (1991) Nitrate-inducible formate dehydrogenase in *Escherichia coli* K-12 I. Nucleotide sequence of the *fdnGHI* operon and evidence that opal (UGA) encodes selenocysteine, *J. Biol. Chem.* **266**, 22380-22385.

21. Berks, B.C., Page, M.D., Richardson, D.J., Reilly, A., Cavill, A., Outen, F., and Ferguson, S.J. (1995) Sequence analysis of subunits of the membrane-bound nitrate reductase from a denitrifying bacterium: the integral membrane subunit provides a prototype for the dihaem-carrying arm of a redox-loop, *Mol. Microbiol.* **15**, 319-331.

22. Kröger, A. and Innerhofer, A. (1976) The function of the b cytochromes in the electron transport from formate to fumarate of *Vibrio succinogenes*, *Eur. J. Biochem.* **69**, 497-506.

23. Kröger, A., Dorrer, E., and Winkler, E. (1980) The orientation of the substrate sites of formate dehydrogenase and fumarate reductase in the membrane of *Vibrio succinogenes*, *Biochim. Biophys. Acta* **589**, 118-136.

24. Zöphel, A., Kennedy, M.C., Beinert, H., and Kroneck, P.M.H. (1991) Investigations on microbial sulfur respiration. 2. Isolation, purification, and characterisation of cellular components from Spirillum 5175, *Eur. J. Biochem.* **195**, 849-856.

25. Page, L., Griffiths, L., and Cole, J.A. (1990) Different physiological roles of two independent pathways for nitrite reduction to ammonia by enteric bacteria, *Arch. Microbiol.* **154**, 349-354.

26. Steenkamp, D.J. and Peck Jr., H.D. (1980) The association of hydrogenase and dithionite reductase activities with the nitrite reductase of *Desulfovibrio desulfuricans*, *Biochem. Biophys. Res. Commun.* **94**, 41-48.

27. Steenkamp, D.J. and Peck Jr., H.D. (1981) Proton translocation associated with nitrite respiration in *Desulfovibrio desulfuricans*, *J. Biol. Chem.* **256**, 5450-5458.

28. de Vries, W., Niekus, H.G.D., van Berchum, H., and Stouthamer, A.H. (1982) Electron transport-linked proton translocation at nitrite reduction in *Campylobacter sputorum* subspecies *bubulus*, *Arch. Microbiol.* **131**, 132-139.

29. Pope, N.R. and Cole, J.A. (1982) Generation of a membrane potential by one of two independent pathways for nitrite reduction by *Escherichia coli*, *J. Gen. Microbiol.* **128**, 319-322.

30. Barton, L.L., LeGall, J., Odom, J.M., and Peck Jr., H.D. (1983) Energy coupling to nitrite respiration in the sulfate-reducing bacterium *Desulfovibrio desulfuricans*, *J. Bacteriol.* **153**, 86-871.

31. Kröger, A. and Winkler, E. (1981) Phosphorylative fumarate reduction in *Vibrio succinogenes*: stoichiometry of ATP synthesis, *Arch. Microbiol.* **129**, 100-104.

32. Odom, J.M. and Peck Jr., H.D. (1981) Localization of dehydrogenases, reductases and electron transfer components in the sulfate-reducing bacterium *Desulfovibrio gigas*, *J. Bacteriol.* **147**, 161-169.

33. Bokranz, M., Gutmann, M., Körtner, C., Kojro, E., Fahrenholz, F., Lauterbach, F., and Kröger, A. (1991) Cloning and nucleotide sequence of the structural genes encoding the formate dehydrogenase of *Wolinella succinogenes*, *Arch. Microbiol.* **156**, 119-128.

34. Schröder, I., Roberton, A.M., Bokranz, M., Unden, G., Böcher, R. and Kröger, A. (1985) The membraneous nitrite reductase involved in the electron transport of *Wolinella succinogenes*. *Arch. Microbiol.*, **140**, 380-386.

35. Fujita, T.and Sato, R. (1966) Studies on soluble cytochromes in Enterobacteriaceae. III. Localization of cytochrome c_{552} in the surface layer of cells, *J. Biochem. (Tokyo)* **60**, 568-577.

36. Darwin, A., Hussain, H., Griffiths, L., Grove, J., Sambongi, Y., Busby, S., and Cole, J. (1993) Regulation and sequence of the structural gene for cytochrome c_{552} from *Escherichia coli*, not a hexaheme but a 50 kDa tetraheme nitrite reductase, *Mol. Microbiol.* **9**, 1255-1265.

37. Collins, M.D. and Widdel, F. (1986) Respiratory quinones of sulphate-reducing and sulphur-reducing bacteria: a systematic investigation, *System. Appl. Microbiol.* **8**, 8-18.

38. Hussain, H., Grove, J., Griffiths, L., Busby, S., and Cole, J. (1994) A seven-gene operon essential for formate-dependent nitrite reduction to ammonium by enteric bacteria, *Mol. Microbiol.* **12**, 153-163.

39. Geisler , V., Ullmann, R., and Kröger, A. (1994) The direction of proton exchange associated with the redox reactions of menaquinone during electron transport in *Wolinella succinogenes*, *Biochim. Biophys. Acta* **1184**, 219-226.

40. Fujita, T. (1966) Studies on soluble cytochromes in Enterobacteriaceae. I. Detection, purification and properties of cytochrome c_{552} in anaerobically grown cells, *J. Biochem. (Tokyo)* **60**, 204-215.

41. Kajie, S.-I. and Anraku, Y. (1986) Purification of a hexaheme cytochrome c_{552} from *Escherichia coli* K12 and its properties as a nitrite reductase, *Eur. J. Biochem.* **154**, 457-463.53.

42. Prakash, O. and Sadana, J.C. (1972) Purification, characterisation and properties of nitrite reductase of *Achromobacter fischeri*, *Arch. Biochem. Biophys.* **148**, 614-632.

43. Liu, M.-C., Bakel, B.W., Liu, M.-Y., and Dao, T.N. (1988) Purification of *Vibrio fischeri* nitrite reductase and its characterisation as a hexaheme c-type cytochrome, *Arch. Biochem. Biophys.* **262**, 259-265.

44. Liu, M.-C. and Peck Jr., H.D. (1981) The isolation of a hexaheme cytochrome from *Desulfovibrio desulfuricans* and its identification as a new type of nitrite reductase, *J. Biol. Chem.* **256**, 13159-13164.

45. Liu, M.-C., Liu, M.-Y., Payne, W.J., Peck Jr., H.D., and LeGall, J. (1983) *Wolinella succinogenes* nitrite reductase: purification and properties, *FEMS Microbiol. Lett.* **19**, 201-206.

46. Blackmore, R., Roberton, A.M., and Brittain, T. (1986) The purification and some equilibrium properties of nitrite reductase of the bacterium *Wolinella succinogenes*, *Biochem. J.* **323**, 547-552.

47. Rehr, B. and Klemme, J.H. (1986) Metabolic role and properties of nitrite reductase of nitrate-ammonifying marine *Vibrio* species, *FEMS Microbiol. Lett.* **35**, 325-328.

48. Schumacher, W. and Kroneck, P.M.H. (1991) Dissimilatory hexaheme c nitrite reductase of "Spirillum" strain 5175: purification and properties, *Arch. Microbiol.* **156**, 70-74.

49. Schumacher, W., Hole, U.H., and Kroneck, P.M.H. (1994) Ammonia-forming cytochrome c nitrite reductase from *Sulfurospirillum deleyianum* is a tetraheme protein: new aspects of the molecular composition and spectroscopic properties, *Biochem. Biophys. Res. Commun.* **205**, 911-916.

50. Pereira, I.C., Abreu, I.A., Xavier, A.V., LeGall, J., and Teixeira, M. (1996) Nitrite reductase from Desulfovibrio desulfuricans (ATCC 27774). A heterooligomer heme protein with sulfite reductase activity, *Biochem. Biophys. Res. Commun.* **224**, 611-618.

51. Fleischmann, R.D. *et al.* (1995) Whole genome random sequencing and assembly of *Haemophilus influenzae* Rd, *Science* **269**, 496-512.

52. Moreno, C., Costa, C., Moura, I., LeGall, J., Liu, M.Y., Payne, W.J., van Duk, C., and Moura, J.J.G. (1993) Electrochemical studies of the hexaheme nitrite reductase from *Desulfovibrio desulfuricans* ATCC 27774, *Eur. J. Biochem.* **212**, 79-86.

53. Strehlitz, B., Gründig, B., Schumacher, W. Kroneck, P.M.H., Vorlop, K.-D., and Kotte, H. (1996) A nitrite sensor based on a highly sensitive nitrite reductase mediator-coupled amperometric detection. *Anal. Chem.* **68**, 807-816.

54. Blackmore, R.S., Gadsby, P.M.A., Greenwood, C., and Thomson, A.J. (1990) Spectroscopic studies of partially reduced forms of *Wolinella succinogenes* nitrite reductase, *FEBS Lett.* **264**, 257-262.

55. Costa, C., Moura, J.J.G., Moura, I., Liu, M.-Y., Peck Jr., H.D., LeGall, J., Wang, Y., and Huynh, B.H. (1990) Hexaheme nitrite reductase from *Desulfovibrio desulfuricans*. Mössbauer and EPR characterisation of the heme groups, *J. Biol. Chem.* **265**, 14382-14387.

56. Costa, C., Moura, J.J.G., Moura, I., Wang, Y., and Huynh, B.H. (1996) Redox properties of cytochrome c nitrite reductase from *Desulfovibrio desulfuricans* ATCC 27774, *J. Biol. Chem.* **271**, 23191-23196.

57. Blackmore, R.S., Brittain, T., Gadsby, P.M.A., Greenwood, C., and Thomson, A.J. (1987) Electron paramagnetic resonance and magnetic circular dichroism studies of a hexa-heme nitrite reductase from *Wolinella succinogenes*, *FEBS Lett.* **219**, 244-248.

58. Margoliash, E. and Frohwirt, N. (1959) Spectrum of horse heart cytochrome c, *Biochem. J.* **71**, 570-572.

59. Bruschi, M., Hatchikian, C.E., Golovleva, L.A., and LeGall, J. (1977) Purification and characterization of cytochrome c_3, ferredoxin, and rubredoxin isolated from *Desulfovibrio desulfuricans* Norway, *J. Bacteriol.* **129**, 30-38.

60. Andersson, K.K., Lipscomb, J.D., Valentine, M., Münck, E., and Hooper, A.B. (1986) Tetraheme cytochrome c_{554} from *Nitrosomonas europaea*. Heme-heme interactions and ligand binding, *J. Biol. Chem.* **261**, 1126-1138.

61. Hendrich, M.P., Logan, M., Andersson, K.K., Arciero, D.M., Lipscomb, J.D., and Hooper, A.B. (1994) The active site of hydroxylamine oxidoreductase from *Nitrosomonas*: evidence for a new metal cluster in enzymes, *J. Am. Chem. Soc.* **116**, 11961-11968.

62. Liu, M.-C., Liu, M.-Y., Payne, W.J., Peck Jr., H.D., LeGall, J., and DerVartanian, D.V. (1987) Comparative EPR-sudies on the nitrite reductases from *Escherichia coli* and *Wolinella succinogenes*, *FEBS Lett.* **218**, 227-230.

63. Kastrau, D.H.W., Heiss, B., Kroneck, P.M.H., and Zumft, W.G. (1994) Nitric oxide reductase from *Pseudomonas stutzeri*, a novel cytochrome *bc* complex. Phospholipid requirement, electron paramagnetic resonance and redox properties, *Eur. J. Biochem.* **222**, 293-303.

MOLECULAR BASIS FOR ENERGY TRANSDUCTION: MECHANISMS OF COOPERATIVITY IN MULTIHAEM CYTOCHROMES

R.O. LOURO[1], T. CATARINO[1], C.A. SALGUEIRO[1], J. LEGALL[1,2], D.L. TURNER[1,3] and A.V. XAVIER[1]

1. Instituto de Tecnologia Química e Biológica, Universidade Nova de Lisboa, Rua da Quinta Grande, 6, Apt. 127, 2780 Oeiras, Portugal
2. Dept. of Biochemistry and Molecular Biology, University of Georgia, Athens, GA30602, USA
3. Dept. of Chemistry, University of Southampton, Southampton, SO17 1BJ, UK

1. Introduction

Energy transduction through electron/proton cooperativity is at the heart of the metabolism of every living organism. Nonetheless, the search for the structural bases sustaining these phenomena has been hindered by the fact that they are usually associated with complex transmembrane proteins of high molecular weight.

Tetrahaem cytochromes c_3 (from here on designated cytc3) are small (14-15 kDa) proteins found in the periplasmic space of sulphate reducing bacteria [1]. They are soluble proteins containing four haems covalently bound to a single polypeptide chain by thioether bridges to cysteine side chains. Several X-ray structures are reported in the literature [2-9] showing that, despite the low homology in primary structure of these proteins, the haem core and general folding are maintained.

The presence of cytc3 was shown to be constitutive in *Desulfovibrio vulgaris* (Miyazaki) [10] and it was identified as a partner to the periplasmic hydrogenase [11]. This places cytc3 at the centre of the bioenergetic metabolism of sulphate reducing bacteria which perform oxidative phosphorylation [12, 13] using a unique mechanism known as 'hydrogen cycling' [14]. Hydrogen, produced in the cytoplasm during the oxidation of organic substrates, diffuses across the cytoplasmic membrane to the periplasm where it is oxidised and, while the electrons are used for the reduction of sulphite, the protons are used to drive the H^+-ATPases. Although this mechanism is still controversial, there are several lines of evidence that support it [10,15,16].

In addition to its central role in *Desulfovibrio* metabolism, the small size of this globular protein implies that the haems in the cluster are placed at a very short distance

209

G.W. Canters and E. Vijgenboom (eds.),
Biological Electron Transfer Chains: Genetics, Composition and Mode of Operation, 209-223.
© 1998 *Kluwer Academic Publishers.*

from each other (with porphyrin atom distances as short as 5.4 nm [6]), opening the possibility for haem-haem interactions [17]. The small size, together with the fact that the haems are low-spin both in the reduced (diamagnetic) and in the oxidised (paramagnetic) states, make NMR a particularly well suited technique to characterise this protein from the structural and thermodynamic point of view [18-34], and determine the interactions between the haems (homotropic electron/electron cooperativities), and between the haems and ionisable groups (heterotropic electron/proton cooperativities), known as redox-Bohr effect [35, 36] by analogy with the Bohr effect described for hemoglobin [37].

The studies performed on cytc3 from *D. vulgaris* (Hildenborough) will be discussed in this article.

2. Thermodynamic Characterisation

2.1. NMR REDOX TITRATIONS

The ^{1}H-NMR spectra of oxidised haem proteins are characterised by having haem methyl substituent signals shifted to low field, so that several of them appear isolated from the protein spectral envelope. For this reason these signals can be easily followed when the proteins are poised in different solution conditions, providing information on the properties of the haems to which they belong and also on the surrounding environment.

Careful analysis of the NMR redox patterns show that: *i*) the intramolecular electron exchange is very fast, and *ii*) the intermolecular electron exchange is slow. As a consequence, each of the haem methyls has a maximum of five signals, one for each oxidation stage. The signals corresponding to the same methyl in different oxidation stages can be cross-correlated by performing Exchange Spectroscopy experiments (EXSY) [38] in partially reduced samples (Fig. 1), and identified by correlation with specifically assigned signals in the reduced form [23, 24, 26, 27, 32]. The NMR experiments confirmed the existence of haem-haem interactions and, when performed at various solution pHs, showed that the haem midpoint redox potentials are also pH dependent [18-20, 28, 31, 33] with a protonation equilibrium responsible for the redox-Bohr effect which is fast on the NMR time scale [33].

Several approaches have been used to interpret the results but it is now clear [34] that the redox interactions cannot be separated from the redox-Bohr interactions by using a single pH or a series of pH's treated independently, because the pK_a changes between oxidation stages [19-21, 28, 33]. This inevitably leads to the conclusion that the redox interaction potentials would themselves have to be pH dependent [19, 31].

If instead the system is approached in a global way, the results can be described without discontinuities in the pH scale. The simplest model capable of describing the data requires five charged centres (four haems and one ionisable centre) involving only two-site interactions, so that the pH dependence of the redox potentials (and thus of the NMR signals) is a result of the interaction between each haem and the ionisable centre.

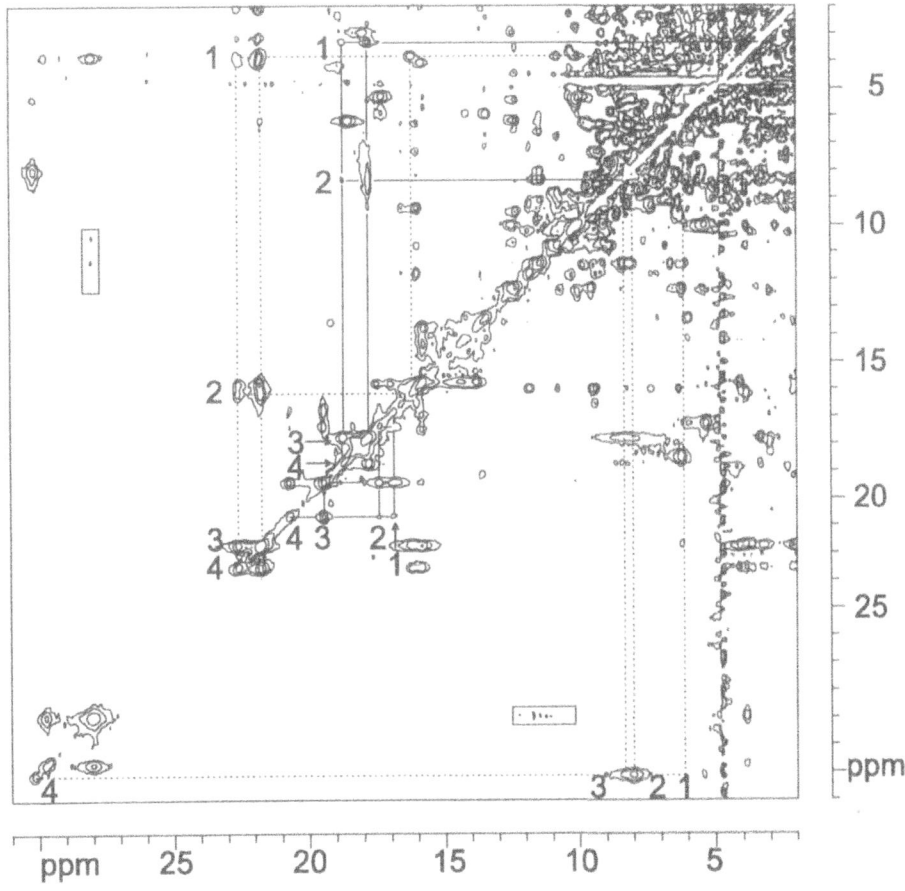

Figure 1. Two-dimensional EXSY spectrum of partially reduced cytc3 from *D. vulgaris* (Hildenborough) at pH 6.0, showing correlations between signals of the same methyl group at different oxidation stages, which are indicated by arabic numbers [33]. Boxed regions indicate broad signals that are not easily seen at this level of intensity cut-off.

Such a model defines 15 thermodynamic parameters (i.e. four energies of oxidation of the haems when the protein is fully reduced and protonated, one energy of deprotonation of the ionisable centre in the fully reduced protein, six haem-haem interactions, four haem-ionisable centre interactions, known as redox-Bohr) that describe a total of 32 distinct populations (Fig. 2) [28, 33].

In order to deconvolute these parameters, the ^{1}H-NMR resonance of one methyl group of each haem was followed at the various oxidation stages and pH values (Fig.3). The chosen methyl groups point towards the protein surface and have large paramagnetic chemical shifts. These conditions reduce the influence of the dipolar

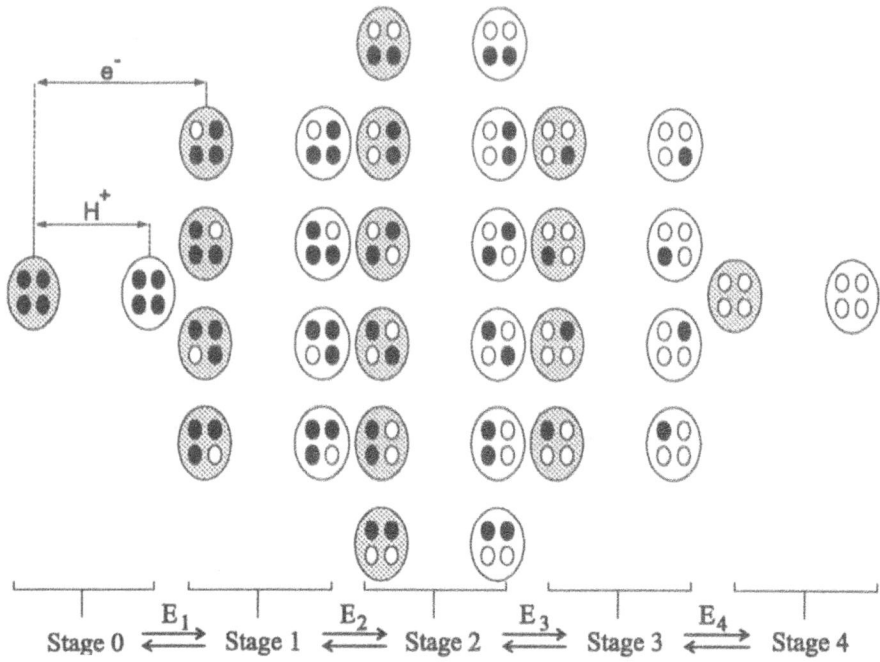

Figure 2. Redox equilibria for a tetrahaem cytochrome with a proton-linked equilibrium showing the 32 possible microstates. The four inner elipses represent the haems which can be either reduced (black) or oxidised (white), and the larger elipses represent the protein which can be either protonated (shaded) or deprotonated (white). Cytc3 from *D. vulgaris* Hildenborough displays fast proton exchange and fast intramolecular electron exchange (within each stage, i.e., molecules with the same number of haems oxidised) [33]. Equilibrium between stages is characterised by the macroscopic redox potentials of the cytochrome (E_1-E_4) and is slow on the NMR time scale.

shifts induced by the oxidation of other haems in the paramagnetic shift of the resonances of each particular haem, which are a consequnce of the short haem-haem distances in this molecule.

The values for the oxidation energies of the haems and the interaction energies between the haems are relative because the redox potential of the solution was not measured inside the NMR tube. Thus, one oxidation energy and one interaction are arbitrarily fixed as standards which are subsequently calibrated by fitting the parameters obtained from the NMR data to redox titrations followed by visible spectroscopy and allowing these arbitrary standards to vary. Since the contribution of each haem to the visible spectra cannot be distinguished, the calibration was performed by fitting the macroscopic redox potentials of the cytochrome (E_i) to the visible data [28]. The energy of deprotonation of the ionisable centre and the interactions between this centre and the haems do not require such a calibration because the pH of the samples was measured inside the NMR tube [28, 33].

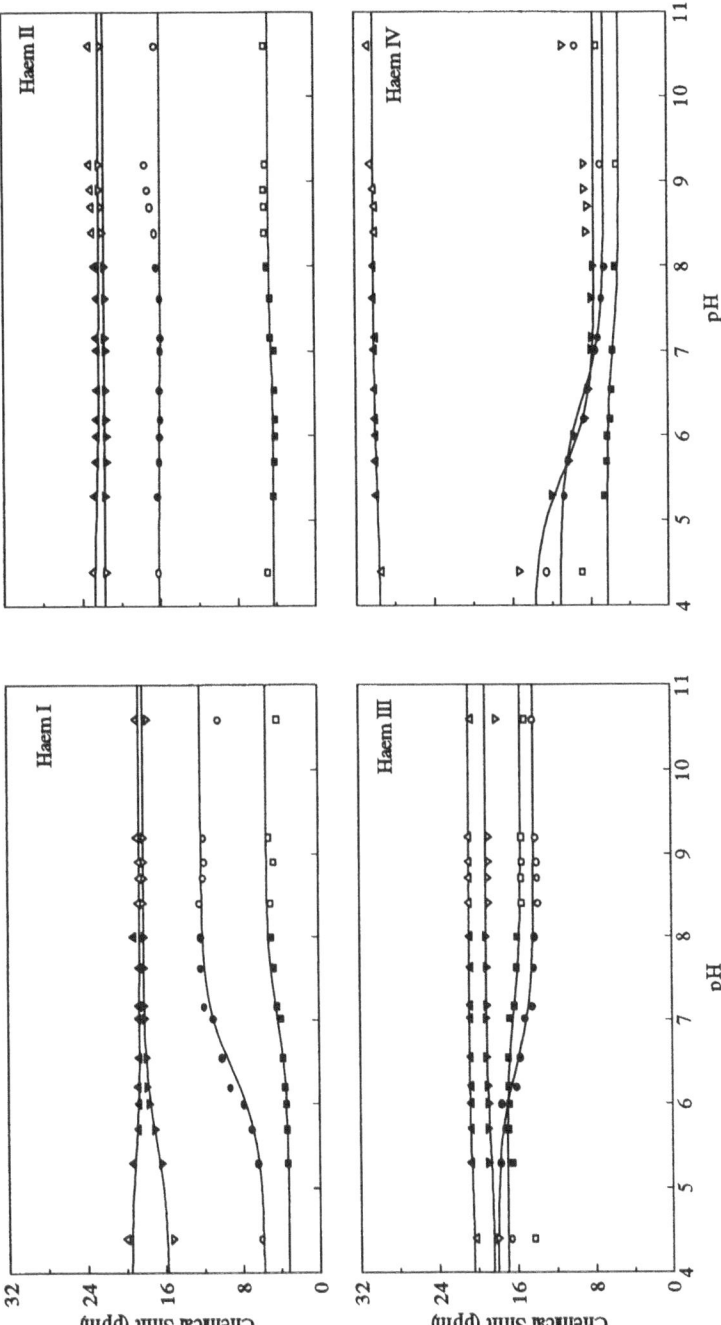

Figure 3. Chemical shifts of the haem methyl group resonances $2^1 CH_3^I$, $18^1 CH_3^I$, $18^1 CH_3^{II}$, $12^1 CH_3^{II}$, $18^1 CH_3^{III}$, $18^1 CH_3^{IV}$ of Dvc_3. Roman numbers refer to the haems in order of attachment to the polypeptide sequence in the oxidation stages 1(■), 2(●), 3(▼) and 4(▲), at different pHs. The full lines represent the best fit (rmsd = 0.25 ppm) of the filled symbols to the model of five interacting centres [33].

Table 1 shows the results obtained for cytc3 from *D. vulgaris* (Hildenborough) [33]. All energies are represented in meV and thus the values are numerically equal to the redox potentials and redox interactions in mV. The pK_a of the completely reduced protein can be calculated by multiplying the energy of deprotonation of the ionisable centre by $F/(2RT)$ [33]. It is observed that there is a large electron-electron (redox) positive cooperativity between two of the haems (I and II, i.e., the second and third haems to be reduced [28, 33]). Such occurrence can not be due to pure electrostatic interactions and requires some structural modification of the protein. It should be stressed that these structural predictions were made based on the results obtained from a theoretical model which does not make any structural assumptions *a priori*. In addition, the large positive cooperativities between the ionisable centre and these two haems enhance the global cooperativity of the system resulting in a proton(s)-assisted two-electron step, which results in an inversion of the macroscopic redox potentials of the cytochrome for a narrow pH range, that actually corresponds to the physiological pH for these organisms (Fig. 4).

TABLE 1: Energy parameters (meV) for the five interacting centres in cytc3 *D. vulgaris* (Hildenborough).

centre	I	II	III	IV	ionisable centre	pKa/ΔpKa
I	**-245**	-43	20	-4	-70	1.2
II		**-267**	-8	8	-30	0.5
III			**-334**	32	-18	0.3
IV				**-284**	-6	0.1
ionisable centre					**439**	7.4

Diagonal elements (boldface) represent the energy for oxidising haems (i.e. reduction potentials in mV) and deprotonating the ionisable group (bottom line) in the fully reduced molecule. The off-diagonal elements represent the energy of interaction between the oxidised haems and between these haems and the ionisable group. Note that the energy of each oxidised haem is lowered by deprotonation (positive cooperativity), as expected on electrostatic grounds, and the substantial negative redox interaction between haems I and II represents positive cooperativity.

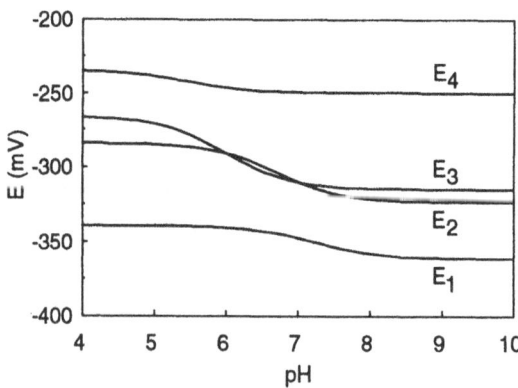

Figure 4. Macroscopic reduction potentials of cytc3 from *D. vulgaris* (Hildenborough) calculated as a function of pH [33]. Between pH 6 and 7 the reduction steps E_2 and E_3 (i.e. between stages 1 and 2 and between stages 2 and 3) show inversion of their redox potentials which ensures the simultaneous transfer of two electrons.

2.2. POTENTIOMETRIC TITRATIONS

To determine the number of protons involved in the redox-Bohr effect, and thus obtain a complete description of this phenomenon, potentiometric titrations of the cytc3 from *D. vulgaris* (Hildenborough) were performed in the oxidised and reduced forms [36]. In both redox states the titrations were performed with addition of acid and base to ensure that there is no hysteresis. The results are plotted in Fig. 5 and show that the titration curves are different in the oxidised and reduced forms. The difference between the two curves is a measure of the number of protons coupled to the redox process showing that two protons participate in the redox-Bohr effect.

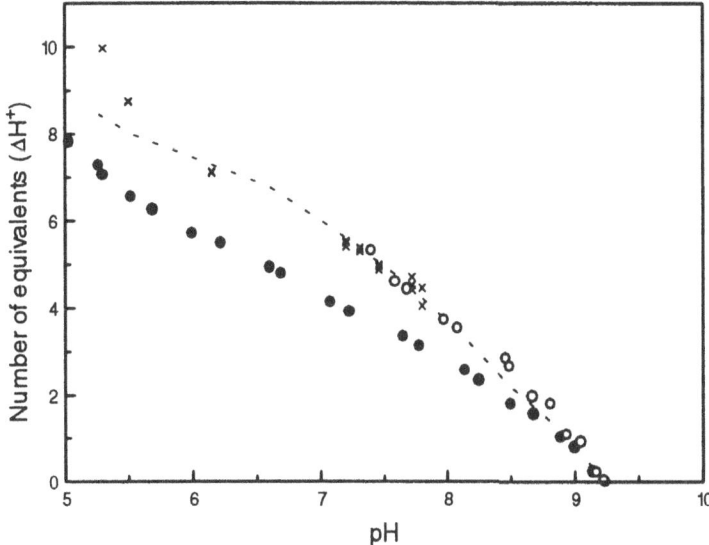

Figure 5. Potentiometric titrations of ferri- and ferro- cytc3 from *D. vulgaris* (Hildenborough) performed as previously described [36]. Data points (●) are from potentiometric titrations of ferricytochrome. In ferrocytochrome, points (O) were obtained from titrations and (X) from single point determinations. The dashed line was calculated for the participation of two protons in the redox-Bohr effect [36].

3. Functional Mechanism

3.1. THERMODYNAMIC BASIS FOR ENERGY TRANSDUCTION

The thermodynamic parameters discussed above are of utmost importance in the light of the role of cytc3 in the bioenergetic metabolism of *Desulfovibrio* cells. Indeed, they show that cytc3 has the necessary properties to perform electron to proton energy transduction by driving a one-way thermodynamic cycle where high-energy electrons (low redox potential) and low-energy protons (high pKa) received from the periplasmic hydrogenase, can be subsequently released as lower energy electrons for the reduction of sulphite, and higher energy protons for the production of ATP. Such a mechanism, is phenomenologically similar to that proposed for energy transduction performed by the transmembrane cytochrome c oxidases [39, 40], and amenable to an equivalent description through the cubane scheme proposed by Wyman [36, 41] (Fig. 6).

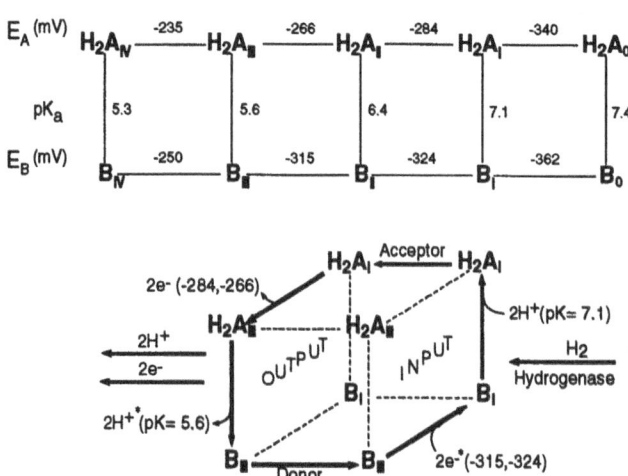

Figure 6. Thermodynamic and mechanistic bases for energy transduction by cytc3. The top diagram gives the values for the macroscopic pK$_a$'s for each of the oxidation stages and the midpoint redox potentials for the transitions between stages in the protonated and deprotonated protein. H$_2$A and B represent the acidic and basic states, respectively and Roman numbers indicate the oxidation stage of the cytochrome [36].The cubane diagram for the charge separation mechanism achieved by cycling between oxidation stages I (one haem oxidised molecule) and III (three haems oxidised molecule) is shown at the bottom. The functional pathway is drawn with arrows. Hydrogenase supplies two protons and two electrons to the cytochrome, which upon diffusion to the acceptor transfers the electrons. Due to the lowering of the pK$_a$ the protons can be released When the cytochrome encounters a new hydrogenase the cycle is resumed. The input and output faces of the cube are indicated to show the states where the electrons and protons are uptake (coupled to hydrogenase) and released (coupled to the transmembrane electron and proton carriers [42]). Energised protons and electrons are indicated with an asterisk.

3.2. PHYSIOLOGICAL IMPORTANCE OF THE ENERGY TRANSDUCTION MECHANISM

The physiological importance of these properties was demonstrated in a study of the pH dependence of the hydrogenase activity for hydrogen uptake [42]. It is known that, despite the fact that intrinsic activity of hydrogenase (as measured by proton-deuterium exchange) is optimum at pH 4.5-5.0 [43], the hydrogen uptake activity is maximum at pH 8.5, in agreement with the fact that increasing the proton concentration favours the reaction in the opposite direction. However, when oxidised cytc3 in buffered solutions is used as electron acceptor, an activity close to maximum is maintained at physiological pH (Fig. 7), thus showing that:

i) the concerted capture of two electrons and two protons enables cytc3 to modify the pH dependence of the hydrogenase activity for hydrogen uptake, maintaining a high rate at physiological pH.

ii) since the experiments were performed in buffered solutions the protons captured and energised by cytc3 are those that are produced upon hydrogen oxidation.

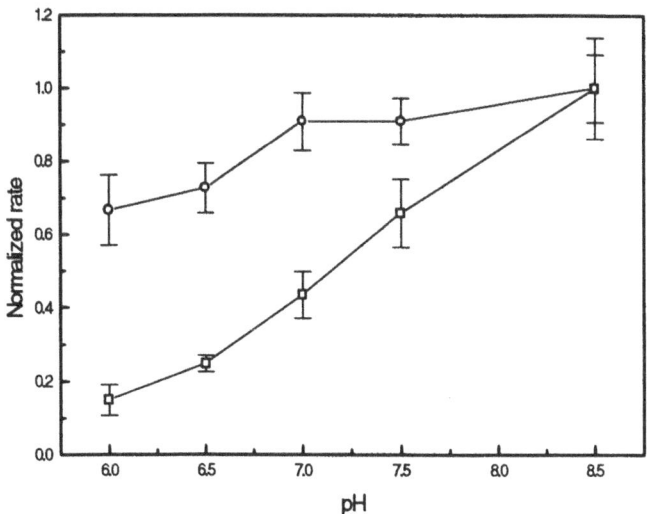

Figure 7. Dependence of hydrogenase activity for hydrogen uptake with pH, using either oxidised benzyl-viologen (□) or ferri-cytc3 (O) [42].

The physiological consequence of these findings is the demonstration of the role of cytc3 in maintaining a high rate of hydrogen uptake by hydrogenase at physiological pH, thus avoiding the leakage of the hydrogen generated in the cytoplasm through the oxidation of organic substrates. The efficient capture and oxidation of this hydrogen and subsequent use of the electrons for sulphite reduction and the protons for producing ATP is crucial to maintaining the metabolism of these organisms and thus sustaining cell growth.

Since both cytc3 and hydrogenase are soluble proteins present in the periplasmic space, the experimental results described above demonstrate the physiological importance of an energy transduction event performed in aqueous solution, in agreement with the Williams localised theory [44]. In the case of transmembrane '*proton pumps*', the electron to proton energy transduction is performed in a three phase system, with the energy of the electrons used to generate a proton gradient across a membrane, whereas the hydrogenase/cytc3 system works as a soluble (two phase) '*proton thruster*', pushing the protons forward in the metabolic pathway [42].

4. Structure/Function Studies

4.1. NMR SOLUTION STRUCTURE STUDIES OF THE REDUCED CYTOCHROME

To unravel the structural basis for these important functional properties, accurate structural models of the protein are required. Since all X-ray structures available were obtained in the oxidised state, information is lacking for comparison with the reduced state. Through the use of solution NMR techniques, the 3D structure has already been obtained for the reduced cytc3 from *D. vulgaris* (Hildenborough) [45], showing that the overall structure is maintained. However, the resolution attainable by NMR, using the currently available methods, may not be enough to observe small structural modifications.

4.2. NMR SOLUTION STRUCTURES OF THE OXIDISED CYTOCHROME

The use of paramagnetic probes in NMR to obtain structural information in proteins is an established technique but, with both theoretical and technological advances made in the field in recent years, there has also been success in the determination of 3D-structures of paramagnetic proteins [46]. The ideal situation would be to obtain a solution structure both of the oxidised and the reduced form of cytc3 by NMR. Then, the paramagnetic probes, which are intrinsic in oxidised haem proteins, will provide stringent geometrical information which should be able to detect the small structural changes which are expected on the basis of the fast intramolecular electron exchange and fast exchange between protonated and deprotonated forms [33].

Using inverse ^{1}H-^{13}C correlated experiments, it is possible to determine the paramagnetic ^{13}C shifts of haem substituents in positions β relative to the haem for natural abundance samples. These shifts can be analysed within the framework of a theoretical description of the electronic structure of the oxidised low-spin haem [30, 47], even when included in π delocalised systems as in the case of *b* type cytochromes [48]. This theoretical background allows the positions of the axial Fe ligands to be determined, and is a method of choice to characterise the active centres of oxidised low-spin haem proteins when complete structural information is either not available [30], or it is not feasible to obtain, as in the case of peroxidases [49]. In addition, the orientation of the magnetic y axis is defined, as expected from theory [47, 50]. By considering the simple aproximation that the z axis is normal to the haem plane, which is reasonable in the light of the data available in the literature [51], and using temperature corrected [52, 53] EPR g-values, the dipolar field generated by the unpaired electron can be described and the effect on neighbouring atoms calculated [34]. Once the magnetic axes are positioned, precise pseudocontact shifts can be calculated and, when compared with those observed for the molecule in different conditions, they can be used to detect subtle structural or chemical modifications.

Such a procedure has been tested with success in the case of oxidised cyt c3 from *D. vulgaris* (Hildenborough) [34]. The pseudocontact shifts calculated by this method and using the X-ray crystal structure were in good agreement with those observed for

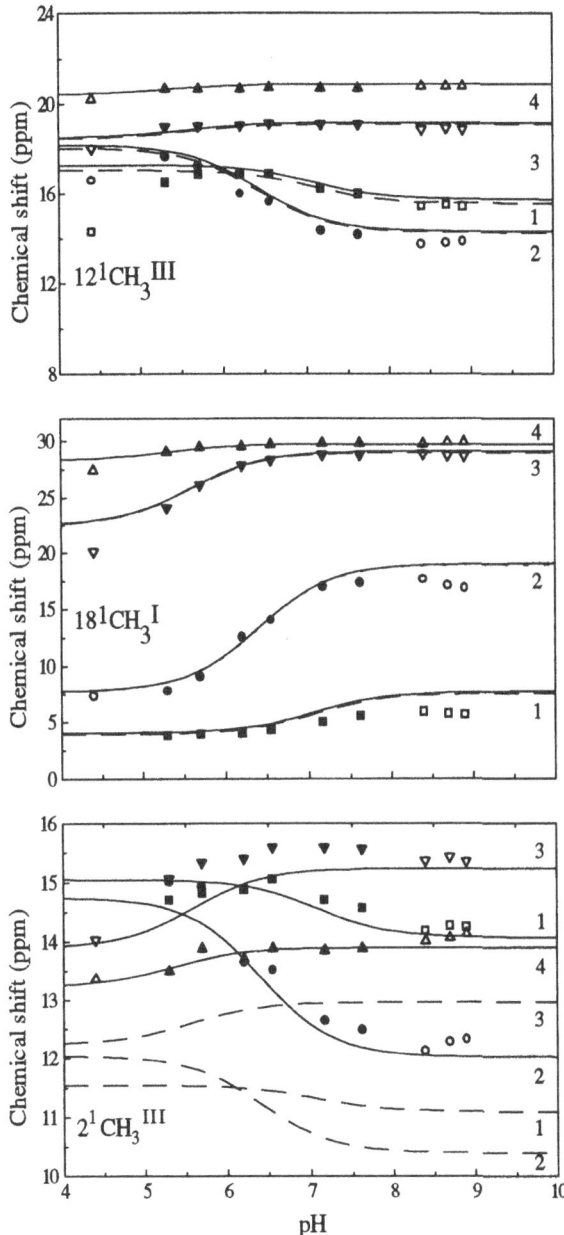

Figure 8. Chemical shift of the haem methyl resonances $12^1CH_3^{III}$, $18^1CH_3^{I}$ and $2^1CH_3^{III}$ in the oxidation stages 1 (■), 2 (●), 3 (▼) and 4 (▲). Dashed lines represent the predicted chemical shifts considering that the the fully oxidised shifts are purely intrinsic, while the continuous lines are the predictions considering the extrinsic paramagnetic shift contribution. Open symbols represent data point outside of the pH range used in the original fit of the NMR data [33]. Arabic numbers near to each line represent the stage of oxidation [34].

resonances of selected residues, and the complicated pH dependent redox pattern of some haem methyl resonances subject to pseudocontact shifts [31, 34] accurately described (Fig. 8).

Due to a favourable distribution of resonances in the NMR spectra, part of the structural basis for the redox-Bohr effect has indeed been unravelled. The resonances attributed to propionate 13^2 in the oxidised from were shown to have a pH dependent behaviour with a pKa similar to the redox-Bohr pKa [31, 33]. This observation confirms the earlier thermodynamic measurements reported in Table 1 showing that this effect is strongest for haem I [28, 33], which locates the ionisable centre near this haem.

4.3. THERMODYNAMIC AND STRUCTURAL BASIS FOR THE COOPERATIVITY

The thermodynamic and structural characterization of cytc3 from *D. vulgaris* (Hildenborough) enabled us to propose the mechanism and structural basis for the electron-electron and electron-proton (redox-Bohr) cooperativities.

Several conclusions can be drawn:

1) Due to the fast intramolecular electron exchange, the system is under thermodynamic control.

2) The two intermediate electrons are received or donated in a concerted step.

3) At physiological pH two protons in fast exchange, assist the uptake and release of the electrons through the redox-Bohr effect (Fig. 6).

Putting this information together with that obtained from the crystal structure it is proposed that the structural motif for the two-proton redox-Bohr effect is based on the architecture of haem 1 ionisable propionic substituents and their interaction with neighbouring residues: propionate 13^2 carboxylate makes a hydrogen bond with the peptidyl NH of C46 and the carboxylate of propionate 17^2 is stabilized by the positive charge of K45. Uptake of a proton by propionate 13^2 breaks the H-bond, moving away the positive charge of K45 and enabling the simultaneous protonation of propionate 17^2. Furthermore, since C46 is covalently bound to haem 2, rearrangements of this region can also be responsible for the positive redox cooperativity between haems I and II.

Future work, namely site-directed mutagenesis [54], will provide confirmation or rebuttal of the hypothesis presented.

5. References

1. LeGall, J. and Peck, Jr., H.D. (1987) Amino-terminal amino acid sequences of electron transfer proteins from Gram-negative bacteria as indicators fot their cellular localization: the sulfate-reducing bacteria, *FEMS Microbiol. Rev.*, **46**, 35-40.

2. Haser, R., Pierrot, M., Frey, M., Payan, F., Astier, J.-P., Bruschi, M. and LeGall, J. (1979) Structure and sequence of multiheme cytochrome c_3, *Nature*, **282**, 806-810.

3. Higuschi, Y., Kusonoki, M., Matsuura, Y., Yasuoka, N. and Kakudo, M. (1984) Refined structure of cytochrome c_3 at 1.8 Å resolution, *J. Biol. Chem.*, **172**, 109-139.

4. Morimoto, Y., Tani, T., Okumura, H., Higuschi, Y. and Yasuoka, N. (1991) Effects of aminoacid substitution on three-dimensional structure: an X-ray analysis of cytochrome c_3 from *Desulfovibrio vulgaris* at 2 Å resolution, *J. Biochem. (Tokyo)*, **110**, 532-540.

5. Coutinho I.B., Turner, D.L., LeGall, J., Xavier, A.V. (1992) Revision of the haem-core architecture in the tetrahaem cytochrome c_3 from *Desulfovibrio baculatus* by two-dimensional ^{1}H NMR, *Eur.J. Biochem.*, **209**, 329-333.

6. Matias, P.M., Frazão, C., Morais, J., Coll, M., and Carrondo, M.A. (1993) Structure analysis of cytochrome c_3 from *Desulfovibrio vulgaris* Hildenborough at 1.9Å resolution, *J. Mol. Biol.*, **234**, 680-699.

7. Czjzek, M., Payan, F., Guerlesquin, F., Bruschi, M. and Haser, R. (1994) Crystal structure of cytochrome c_3 from *Desulfovibrio desulfuricans* Norway at 1.7 Å resolution, *J. Mol Biol*, **243**, 653-667.

8. Morais, J., Palma, P.N., Frazão, C., Caldeira, J., LeGall, J., Moura, I., Moura, J.J.G. and Carrondo, M.A. (1995) Structure of the tetraheme cytochrome from *Desulfovibrio desulfuricans* ATCC 27774: X-ray diffraction and electron paramagnetic resonance studies, *Biochemistry*, **34**, 12830-12841.

9. Matias, P.M., Morais, J., Coelho, R., Carrondo, M.A., Wilson, K., Dauter, Z., Sieker, L. (1996) Cytochrome c_3 from *Desulfovobrio gigas*: crystal structure at 1.8 A resolution and evidence for a specific calcium-binding site, *Protein Sci.*, **5**, 1342-1354.

10. Tsuji, K., Yagi, T. (1980) Significance of hydrogen burst from growing cultures of *Desulfovibrio vulgaris* Miyazaki, and the role of hydrogenase and cytochrome c_3 in energy production system, *Arch. Microbiol.*, **125**, 35-42.

11. Yagi, T.,Honya, M. and Tamiya N. (1968) Purification and properties of hydrogenases of different orogins, *Biochim. Biophys. Acta* **153**, 699-705.

12. Peck Jr., H.D. (1960) Evidence for oxidative phosphorylation during reduction of sulfate with hydrogen by *Desulfovibrio desulfuricans*, *J. Biol. Chem.*, **235**, 2734-2738.

13. Peck Jr, H.D. (1966) Phosphorylation coupled with electron transfer in extracts of the sulfate reducing bacterium *Desulfovibrio gigas*, *Biochem. Biophys. Res. Commun.* **22**, 112-118.

14. Odom, J.M. and Peck Jr., H.D. (1981) Hydrogen cycling as a general mechanism for energy coupling in the sulfate-reducing bacteria *Desulfovibrio sp.*, *FEMS Microbiol. Lett.* **12**, 47-50.

15. Badziong, W., Thauer, R.K. and Zeikus, J.G. (1978) Isolation and characterization of *Desulfovibrio* growing on hydrogen plus sulfate as the sole energy source, *Arch. Microbiol.* **116**, 41-49.

16. Badziong, W., and Thauer, R.K. (1980) Vectorial electron transport in *Desulfovibrio vulgaris* (Marburg) growing on hydrogen plus sulfate as sole energy source, *Arch. Microbiol.*, **125**, 167-174.

17. Dobson, C.M., Hoyle, N.J., Geraldes, C.F., Bruschi, M.,LeGall, J., Wright, P.E. and Williams, R.J.P. (1974) Outline structure of cytochrome c_3 and consideration of its properties, *Nature*, **249**, 425-429.

18. Moura, J.J.G., Santos, H., Moura, I., LeGall, J., Moore, G.R., Williams, R.J.P. and Xavier, A.V. (1982) NMR redox studies of *Desulfovibrio vulgaris* cytochrome c_3, electron transfer mechanisms *Eur. J. Biochem.* **127**, 151-155.

19. Santos, H., Moura, J.J.G., Moura, I., LeGall, J. and Xavier, A.V. (1984) NMR studies of electron transfer mechanisms in a protein with interacting redox centres: *Desulfovibrio gigas* cytochrome c_3, *Eur. J. Biochem.*, **141**, 283-296.

20. Fan, K., Akutsu, H., Kyoguku, Y. and Niki, K. (1990) Estimation of microscopic redox potentials of a tetraheme protein cytochrome c_3 of *Desulfovibrio vulgaris*, Miyazaki F, and partial assignments of heme groups, *Biochemistry*, **29**, 2257-2263.

21. Coletta M., Catarino, T., LeGall, J. and Xavier, A.V. (1991) A thermodynamic model for the cooperative functional properties of the tetraheme cytochrome c_3 from *Desulfovibrio gigas*, *Eur. J. Biochem.*, **202**, 1101-1106.

22. Park, J.-S., Kano, K., Niki, K. and Akutsu, H. (1991) Full assignment of heme redox potentials of cytochrome c_3 of *Desulfovibrio vulgaris*, Miyazaki F by ^{1}H-NMR, *FEBS Lett.*, **285**, 149-151.

23. Turner, D.L., Salgueiro, C.A., LeGall, J.and Xavier, A.V. (1992) Structural studies of *Desulfovibrio vulgaris* ferrocytochrome c_3 by two-dimensional NMR, *Eur. J. Biochem.*, **210**, 931-936.

24. Salgueiro, C.A., Turner, D.L., Santos, H., LeGall, J.and Xavier, A.V. (1992) Assignment of the redox potentials to the four haems in *Desulfovibrio vulgaris* cytochrome c_3 by 2D-NMR, *FEBS Lett.*, **314**, 155-158.

25. Sola, M. and Cowan, J.A. (1992) Assignment of heme resonances in the ^{1}H NMR spectrum of oxidized *Desulfovibrio vulgaris* (Hildenborough) cytochrome c_3, *Inorg. Chim. Acta*, **202**, 241-251.

26. Coutinho, I.B.,Turner, D.L., LeGall, J. and Xavier, A.V. (1993) Characterization of the structure and redox behaviour of cytochrome c_3 from *Desulfovibrio baculatus* by [1]H- nuclear-magnetic-resonance spectroscopy, *Biochem J.*, **294**, 899-908.

27. Piçarra-Pereira, M.A., Turner, D.L., LeGall, J. and Xavier, A.V. (1993) Structural studies on *Desulfovibrio gigas* cytochrome c_3 by two-dimensional [1]H-nuclear-magnetic-resonance spectroscopy, *Biochem. J.*, **294**, 909-915.

28. Turner, D.L., Salgueiro, C.A., Catarino, T., LeGall, J. and Xavier, A.V. (1994) Homotropic and heterotropic cooperativity in the tetrahaem cytochrome c_3 from *Desulfovibrio vulgaris*, *Biochim. Biophys. Acta*, **1187**, 232-235.

29. Coutinho, I.B.,Turner, D.L., LeGall, J. and Xavier, A.V. (1995) NMR studies and redox titration of the tetraheme cytochrome c_3 from *Desulfomicrobium baculatum*, *Eur. J. Biochem.*, **230**, 1007-1013.

30. Turner, D.L., Salgueiro, C.A., Schenkels, P., LeGall, J. and Xavier, A.V. (1995) Carbon-13 NMR studies of the influence of axial ligand orientation on haem electronic structure, *Biochim. Biophys. Acta*, **1246**, 24-28.

31. Park, J.-S., Ohmura, T., Kano, K., Sagara, T., Niki, K., Kyogoku, Y. and Akutsu, H. (1996) Regulation of the redox order of four hemes by pH in cytochrome c_3 from *D. vulgaris* Miyazaki F, *Biochim. Biophys. Acta*, **1293**, 45-54.

32. Louro, R.O., Pacheco, I.P., Turner, D.L., LeGall, J. and Xavier, A.V. (1996) Structural and functional characterization of cytochrome c_3 from *D. desulfuricans* ATCC 27774 by ^{1}H-NMR, *FEBS Letts.*, **390**, 59-62.

33. Turner, D.L., Salgueiro, C.A., Catarino, T., LeGall, J. and Xavier, A.V. (1996) NMR studies of cooperativity in the tetrahaem cytochrome c_3 from *Desulfovibrio vulgaris*, *Eur. J. Biochem.*, **241**, 232-235.

34. Salgueiro, C.A., Turner, D.L. and Xavier, A.V. (1997) Use of paramagnetic NMR probes for structural analysis in cytochrome c_3 from *Desulfovibrio vulgaris*, *Eur. J. Biochem.*, **244**, 721-734.

35. Papa, S., Guerrieri, F. and Izzo, G. (1979) Redox Bohr-effects in the Cytochrome System of Mitochondria, *FEBS Lett.*, **105**, 213-216.

36. Louro, R.O., Catarino, T., Salgueiro, C.A., LeGall, J. and Xavier, A.V. (1996) Redox-Bohr effect in the tetrahaem cytochrome c_3 from *Desulfovibrio vulgaris*: a model for energy transduction mechanisms *JBIC*, **1**, 34-38.

37. Bohr, C., Hasselbach, K.A. and Krogh, A. (1904) Ueber einen in biologischer Beziehung wichtigen Einfluss, den die Kohllensäurespannung des Blutes auf dessen Sauerstoffbindung übt, *Skand. Arch. Physiol.*, **16**, 402-412.

38. Xavier, A.V., Turner, D.L. and Santos, H. (1993) Two-dimensional nuclear magnetic resonance of paramagnetic metalloproteins, *Methods Enzymol.*, **227**, 1-16.

39. Babcock, G.T. and Wikström M. (1992) Oxygen activation and the conservation of energy in cell respiration, *Nature* **356**, 301-309.

40. Ädelroth P., Sigurdson H., Hallén S. and Brzezinski P. (1996) Kinetic coupling between electron and proton transfer in cytochrome c oxidase: simultaneous measurements of conductance and absorbance changes, *Proc. Nat. Acad. Sci.*, **93**, 12292-12297.

41. Wyman, J. (1975) The turning wheel: a study in steady states, *Proc. Natl. Acad. Sci. USA* **72**, 3983-3987.

42. Louro, R.O., Catarino, T., LeGall, J. and Xavier, A.V. (1997) Redox-Bohr Effect in Electron/Proton Energy Transduction: Cytochrome c_3 Coupled to Hydrogenase Works as a 'Proton Thruster' in *Desulfovibrio vulgaris.*, *JBIC*, **2**, 488-491.

43. Fauque, G.D., Berlier, Y.M., Czechowski, M.H., Dimon, B., Lespinat, P.A. and LeGall, J. (1987) A proton-deuterium exchange study of three types of *Desulfovibrio* hydrogenases *J. Ind. Microbiol.*, **1**, 1-9.

44. Williams, R.J.P. (1978) The history and the hypotheses concerning ATP-formation by energised protons, *FEBS Lett.* **85**, 9-19.

45. Messias, A.C., Kastrau, D.H.W., Costa, H.S., LeGall, J., Banci, L., Turner, D.L., Santos, H. and Xavier, A.V. (1996) Solution Structure of *Desulfovibrio vulgaris* (Hildenborough) ferrocytochrome c_3 by 2D ^{1}H-NMR and distance geometry calculations. *Proceedings of the XVII Conference on Magnetic Resonance in Biological Systems, Keystone, Colorado, USA.*

46. Banci, L and Pierattelli, R. (1995) 3D structure of HIPIP's in solution through NMR and molecular dynamics studies, in G.N. La Mar (ed.), *Nuclear magnetic resonance of paramagnetic molecules*, Kluwer Academic Publishers, Dordrecht, pp. 281-296.

47. Turner, D.L. (1995) Determination of haem electronic structure in His-Met cytochromes c by ^{13}C NMR. The effect of the axial ligands, *Eur. J. Biochem.*, **227**, 829-837.

48. Banci, L., Pierattelli, R. and Turner, D.L. (1995) Determination of haem electronic structure in cytochrome b_5 and metcyanomyoglobin, *Eur. J. Biochem.*, **232**, 522-527.

49. Pierattelli, R., Banci, L. and Turner, D.L. (1996) Indirect determination of magnetic susceptibility tensors in peroxidases: a novel approach to structure elucidation by NMR, *JBIC*, **1**, 320-329.

50. Oosterhuis, W.T. and Lang, G. (1969) Mössbauer effect in $K_3Fe(CN)_6$, *Phys. Rev.*, **178**, 439-456.

51. Moore, G.R. and Pettigrew, G.W. (1990) *Cytochromes c: Evolutionary, Structural and physicochemical aspects*, Springer-Verlag Berlin.
52. Kurland, R.J. and McGarvey, B.R.J. (1970) Isotropic NMR shifts in transition metal complexes: the calculation of the Fermi contact and pseudocontact terms, *J. Magn. Reson.*, **2**, 286-301.
53. Horrocks, W. and Greenberg, E.S. (1973) Evaluation of dipolar nuclear magnetic resonance shifts in low-spin hemin systems: ferricytochrome *c* and metmyoglobin cyanide, *Biochim. Biophys. Acta*, **322**, 38-44.
54. Saraiva, L.M., Salgueiro, C.A., LeGall, J., van Dongen, W.M.A.M. and Xavier, A.V. (1996) Site-directed mutagenesis of a phenylalanine residue strictly conserved in cytochromes c_3, *JBIC*, **1**, 542-550.

THE SOLUTION STRUCTURE OF REDOX PROTEINS AND BEYOND

L. BIANCI, I. BERTINI AND P. TURANO
Department of Chemistry, University of Florence,
Via Gino Capponi 7, Florence, Italy

and

C. LUCHINAT
Department of Soil Science and Plant Nutrition, University of Florence
P. le delle Cascine 28, Florence, Italy

1. Solution Structures of Paramagnetic Metalloproteins

The first solution structure determined by NMR has been solved by Wüthrich and coworkers [1] in 1984. Since that time , the number of solution structures of proteins determined by NMR has grown very fast [2-5]. However, proteins containing paramagnetic metal ions represent an exception due to the experimental difficulties connected to the presence of one or more paramagnetic centres. The effect of the coupling between the spins of the unpaired electron(s) and the nuclei cause broadening of the NMR signals making there detection more difficult [6-9]. Moreover, the strong reduction in the relaxation times of the nuclear spins represents a severe drawback with respect to the observation of nucleus-nucleus connectivities [9-11]. Only ten years later the first solution structure of a paramagnetic metalloprotein has been published by our group [12]. The way that allowed us to reach this goal went first of all through the optimization of pulse sequences to detect dipolar connectivities between protons interacting with the paramagnetic centre. These pulse sequences included 1D NOEs experiments with special tricks to avoid the strong off-resonance effects arising from the high power level needed for the saturation of fast relaxing resonances [13,14]; 2D NOESY optimized to detect cross-peaks involving fast relaxing signals (in particular the mixing time should be chosen on the basis of the T_1 values of the two signals between which the cross peak is expected [10]); SUWEFT-NOESY, i.e. NOESY experiments preceded by a WEFT pulse sequence to eliminate the resonances from the diamagnetic envelope thus allowing the detection of cross peaks between fast relaxing resonances whose shifts fall (accidentally) in the diamagnetic region; NOE-NOESY experiments are also feasible thanks to the presence of well-resolved hyperfine shifted resonances that can easily be saturated [15]. The latter pulse sequence is useful to selectively observe cross-peaks between protons dipolarly coupled with the saturated one. These experiments allow us to exhaustively assign the NMR spectrum and to obtain a

G.W. Canters and E. Vijgenboom (eds.),
Biological Electron Transfer Chains: Genetics, Composition and Mode of Operation, 225-238.
© 1998 *Kluwer Academic Publishers.*

reasonable number of interproton distance constraints also for those residues close to the paramagnetic centre, and therefore to the determine the NMR solution structure to a quite refined level. Still, the number of NOEs may be smaller than in the analogous diamagnetic molecules, especially around the metal centre.

However, thanks to the nature of the hyperfine coupling between the resonating nucleus and the unpaired electron(s), one can take advantage of the presence of the paramagnetic centre to obtain new structural constraints. The most obvious of these constraints is derived from non-selective protons T_1 measurements. The nuclear relaxation rate enhancement (ρ_i^{para}) due to the presence of a paramagnetic centre in a metalloprotein can be expressed by the following equation [7,16]:

$$\rho_i^{para} = \frac{2}{15}(\frac{\mu_0}{4\pi})\frac{\gamma_I^2 g_e^2 \mu_B^2 S(S+1)}{r_i^6}(\frac{7\tau_s}{1+\omega_s^2\tau_s^2}+\frac{3\tau_s}{1+\omega_I^2\tau_s^2}) \qquad (1)$$

where γ_I is the nuclear magnetogiric ratio, ω_I is the nuclear Larmor frequency, r_i is the metal to nucleus distance, τ_s is the correlation time for the electron spin relaxation, and all other symbols have their usual meaning.

According to the above reported equation (1) the metal-to-nucleus distance can be derived from the T_1 values, once the diamagnetic contribution to the relaxation rate is known. In practice the latter term is always of the order of 4-5 s^{-1} [12]. As an upper proton-proton distance limit is obtainable from the extent of the NOE effect [17,18], from T_1 measurements an upper limit for the nucleus-metal distance can be evaluated from equation (1). In both cases the effect depends on the reciprocal of the sixth power of the distance.

These T_1 constraints have been successfully applied to refine the solution structures of some iron-sulfur proteins that are characterized by T_1 values of the order of 20 ms for protons at about 5 Å from the Fe-S cluster [19,20].

The dipolar coupling between a proton and the unpaired electron also yields a contribution to the hyperfine shift in solution, which is called pseudocontact shift [9,21,22]:

$$\delta_i^{pc} = \frac{1}{12\pi r_i^3}[\Delta\chi_{ax}(3n_i^2-1)+\frac{3}{2}\Delta\chi_{rh}(l_i^2-m_i^2)] \qquad (2)$$

where $\Delta\chi_{ax}$ and $\Delta\chi_{rh}$ are the axial and rhombic anisotropy of the magnetic susceptibility induced by the paramagnetic ion, respectively, r_i is defined as in equation (1), and l_i, m_i, and n_i are the direction cosines of the position vector of atom i with respect to the orthogonal reference system formed by the principle axes of the magnetic susceptibility tensor.

In pseudocontact shifts the dependence on the metal-to-proton distance is longer range with respect to that of relaxation rates (as it decreases with the reciprocal of the third power instead of the sixth) and therefore more effective also at larger distances from the metal ion. The nature of the structural constraints is complicated by the angular dependence shown in equation (2) and from the fact that the extent of the magnetic susceptibility anisotropy should be determined independently. This type of constraints has been successfully used to refine the solution structure of cytochromes, i.e. of systems containing a low spin iron(III) characterized by a sizable magnetic anisotropy [23-26].

For systems with negligible pseudocontact shift contribution to the hyperfine shift, or where it has been possible to factorize out the pseudocontact contribution, the contact term can also be used as further structural constraint. For instance, in $[Fe_4S_4]^{2+}$-containing iron-sulfur proteins if has been found that the contact hyperfine shift values for the β-CH_2 of the coordinated cysteines depends on the dihedral angle between the plane formed by the S_γ, the C_β and the observed $C\alpha$ or H_β nuclei and the C_β-S_γ-Fe plane [27,28]. The type of angular dependence is not known a priori and can only be determined when several examples for a certain system are already available. However, once this kind of dependence is found, it could also be used as further structural constraints in new proteins containing the same prosthetic group. Recently, the availability of the solution structure and of the magnetic anisotropy tensor for three c-type cytochromes allowed us to find a similar angular dependence for the contact shift of the α-CH_2 protons of propionates (unpublished results from this lab).

The above described non-conventional structural constraints are important for structure determination from two different points of view: i) they allow us to safely position the metal ion within the protein frame; ii) they provide a better definition of the residues around the paramagnetic centre, where the number of observed NOEs is usually lower.

2. Electron Transfer Proteins in Solution

The possibility to solve the solution structure of paramagnetic systems allowed us to structurally characterize electron transfer proteins in both the oxidation states relevant for their biological function.

The first electron transfer proteins we studied are the HiPIPs from *Ectothiorhodospira halophila* and *Chromatium vinosum* [12,19,29-32]. a pairwise comparison of the solution structures in the two redox states is reported in Tables 1 and 2, together with a comparison with the available crystal structures [33,34]. For both proteins the RMSD values between the reduced and oxidized form indicate that the protein conformation is not exactly the same. However, it was not possible to identify detailed conformational differences well beyond the determination of the structure definition. The major difference found between the two redox forms is related to the structural stability towards denaturing agents, the oxidized protein being less stable than the reduced one [35].

TABLE 1. Pairwise RMSD values (Å) for the backbone (above the diagonal) and all heavy atoms (below the diagonal) for the solution structures of oxidized and reduced HiPIP from *Ectothiorhodospira halophila* refined by restrained molecular dynamics in water (RMDw), and for the two independent molecules A and B observed in the X-ray structure [13].

	X-ray A	X-ray B	RMDw ox	RMDw red
X-ray A	-	0.50	0.97	0.72
X-ray B	1.18	-	0.87	0.67
RMDw ox	1.68	1.66	-	0.77
RMDw red	1.53	1.46	1.37	-

TABLE 2. Pairwise RMSD values (Å) for the backbone (above the diagonal) and all heavy atoms (below the diagonal) for the solution structures of oxidized and reduced HiPIP from *Chromatium vinosum* refined by restrained molecular dynamics in water (RMDw) [31,32], and for the X-ray structure of the oxidized protein [34].

	X-ray ox	RMDw ox	RMDw red
X-ray ox	-	0.76	0.61
RMDw ox	1.25	-	0.65
RMDw red	1.19	1.14	-

The cytochromes, at least apparently, have a different behaviour. Both X-ray on the crystal and NMR in solution have pointed out several small structural rearrangements upon oxidation. We have solved the solution structure of three c-type cytochromes [24,25,36-38]. In Table 3-5 the constraints used to determine the solution structures of the oxidized forms of *Saccharomyces cerevisiae* and horse heart cytochrome c, and of *Monoraphidium braunii* cytochrome c_6 are reported together with some statistics on the resulting structures. Analogous information for the reduced forms of the same three cytochromes is reported in Tables 6-8. All the structures appear relatively well defined in both oxidation states. In the oxidized paramagnetic states the definition of the haem substituents and surrounding residues is comparable with that obtained in the reduced diamagnetic form.

TABLE 3. Constraints and final data relative to *Saccharomyces cerevisiae* iso-1-ferricytochrome *c* [25]

Number of amino acids	108
Metal cofactor	one *c*-type haem
Number of assigned amino acids	105
% assignment of ^{1}H resonances	77
Type of constraint:	
2D NOESY	1671 (1356)
1D NOE	5
Pseudocontact shifts	117
H-bonds	14
Refinement procedure	PSEUDO-REM*
Number of structures	20
RMSD (Å)	backbone: 0.58±0.08
(from the mean structure, residues 1-102)	heavy atoms: 1.05±0.10
NOEs target function (Å^2)	≤ 0.87
δ_{pc} target function (Å^2)	≤ 3.42

* Restrained Energy Minimization with the inclusion of pseudocontact shifts

TABLE 4. Constraints and final data relative to horse heart ferricytochrome *c* [24].

Number of amino acids	103
Metal cofactor	one *c*-type haem
Number of assigned amino acids	103
% assignment of ^{1}H resonances	75
Type of constraint:	
2D NOESY	2250 (1729)
1D NOE	28
Pseudocontact shifts	241
H-bonds	14
Refinement procedure	PSEUDO-REM
Number of structures	35
RMSD (Å)	backbone: 0.70±0.11
(from the mean structure, residues 5-100)	heavy atoms: 1.21±0.14
NOEs target function (Å^2)	0.97 (average value)
δ_{pc} target function (Å^2)	0.20 (average value)

TABLE 5. Constraints and final data relative to *Monoraphidium braunii* ferricytochrome c_6 [37].

Number of amino acids	89
Metal cofactor	one *c*-type haem
Number of assigned amino acids	88
% assignment of ^{1}H resonances	84
Type of constraint:	
2D NOESY	1657 (1100)
1D NOE	11
Pseudocontact shifts	288
H-bonds	-
Refinement procedure	PSEUDO-REM
Number of structures	40
RMSD (Å)	backbone: 0.67±0.09
(from the mean structure, residues 3-87)	heavy atoms: 1.95±0.11
NOEs target function (Å^2)	n.r.
δ_{pc} target function (Å^2)	n.r.

TABLE 6. Constraints and final data relative to *Saccharomyces cerevisiae* iso-1-ferricytochrome *c* [36].

Number of amino acids	108
Metal cofactor	one *c*-type haem
Number of assigned amino acids	105
% assignment of ^{1}H resonances	77
Type of constraint:	
2D NOESY	1702 (1442)
1D NOE	-
Pseudocontact shifts	-
H-bonds	14
Refinement procedure	DG[a]
Number of structures	20
RMSD (Å)	backbone: 0.61±0.09
(from the mean structure, residues 6-100)	heavy atoms: 0.98±0.09
NOEs target function (Å^2)	≤ 0.82
δ_{pc} target function (Å^2)	-

[a] Distance Geometry calculations

TABLE 7. Constraints and final data relative to horse heart ferrocytochrome c [39].

Number of amino acids	103
Metal cofactor	one c-type haem
Number of assigned amino acids	-
% assignment of ^{1}H resonances	81
Type of constraint:	
2D NOESY	1940
1D NOE	-
Pseudocontact shifts	-
H-bonds	26
Refinement procedure	SA_{rw}[a]
Number of structures	44
RMSD (Å)	backbone: 0.47±0.09
(from the mean structure, residues 5-100)	heavy atoms: 0.91±0.07
NOEs target function (Å^2)	n.r.
δ_{pc} target function (Å^2)	-

[a] Restrained Simulated Annealing with the inclusion of water molecules.

TABLE 8. Constraints and final data relative to *Monoraphidium braunii* ferrocytochrome c_6 [38].

Number of amino acids	89
Metal cofactor	one c-type haem
Number of assigned amino acids	89
% assignment of ^{1}H resonances	96
Type of constraint:	
2D NOESY	1776 (1278)
1D NOE	-
Pseudocontact shifts	-
H-bonds	15
Refinement procedure	RMD[a]
Number of structures	20
RMSD (Å)	backbone: 0.34±0.06
(from the mean structure, residues 5-100)	heavy atoms: 0.67±0.06
NOEs target function (Å^2)	≤ 0.54
δ_{pc} target function (Å^2)	-

[a] Restrained Molecular Dynamics

For each cytochrome the average solution structures obtained for the oxidized form have been compared with the corresponding average solution structure of the reduced protein and with all the available crystal structures. The results are summarized in Tables 9-11. Five structures for *S. cerevisiae* iso-1-cytochrome *c* are now available; these are the solution structures of the reduced [36] and oxidized [25] proteins, the X-ray crystal structures of reduced [40] and oxidized [41] uncomplexed species, and the X-ray structure of the oxidized protein in the cytochrome *c*-cytochrome *c* peroxidase complex [42] (Table 9). The extremely low RMSD values for the backbone atoms of the reduced and oxidized cytochrome *c* in the crystal clearly point out that no differences in the folding are observed. The slightly larger RMSD values for the side chains can be ascribed to the following small but probably meaningful differences: haem distortion; rearrangement of the complicated H-bond network involving propionate-7, Gly41, Asn52, Trp59, Tyr67 and a water molecule on the distal side of the haem; reorientation of the proximal histidine side chain. Within the resolution of the NMR data the rearrangement of the side chain of propionate-7 upon oxidation can be detected also in solution. A change in conformation of propionate-7 upon oxidation has also been observed for the solution structures of the two other cytochromes. The movement of this group is not the same in the three proteins, but is always clearly detected (Figure 1). In cytochrome c_6 from *M. braunii* the movement of propionate-7 is accompanied by a rearrangement of the ring of His30: in the oxidized solution structure the Oδ1 of the carboxylate of propionate-7 forms a hydrogen bond with the Hδ1 of the histidine [37]; in the reduced solution structure both these groups have a different conformation but the hydrogen bond is still maintained, even if with Oδ2 of propionate-7 [38]. The structural features of the reduced form in the crystal [43] are intermediate between those observed in the two solution structures: propionate-7 has the same conformation as in the reduced solution structure., the His30 ring has the conformation observed in the oxidized solution structure. However, as a consequence of the increased distance between the two groups the hydrogen bond is absent.

TABLE 9. RMSD values (Å) for the backbone (above the diagonal) and all heavy atoms (below the diagonal) of the energy minimized average solution structures of the reduced and oxidized *Saccharomyces cerevisiae* iso-1-cytochrome *c* relative to all the other available structures of the same protein [25,36,40-42].

	sol ox	sol red	X-ray ox	X-ray red	cyt *c*-CcP
sol ox	-	0.97	0.88	0.86	0.80
sol red	1.72	-	0.88	0.87	0.90
X-ray ox	1.56	1.68	-	0.28	0.44
X-ray red	1.52	1.62	0.70	-	0.40
cyt *c*-Ccp	2.25	2.46	2.07	2.00	-

TABLE 10. RMSD values (Å) for the backbone (above the diagonal) and all heavy atoms (below the diagonal) of the energy minimized average solution structures of the reduced and oxidized horse heart cytochrome c relative to the solution structure of the reduced form and the crystal structure of the oxidized form [24,39,40].

	sol ox	sol red	X-ray ox
sol ox	-	1.85	1.51
sol rad	2.81	-	2.25
X-ray ox	2.33	3.18	-

TABLE 11. RMSD values (Å) for the backbone (above the diagonal) and all heavy atoms (below the diagonal) of the energy minimized average solution structures of the reduced and oxidized cytochrome c_6 from *Monoraphidium braunii* relative to the solution and crystal structures of the reduced form [37,38,43].

	sol ox	sol red	X-ray ox
sol ox	-	0.87	1.14
sol rad	1.46	-	0.89
X-ray ox	1.66	1.38	-

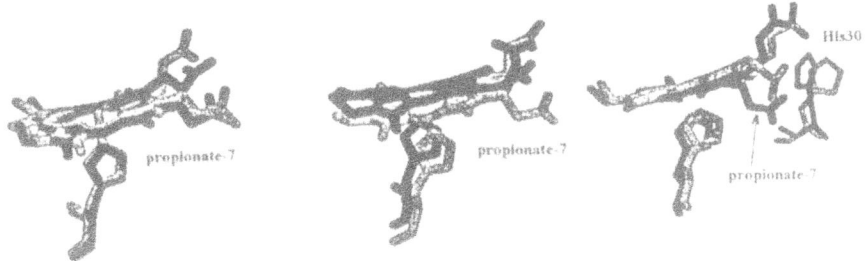

Figure 1. The conformation of propionate-7 in the reduced and oxidized forms of *S. cerevisiae* iso-1-cytochrome c (left), horse heart cytochrome c (middle), and *M. braunii* cytochrome c_6. The reduced form is represented in black, the oxidized form in grey.

Beside the movement of propionate-7 and surrounding residues other differences are observed. For horse heart cytochrome c the most noticeable difference appears the rotation of about 17° of the His18 plane [24], however the functional relevance of a movement of this type is not clear at the moment. More interesting appears the case of cytochrome c_6, where structural rearrangements for some side chains upon oxidation have been observed in the acidic patch involving residues 69-72 [37], which have been proposed to be involved in the interaction with redox partners [44,45]. This region also shows a larger mobility in the oxidized protein [37].

3. What is Beyond the Structure?

Are the above described differences functionally meaningful?

For yeast cytochrome c it has been suggested that the Gly41/propionate-7 hydrogen bond is part of a direct tunnelling pathway that is responsible for the relatively strong coupling between the haem and a Ru complex bonded to His39 in the protein [46]. So, it seems reasonable to assume that this propionate adjusts itself to the best conformation to favour the electron uptake and release in the various proteins. The changes in the His30/propionate-7 region in $M.$ $braunii$ cytochrome c_6 have also been proposed to be relevant for protein-protein recognition [37]. Kinetic analyses suggested that this cytochrome forms electrostatic transient complexes with redox partners [44,45]. So, it is tempting to propose that the structural differences between the two redox forms in solution may at least partially account for a change in affinity for the redox partners. However, how can we explain the fact that in cytochrome c_6 the conformation of His30 in the solid state of the reduced form is similar to that of the oxidized solution structure?

Crystallization of proteins requires protein-protein interactions. This could somehow affect the conformation of surface residues with respect to the "free" state. Therefore, the differences between the solution and the X-ray structures could arise from the protein-protein interactions in the crystal. Indeed, in solution this kind of interactions is reduced because of dilution. A good example for this effect could be taken from the behaviour of Arg13 and Gln16 in $S.$ $cerevisiae$ cytochrome c: from Figure 2 it clearly appears that the crystal packing effects force these two side chains to assume a conformation different from that experimentally found in solution by NMR and obtained through MD simulations in water starting from the X-ray structures.

However, some conformational effects from the interactions with the buffer cannot be ruled out. A good example for the above consideration is represented by cytochrome c where increasing the ionic strength of the solution produces a stabilization of some of the possible conformations for certain a number of residues (unpublished results of this lab).

The energy associated to a change in one or more torsional angles that accompanies the reorientation of a surface residue may be very small with respect to kT. However, this conformational change is not an isolated event but is somehow transmitted into the interior of the protein through a series of small readjustments of several torsion angles. All together these movements can account for an energy

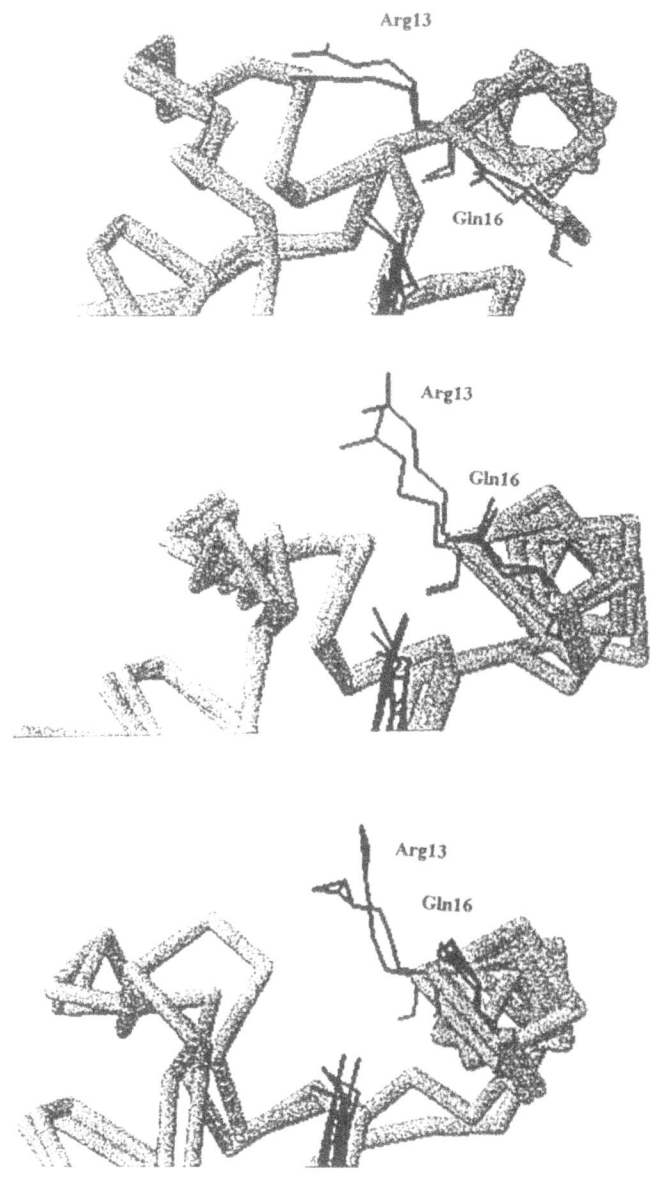

Figure 2. The conformation of Arg13 and Gln16 in the oxidized and reduced forms of *S. cerevisiae* iso-1-cytochrome *c* as it results from the X-ray crystal structures (top panel), NMR solution structures (middle panel), and the structures calculated through MD simulations (lower panel).

difference of the order 3-4 times kT. This picture in which the reorganization of a patch of surface residues propagates inside the protein so that at the end the overall structure of the molecule is perturbed, can be important to understand the basis of protein-protein recognition. At the present stage, there is no clear evidence for the above ideas, but it is a good working hypothesis as many experimental observations could be framed within this description. Going back to electron transfer proteins, it is tempting to describe the process through the formation of a complex between the two redox partner proteins, occurring only when the two molecules are in the correct redox state. This requires, for instance, that the acceptor protein has the correct conformation to interact with the reduced donor proteins only when it is in the oxidized form. In this case we should think that a change in the oxidation state of the metal ion in the interior of the protein changes the conformation of a few residues surrounding it and this movement propagates towards the surface amino acids, which therefore adopt the optimum conformation for the interaction with the partner molecule.

4. References

1. Williamson, M.P., Havel, T.F. and Wüthrich, K. (1985) Solution conformation of proteinase inhibitor IIA from bull seminal plasma by 1H nuclear magnetic resonance and distance geometry, *J. Mol. Biol.* **185**, 295-315.

2. Wayne, A., Hendrickson, Wüthrich, K. (eds.) (1994) *Macrom. Struct.* 1, Current Biology Ltd.

3. Wayne, A., Hendrickson, Wüthrich, K. (eds.) (1993) *Macrom. Struct.* 1, Current Biology Ltd.

4. Wayne, A., Hendrickson, Wüthrich, K. (eds.) (1992) *Macrom. Struct.* 1, Current Biology Ltd.

5. Wayne, A., Hendrickson, Wüthrich, K. (eds.) (1991)*Macrom. Struct.* 1, Current Biology Ltd.

6. Bertini, I. and Luchinat, C. (1986) *NMR of paramagnetic molecules in biological systems*, Benjamin/Cummings, Menlo Park, CA.

7. Banci, L., Bertini, I. and Luchinat, C. (1991) *Nuclear and electron relaxation. The magnetic nucleus-unpaired electron coupling in solution*, VCH, Weinheim.

8. Bertini, I., Turano, P. and Vila, A.J. (1993) NMR of paramagnetic metalloproteins, *Chem. Rev.* **93**, 2833-2932.

9. Bertini, I. and Luchinat, C. (1996) *NMR of paramagnetic substances*, Coord. Chem. Rev. **150**, Elsevier, Amsterdam.

10. Banci, L., Bertini, I. and Luchinat, C. (1994) 2D NMR spectra of paramagnetic systems, in James, T.L. and Oppenheimer, N.J. (eds.) *Methods in Enzymology*, Academic press, London, **239**, 485-514.

11. Bertini, I., Luchinat, C. and Rosato, A. (1996) The solution structure of paramagnetic metalloproteins, Prog. Biophys. Mol. Biol. **66**, 43-80.

12. Banci, L., Bertini, I., Eltis, L.D., Felli, I.C., Kastrau, D.H.W., Luchinat, C., Piccioli, M., Pierattelli, R. and Smith, M. (1994) The three dimensional structure in solution of the paramagnetic protein high-potential iron-sulfur protein I from *Ectothiorhodospira halophila* through nuclear magnetic resonance, *Eur. J. Biochem.* **225**, 715-725.

13. Banci, L., Bertini, I., Luchinat, C., Piccioli, M., Scozzafava, A. and Turano, P. (1989) ^{1}H NOE studies on dicopper(II)dicobalt(II)superoxide dismutase, *Inorg. Chem.* **28**, 4650-4656.

14. Banci, L., Bertini, I., Luchinat, C. and Piccioli, M. (1990) Transient versus steady state NOE in paramagnetic molecules. Cu_2Co_2SOD as an example, *FEBS Lett.* **272**, 175-180.

15. Bertini, I., Dikiy, A., Luchinat, C., Piccioli, M. and Tarchi, D. (1994) NOE-NOESY, a further tool in NMR of paramagnetic metalloproteins, *J. Magn. Reson., Ser. B* 103, 278-283.

16. Solomon, I. (1955) Relaxation Processes in a system of two spins, *Phys. Rev.* **99**, 559-565.

17. Noggle, J.H. and Schirmer, R.E. (1971) *The nuclear Overhauser effect*, Academic Press, New York.

18. Neuhaus, D. and Williamson, M. (1989) *The nuclear Overhauser effect in structural and conformational analysis*, VCH, New York.

19. Bertini, I., Couture, M.M.J., Donaire, A., Eltis, L.D., Felli, I.C., Luchinat, C., Piccioli, M. and Rosato, A. (1996) The solution structure refinement of the paramagnetic reduced HiPIP from *Ectothiorhodospira halophila* by using stable isotope labelling and nuclear relaxation, *Eur. J. Biochem.* **241**, 440-452.

20. Bertini, I., Donaire, A., Feinberg, B.A., Luchinat, C., Piccioli, M. and Yuan, H. (1995) Solution structure of the oxidized 2[Fe$_4$S$_4$] ferredoxin from *Clostridium pasteurianum*, *Eur. J. Biochem.* **232**, 192-205.

21. La Mar, G.N., Horrocks, W.D.W., and Holm, R.H. (Eds.) (1973) *NMR of paramagnetic molecules*, Academic Press, New York.

22. Shin, K. and Gaff, H.M. (1990) Iron(III) porphyrin promoted aerobic oxidation of sulfur dioxide, *J. Chem. Soc., Chem. Commun.* 461-462.

23. Banci, L., Bertini, I., Bren, K.L., Cremonini, M.A., Gray, H.B., Luchinat, C. and Turano, P. (1996) The use of pseudocontact shifts to refine solution structures of paramagnetic metalloproteins: Met80Ala cyano-cytochrome *c* as an example, *JBIC* **1**, 117-126.

24. Banci, L., Bertini, I., Gray, H.B., Luchinat, C., Reddig, T., Rosato, A. and Turano, P. (1997) Solution structure of oxidized horse heart cytochrome *c*, *Biochemistry* **36**, 9867-9877.

25. Banci, L., Bertini, I., Bren, K.L., Gray, H.B., Sompornpisut, P. and Turano, P. (1997) The solution structure of oxidized cytochrome *c*: hints to understand its function and folding, *Biochemistry* **36**, 8992-9001.

26. Banci, L., Bertini, I., Gori Savellini, G., Romagnoli, A., Turano, P., Cremonini, M.A., Luchinat, C. and Gray, H.B. (1997) The pseudocontact shifts as constraints for energy minimization and molecular dynamic calculations on solution structures of paramagnetic metalloproteins, *Proteins: Structure, Function and Genetics* **29**, 68-76.

27. Bertini, I., Capozzi, F., Luchinat, C., Piccioli, M. and Vila, A.J. (1994) The Fe$_4$S$_4$ centers in ferredoxins studied through proton and carbon hyperfine coupling. sequence specific assignments of cysteines in ferredoxins from *Clostridium acidi urici* and *Clostridium pasteurianum*, *J. Am. Chem. Soc.* **116**, 651-660.

28. Bertini, I., Ciurli, S. and Luchinat, C. (1995) The electronic structure of FeS centers in protein and models. A contribution to the understanding of their electron transfer properties, *Structure and Bonding* **83**, 1-54.

29. Bertini, I., Felli, I.C., Luchinat, C. and Rosato, A. (1996) A complete relaxation matrix refinement of the solution structure of a paramagnetic metalloprotein: reduced HiPIP I from *E. halophila*, *Proteins: Structure, Function and Genetics* **24**, 158-164.

30. Bertini, I., Eltis, L.D., Felli, I.C., Kastrau, D.H.W., Luchinat, C. and Piccioli, M. (1995) The solution structure of oxidized HiPIP I from *Ectothiorhodospira halophila*, can NMR probe rearrangements associated to electron transfer processes? *Chemistry - A European journal* **1**, 598-607.

31. Banci, L., Bertini, I., Dikiy, A., Kastrau, D.H.W., Luchinat, C. and Sompornpisut, P. (1995) The three dimensional solution structure of the reduced high potential iron sulfur protein *Chromatium vinosum* through NMR, *Biochemistry* **34**, 206-219.

32. Bertini, I., Dikiy, A., Kastrau, D.H.W., Luchinat, C. and Sompornpisut, P. (1995) The three dimensional solution structure of oxidized HiPIP from *Chromatium vinosum* through NMR. Comparative analysis with the solution structure of the reduced species, *Biochemistry* **34**, 9851-9858.

33. Breiter, D.R., Meyer, T.E., Rayment, I. and Holden, H.M. (1991) The molecular structure of the high potential iron-sulfur protein isolated from *Ectothiorhodospira halophila* determined at 2.5 Å resolution, *J. Biol. Chem.* **266**, 18660-18667.

34. Carter, C.W.J., Kraut, J., Freer, S.T., Xuong, N.-H., Alden, R.A. and Bartsch, R.G. (1974) Two-angstrom crystal structure of *Chromatium vinosum* high potential iron protein, *J. Biol. Chem.* **249**, 4212-4215.

35. Bertini, I., Cowan, J.A., Luchinat, C., Natarajan, K. and Piccioli, M. (1997) Characterization of a partially unfolded high potential iron protein relevant to the folding pathway and cluster stability, *Biochemistry* **36**, 9332-9339.

36. Baistrocchi, P., Banci, L., Bertini, I., Turano, P., Bren, K.L. and Gray, H.B. (1996) Three-dimensional solution structure of *Saccharomyces cerevisiae* reduced iso-1-cytochrome *c*, *Biochemistry* **35**, 13788-13796.

37. Banci, L., Bertini, I., Quacquarini, G. Walter, O., Diaz, A., Hervás, M. and De la Rosa, M.A. (1996)

The solution structure of cytochrome c_6 from the green alga *Monoraphidium braunii*, *JBIC* 1, 330-340.

38. Banci, L., Bertini, I., De la Rosa, M.A., Koulougliotis, D., Navarro, J.A. and Walter, O. (1998) The solution structure of oxidized cytochrome c_6 from the green alga *Monoraphidium braunii*, submitted.

39. Qi, P.X., Di Stefano, D.L. and Wand, A.J. (1994) Solution structure of horse heart ferrocytochrome c determined by high-resolution NMR and restrained simulated annealing, *Biochemistry* 33, 6408-6417.

40. Louie, G.V. and Brayer, G.D. (1990) High-resolution refinement of yeast iso-1-cytochrome c and comparison with other eukaryotic cytochromes, *J. Mol. Biol.* 214, 527-555.

41. Berghuis, A.M. and Brayer, G.D. (1992) Oxidation state-dependent conformational changes in cytochrome c, *J. Mol. Biol.* 223, 959-976.

42. Pelletier, H. and Kraut, J. (1992) Crystal structure of a complex between electron transfer partners, cytochrome c peroxidase and cytochrome c, *Science* 258, 1748-1755.

43. Frazao, C., Soares, C.M., Carrondo, M.A., Pohl, E., Dauter, Z., Wilson, K.S., Hervás, M., Navarro, J.A., De la Rosa, M.A. and Sheldrick, G. (1995) *Ab initio* determination of the crystal structure of cytochrome c_6 and comparison with plastocyanin, *Structure* 3, 1159-1170.

44. Hervás, M., Navarro, J.A., Diaz, A., Bottin, H. and De la Rosa, M.A. (1995) Laser-flash kinetic analysis of the fast electron transfer from plastocyanin and cytochrome c_6 to photosystem I. Experimental evidence on the evolution of the reaction mechanism, *Biochemistry* 34, 11321-11326.

45. Hervás, M., Navarro, J.A., Diaz, A. and De la Rosa, M.A. (1996) A comparative thermodynamic analysis by laser-flash absorption spectroscopy of photosystem I by plastocyanin and cytochrome c_6 in *Anabaena* PCC7119, *Synechocystis* PCC6803, and spinach, *Biochemistry* 35, 2693-2698.

46. Winkler, J.R. and Gray, H.B. (1992) Electron transfer in ruthenium-modified proteins, *Chem. Rev.* 92, 369-379.

Chapter 4
The Cytochrome *c* Oxidase Family

EXPLORING THE PROTON CHANNELS OF CYTOCHROME OXIDASE

ROBERT B. GENNIS
University of Illinois at Urbana-Champaign
Department of Chemistry
600 S. Mathews Ave.
Urbana, IL 61801

Cytochrome oxidase is a marvelous molecular machine designed to conserve energy by coupling the chemical reduction of dioxygen to the generation of a transmembrane proton electrochemical potential. There is actually a large number of related enzymes, comprising the heme-copper oxidase superfamily, that perform this function in both eukaryotes and prokaryotes [1]. The enzyme isolated from bovine heart mitochondria is the best studied member of this superfamily, but bacterial enzymes from *Escherichia coli*, *Rhodobacter sphaeroides*, and *Paracoccus denitrificans* are also under intense investigation. The use of the prokaryotic oxidases has provided the full array of molecular genetics tools to probe structure/function relationships in this group of enzymes. The high resolution structures of the cytochrome oxidases from both bovine heart [2, 3] and from *Paracoccus denitrificans* [4] have been solved and are stellar achievements. These structures have resolved a number of questions and confirmed the close relationship of the 13-subunit mammalian enzyme and the smaller (3 or 4 subunit) bacterial enzymes. However, many questions concerning the mechanism by which these enzymes work, especially regarding the nature of the proton pump, remain unresolved. The structures provide a rich source of information and clues, but do not define the catalytic mechanism.

The minimum catalytic unit of the cytochrome oxidases consists of subunits I and II [5], which contain the amino acid ligands for all of the redox-active metal centers. Cytochrome c is oxidized initially by Cu_A which is a binuclear copper center located within subunit II [2, 4]. The other redox-active centers are within subunit I, the heart of the heme-copper oxidases. The electron rapidly equilibrates between Cu_A and heme a (<50 μsec), the low spin heme component of the oxidase [6-8]. From heme a, electrons proceed to the heme a_3-Cu_B binuclear center which is where dioxygen binds and is reduced to water. The transmembrane voltage gradient is generated by two different mechanisms. First, the electrons (from cytochrome c) and the protons used in the chemistry to convert dioxygen to water come from opposite sides of the membrane [9]. The heme-copper center is located deep within subunit I, but significantly closer to the outer surface of the membrane where cytochrome c is oxidized [3, 4]. Electron transfer from Cu_A to heme a is electrogenic[6, 7], but the

241

G.W. Canters and E. Vijgenboom (eds.),
Biological Electron Transfer Chains: Genetics, Composition and Mode of Operation, 241-249.
© 1998 *Kluwer Academic Publishers.*

direction of electron transfer from heme a to the heme-copper center is parallel to the plane of the membrane [3, 4]. The protons needed to form water, the so-called chemical protons, are delivered to the active site from the opposite side of the membrane: the bacterial cytoplasm or the eukaryotic mitochondrial matrix (in the case of the eukaryotic enzymes). A pathway, or channel, facilitating the movement of protons to the heme-copper center is a necessary feature of cytochrome oxidase, as was recognized prior to the determination of the actual structures [10, 11]. The combining of the electron from the electropositive side of the membrane and the proton from the electronegative side of the membrane in the chemical reaction at the active site of the enzyme is a voltage-generating process. In addition to this mechanism of generating a voltage, the enzyme functions as an actual proton pump. For each dioxygen reduced to water, 4 protons are pumped across the membrane (1 proton/electron). This generates a pH gradient across the membrane and also is an electrogenic process, contributing to the transmembrane voltage [6, 12]. Clearly, proton pumping also necessitates a pathway for protons to tranverse the membrane and some mechanism to couple the process to the chemistry catalyzed at the active site of the enzyme.

The X-ray structures of cytochrome oxidase from bovine heart [2, 3] and from *P. denitrificans* [4] reveal domains within subunit I that have been interpreted as possible pathways for proton uptake from the negative side of the membrane [3, 4]. These represent pathways of continuous hydrogen bonds between observed water molecules, proposed water molecules within observed cavities, and polar residues from amino acid protein side chains or from the peptide backbone. Several amino acid residues within these putative channels had been predicted on the basis of the site-directed mutagenesis studies [10, 13, 14]. One pathway contains residues in helix VI and helix VIII, including highly conserved polar residues K362, T359 and Y288 (*R. sphaeroides* numbering) and has been speculated [4] to be specifically involved in delivering protons used chemically in the reduction of dioxygen to water ("chemical channel"). This structure will be referred to as the "K-channel".

A second channel involves residues within helix III and IV and has a highly conserved aspartate and asparagine (D132 and N139) near the entrance and a hydrogen bond path can be discerned up to S201 in helix IV. In the *P. denitrificans* oxidase, the pathway leads via water molecules to a highly conserved glutamic acid residue (E286) at which point the pathway is not clear. This structure has been speculated to be a "pumping channel" [4] but will be referred here as the "D-channel" (for the aspartic acid component) until its role is better defined experimentally. In the mammalian enzyme, the pathway of the equivalent channel is proposed to bypass E286 and to proceed through to the electropositive surface.

There is also a third proposed channel that is observed in the bovine heart oxidase [3], but the residues comprising this channel are not highly conserved in the heme-copper oxidases and the equivalent structure does not appear in the oxidase from *Paracoccus*. Our studies have been focused on the roles of the K- and D-channels, and we will consider a two-channel model for proton input for the oxidase,

A Two-channel Model for
Proton Delivery to the Binuclear Center

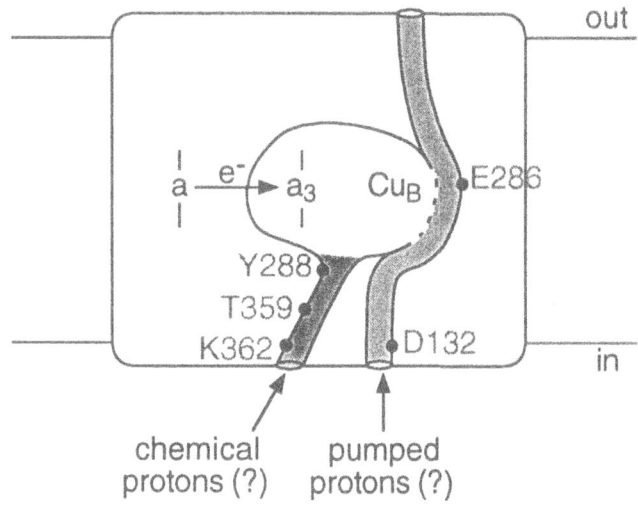

Figure 1. A two-channel model of cytochrome oxidase.

with the recognition that the situation could be even more complex. A schematic of the two-channel model is shown in Figure 1.

Mutations in some residues within either the K-channel or the D-channel result in severe inhibition of the steady state turnover of the oxidases from both *R. sphaeroides* (see Table 1) [13, 15-17] as well as from *E. coli* [11, 14, 18-20]. For example, in the *Rhodobacter* oxidase, both K362M and the E286Q mutants are essentially devoid of cytochrome *c* oxidase activity. The mutagenesis data strongly support an important functional role for E286, although this is not clear in the structure of the mammalian oxidase [3], and the role of this residue is not yet evident. D132N has only 5% of the wild type oxidase activity, and the residual activity is not coupled to proton pumping [13]. This is consistent with the initial finding that the equivalent mutant in cytochrome bo_3 from *E. coli* is deficient in proton pumping, but can still catalyze oxidase activity at a substantial rate; i.e., some mutants in this position result in decoupling the oxidase activity from the proton pump [14, 21]. The proposal that the D-channel is utilized for the translocation of the pumped protons [4], whereas the K-channel is for the chemical protons is logical from observing the structure and from these limited kinetics data. More recent work from our laboratory and

Table 1. "Channel" mutants in the R. *sphaeroides* cyt *c* oxidase

Mutant	Turnover Rate (electrons/sec)	Proton Pumping (H+/e Ratio)
Wildtype	1500	0.6 - 0.8
K-channel: K362M	<5 (<0.3%)	—
K-channel: T359A	250 (17%)	0.7 - 0.8
K-channel: Y288F	<5 (<0.3%)	—
D-channel: D132N	50 (3.3%)	0
D-channel: S201A	550 (37%)	0.4 - 0.7
D-channel: E286Q	<5 (<0.3%)	—

from the laboratories of our collaborators suggests that this is not likely to be the case, however.

Figure 2 illustrates a simplified schematic of the mechanism of cytochrome oxidase in which the status of the heme-copper center is indicated. Starting with the fully oxidized enzyme, reduction by at least two electrons is required before dioxygen binds. The two-electron reduction of the binuclear center has been demonstrated to be associated with the uptake by the enzyme of two protons [22, 23], also shown in the schematic. The binding of dioxygen results in the formation of an intermediate called compound P, or the peroxy form of the enzyme. The actual structure of this species, in particular, whether the O-O bond is broken at this stage, is not clear and is being actively investigated by several laboratories [24-27]. To complete the reduction of dioxygen requires two more electrons (and protons). Compound P is reduced to compound F, the oxoferryl form of the enzyme [8, 24, 28, 29]. At this stage the O-O bond is clearly broken [24]. The fourth electron (F-to-Ox) returns the enzyme to the oxidized (Ox) state, completing the catalytic cycle.

Thermodynamic measurements by the Wikström laboratory [30] have indicated that all the proton pumping is coupled with the delivery of the third and fourth electrons, as indicated in Figure 2. Time-resolved measurements of the generation of the membrane potential in single-turnover experiments support the notion that the P-to-F and F-to-Ox steps are indeed coupled to charge movement across the membrane [6, 7, 12]. Also, in single-turnover experiments proton release from reconstituted vesicles coincident with the F-to-Ox step has been directly observed, though no proton release associated with P-to-F was been observed [31, 32].

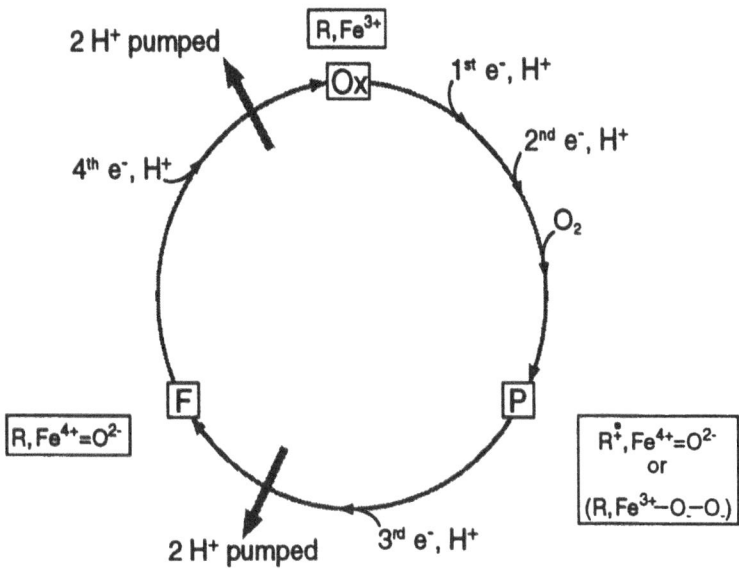

Figure 2. A simple version of the mechanism of cytochrome *c* oxidase. R refers to an amino acid residue that is capable of serving as an electron donor.

The proposition that two protons are pumped coincident with the delivery of each the third and fourth electrons [30] provides constraints in any detailed mechanism of the coupling. The Wikström group proposed the "histidine cycle" [33, 34] to meet this and other considerations, and this was incorporated into the mechanism proposed by Michel and his coworkers [4]. The essence of this is the exchange of one of the Cu_B histidine ligands such that two protons are taken up and released in alternating sequence from the imidazole. Indeed, in the structure of the oxidized *Paracoccus* enzyme with azide bound to the active site, one of the Cu_B histidine ligands is disordered, suggesting that it may occupy more than one position under these conditions [4]. However, much more needs to be done to test whether the proton pumping mechanism involves the Cu_B ligands. The pathway by which protons might get from either the K-channel or the D-channel to the protonatable Cu_B ligand also needs to be explored, but it seems possible, from casual inspection, that neither channel can be excluded as a source for protons for this pathway.

Regardless of the mechanism of coupling, it is clear that the steps following the formation of compound P involve protons involved in both chemistry and pumping. One would expect, therefore, that mutations in either the D-channel or the K-channel would impede these steps. This has been tested experimentally with the *Rhodobacter* oxidase by our group and those of Dr. Alexander Konstantinov (Moscow State University) and Dr. Peter Brzezinski (University of Göteborg). One strategy used by the group of Dr. Konstantinov in collaboration with our group is indicated in Figure 3.

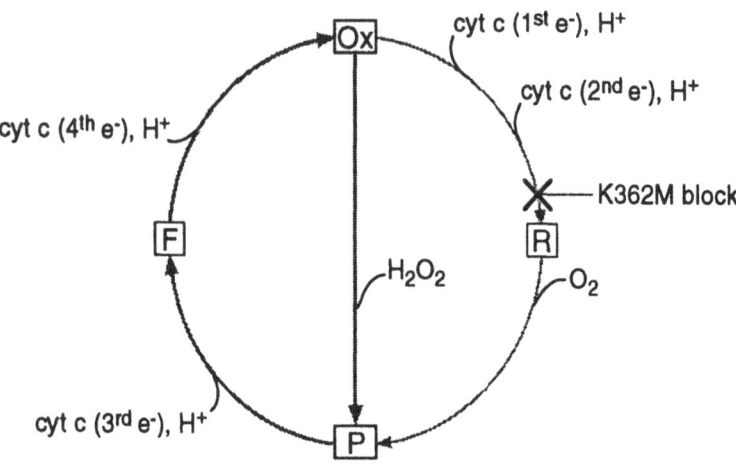

Figure 3. The peroxide shunt. R refers to the form of the enzyme with a reduced binuclear center.

This shows the same catalytic cycle as in Figure 2, but with the addition of the "peroxide shunt". The addition of hydrogen peroxide to the oxidized enzyme results in the formation of compound P, which can be converted to compound F and then back to Ox [35, 36], provided there is not a block in the P-to-F or F-to-Ox steps. Other strategies have been to examine the single turnover of the oxidase species, observing the reaction of dioxygen with the fully reduced enzyme and also the one-electron photo-reduction of enzyme poised in the oxoferryl state, F-to-Ox.

The results from these experiments are as yet unpublished and will only be summarized here. The evidence suggests that the K362M mutation results in a block in the initial electron transfer from heme a to heme a_3, as illustrated in Figure 3. There is no indication, so far, of a block in any step following the formation of compound P in the K362M mutant. Since oxygen will not bind if reduction of the heme-copper center is impeded, the K362M oxidase does not function as an oxidase. The enzyme, however, functions effectively as a cytochrome c peroxidase, which can be explained by the schematic in Figure 3. This result is inconsistent with assigning the K-channel as uniquely required for delivering either the chemical or pumped protons, but requires a different interpretation. In contrast, the data indicate that the E286Q and D132N mutants block the enzyme catalysis in one or more steps following the formation of an oxygenated intermediate, perhaps compound P.

The data suggest a working hypothesis that the D-channel is required for the delivery of protons following the formation of compound P. This would involve protons used both for chemistry at the active site as well as protons that are pumped.

Possibly, E286 acts like a toggle switch and can direct protons either to the heme-copper center, or out, perhaps via a pathway not involving the Cu_B ligands. The K-channel, is suggested to provide protons to one or more groups at the heme-copper center whose protonation is required to stabilize the reduced form of these metals. One of these protonated groups may well be a hydroxide anion associated with the metals. The location of the K-channel is consistent with such a role.

This proposal for the function of the two putative proton channels will require additional experimental tests to either provide needed elaborations or to discard it. Ultimately, it will be important to determine what groups become protonated and deprotonated transiently during the catalytic cycle, to determine the structure or structures referred to as compound P so the chemistry is better defined, and to examine in more detail the nature of the steps in the mechanism associated with proton uptake, internal translocation and release.

This work was supported by grants from the NIH (HL16101) and the Department of Energy (DE-FG02-87ER13716).

References

1. García-Horsman, J. A., Barquera, B., Rumbley, J., Ma, J. and Gennis, R. B. (1994) The Superfamily of Heme-Copper Respiratory Oxidases, *J. Bacteriol.* **176**, 5587-5600.

2. Tsukihara, T., Aoyama, H., Yamashita, E., Tomizaki, T., Yamaguchi, H., Shinzawa-Itoh, K., Nakashima, T., Yaono, R. and Yoshikawa, S. (1995) Structures of Metal Sites of Oxidized Bovine Heart Cytochrome *c* Oxidase at 2.8Å, *Science* **269**, 1069-1074.

3. Tsukihara, T., Aoyama, H., Yamashita, E., Takashi, T., Yamaguichi, H., Shinzawa-Itoh, K., Nakashima, R., Yaono, R. and Yoshikawa, S. (1996) The Whole Structure of the 13-Subunit Oxidized Cytochrome c Oxidase at 2.8 Å, *Science* **272**, 1136-1144.

4. Iwata, S., Ostermeier, C., Ludwig, B. and Michel, H. (1995) Structure at 2.8 Å Resolution of Cytochrome *c* Oxidase from *Paracoccus denitrificans, Nature* **376**, 660-669.

5. Hendler, R. W., Pardhasaradhi, K., Reynafarje, B. and Ludwig, B. (1991) Comparison of Energy-transducing Capabilities of the Two- and Three-subunit Cytochromes aa_3 from *Paracoccus denitrificans* and the 13-subunit Beef Heart Enzyme, *Biophys. J.* **60**, 415-423.

6. Zaslavsky, D. L., Smirnova, I. A., Siletsky, S. A., Kaulen, A. D., Millett, F. and Konstantinov, A. A. (1995) Rapid Kinetics of Membrane Potential Generation by Cytochrome *c* Oxidase with the Photoactive Ru(II)-tris-bipyridyl Derivative of Cytochrome *c* as Electron Donor, *FEBS Lett.* **359**, 27-30.

7. Zaslavsky, D., Kaulen, A. and Smirnova, I. A. (1993) Flash-induced Membrane Potential Generation by Cytochrome *c* Oxidase, *FEBS Lett.* **336**, 389-393.

8. Brzezinski, P. (1996) Internal Electron-Transfer Reactions in Cytochrome *c* Oxidase, *Biochemistry* **35**, 5611-5615.

9. Wikström, M., Krab, K. and Saraste, M. (1981) *Cytochrome Oxidase - A Synthesis*, Academic Press, New York.

10. Hosler, J. P., Ferguson-Miller, S., Calhoun, M. W., Thomas, J. W., Hill, J., Lemieux, L., Ma, J., Georgiou, C., Fetter, J., Shapleigh, J., Tecklenburg, M. M. J., Babcock, G. T. and Gennis, R. B.

(1993) Insight into the Active-Site Structure and Function of Cytochrome Oxidase by Analysis of Site-Directed Mutants of Bacterial Cytochrome aa_3 and Cytochrome bo, *J. Bioenerg. Biomembr.* **25**, 121-136.

11. Thomas, J. W., Lemieux, L. J., Alben, J. O. and Gennis, R. B. (1993) Site-Directed Mutagenesis of Highly Conserved Residues in Helix VIII of Subunit I of the Cytochrome bo Ubiquinol Oxidase from *Escherichia coli*: An Amphipathic Transmembrane Helix That May Be Important in Conveying Protons to the Binuclear Center, *Biochemistry* **32**, 11173-11180.

12. Verkhovsky, M. I., Morgan, J. E., Verkhovskaya, M. L. and Wikström, M. (1997) Translocation of Electrical Charge During a Single Turnover of Cytochrome-c Oxidase, *Biochim. Biophys. Acta* **1318**, 6-10.

13. Fetter, J. R., Qian, J., Shapleigh, J., Thomas, J. W., Garcia-Horsman, A., Schmidt, E., Hosler, J., Babcock, G. T., Gennis, R. B. and Ferguson-Miller, S. (1995) Possible Proton Relay Pathways in Cytochrome c Oxidase, *Proc. Natl. Acad. Sci. USA* **92**, 1604-1608.

14. Thomas, J. W., Puustinen, A., Alben, J. O., Gennis, R. B. and Wikström, M. (1993) Substitution of Asparagine for Aspartate-135 in Subunit I of the Cytochrome bo Ubiquinol Oxidase of *Escherichia coli* Eliminates Proton-Pumping Activity, *Biochemistry* **32**, 10923-10928.

15. Mitchell, D. M., Fetter, J. R., Mills, D. A., Ädelroth, P., Pressler, M. A., Kim, Y., Aasa, R., Brzezinski, P., Malmström, B. G., Alben, J. O., Babcock, G. T., Ferguson-Miller, S. and Gennis, R. B. (1996) Site-Directed Mutagenesis of Residues Lining a Putative Proton Transfer Pathway in Cytochrome c Oxidase from *Rhodobacter sphaeroides*, *Biochemistry* **35**, 13089-13093.

16. Hosler, J. P., Shapleigh, J. P., Mitchell, D. M., Kim, Y., Pressler, M., Georgiou, C., Babcock, G. T., Alben, J. O., Ferguson-Miller, S. and Gennis, R. B. (1996) Polar Residues in Helix VIII of Subunit I of Cytochrome c Oxidase Influence the Activity and the Structure of the Active Site, *Biochemistry* **35**, 10776-10783.

17. Mitchell, D. M., Aasa, R., Ädelroth, P., Brzezinski, P., Gennis, R. B. and Malmström, B. G. (1995) EPR Studies of Wild-type and Several Mutants of Cytochrome c Oxidase from *Rhodobacter sphaeroides*: Glu^{286} is Not a Bridging Ligand in the Cytochrome a_3-Cu$_B$ Center, *FEBS Lett.* **374**, 371-374.

18. Thomas, J. W., Calhoun, M. W., Lemieux, L. J., Puustinen, A., Wikström, M., Alben, J. O. and Gennis, R. B. (1994) Site-Directed Mutagenesis of Residues Within Helix VI in Subunit I of the Cytochrome bo_3 Ubiquinol Oxidase from *Escherichia coli* Suggests That Tyrosine 288 May Be a Cu$_B$ Ligand, *Biochemistry* **33**, 13013-13021.

19. Svensson-Ek, M., Thomas, J. W., Gennis, R. B., Nilsson, T. and Brzezinski, P. (1996) Kinetics of Electron and Proton Transfer during the Reaction of Wild Type and Helix VI Mutants of Cytochrome bo_3 with Oxygen, *Biochemistry* **35**, 13673-13680.

20. Svensson, M., Hallen, S., Thomas, J. W., Lemieux, L., Gennis, R. B. and Nilsson, T. (1995) Oxygen Reaction and Proton Uptake in Helix VIII Mutants of Cytochrome bo_3, *Biochemistry* **34**, 5252-5258.

21. Garcia-Horsman, J. A., Puustinen, A., Gennis, R. B. and Wikström, M. (1995) Proton Transfer in Cytochrome bo_3 Ubiquinol Oxidase of *Escherichia coli*: Second-Site Mutations in Subunit I that Restore Proton Pumping in the Mutant Asp135→Asn, *Biochemistry* **34**, 4428-4433.

22. Rich, P. R., Meunier, B., Mitchell, R. and Moody, A. J. (1996) Coupling of Charge and Proton Movement in Cytochrome c Oxidase, *Biochim. Biophys. Acta* **1275**, 91-95.

23. Mitchell, R., Mitchell, P. and Rich, P. R. (1992) Protonation States of the Catalytic Intermediates of Cytochrome c Oxidase, *Biochim. Biophys. Acta* **1101**, 188-191.

24. Ferguson-Miller, S. and Babcock, G. T. (1996) Heme/Copper Terminal Oxidases, *Chem. Rev.* **7**, 2889-2907.

25. Kitagawa, T. and Ogura, T. (1997) Oxygen Activation Mechanism at the Binuclear Site of Heme-Copper Oxidase Superfamily as Revealed by Time-Resolved Resonance Raman Spectroscopy, *Progress Inorganic Chemistry* **45**, 431-479.

26. Proshlyakov, D. A., Ogura, T., Shinzawa-Itoh, K., Yoshikawa, S. and Kitagawa, T. (1996) Microcirculating System for Simultaneous Determination of Raman and Absorption Spectra of Enzymatic Reaction Intermediates and Its Application to the Reaction of Cytochrome *c* Oxidases With Hydrogen Peroxide, *Biochemistry* **35**, 76-82.

27. Ogura, T., Hirota, S., Proshlyakov, D. A., Shinzawa-Itoh, K., Yoshikawa, S. and Kitagawa, T. (1996) Time-Resolved Resonance Raman Evidence for Tight Coupling Between Electron Transfer and Proton Pumping of Cytochrome *c* Oxidase Upon the Change From the Fe^V Oxidation Level to the Fe^{IV} Oxidation Level, *J. Am. Chem. Soc.* **118**, 5443-5449.

28. Verkhovsky, M. I., Morgan, J. E. and Wikström, M. (1996) Redox Transitions Between Oxygen Intermediates in Cytochrome-*c* Oxidase, *Proc. Natl. Acad. Sci. USA* **93**, 12235-12239.

29. Sucheta, A., Georgiadis, K. E. and Einarsdóttir, O. (1997) Mechanisms of Cytochrome *c* Oxidase-catalyzed Reduction of Dioxygen to Water: Evidence for Peroxy and Ferryl Intermediates at Room Temperature, *Biochemistry* **36**, 554-565.

30. Wikström, M. (1989) Identification of the Electron Transfers in Cytochrome Oxidase that are Coupled to Proton-Pumping, *Nature* **338**, 776-778.

31. Nilsson, T., Hallén, S. and Oliveberg, M. (1990) Rapid Proton Release During Flash-induced Oxidation of Cytochrome *c* Oxidase, *FEBS Lett.* **260**, 45-47.

32. Oliveberg, M., Hallén, S. and Nilsson, T. (1991) Uptake and Relase of Protons During the Reaction Between Cytochrome *c* Oxidase and Molecular Oxygen: A Flow-Flash Investigation, *Biochemistry* **30**, 436-440.

33. Wikström, M., Bogachev, A., Finel, M., Morgan, J. E., Puustinen, A., Raitio, M., Verkhovskaya, M. and Verkhovsky, M. I. (1994) Mechanism of Proton Translocation by the Respiratory Oxidases. The Histidine Cycle, *Biochim. Biophys. Acta* **1187**, 106-111.

34. Morgan, J. E., Verkhovsky, M. I. and Wikström, M. (1994) The Histidine Cycle: A New Model for Proton Translocation in the Respiratory Heme-copper Oxidases, *J. Bioenerg. Biomembr.* **26**, 599-608.

35. Vygodina, T. V. and Konstantinov, A. A. (1988) H_2O_2-Induced Conversion of Cytochrome *c* Oxidase Peroxy Complex to Oxoferryl State, *Ann. NY Acad. Sci.* **550**, 124-138.

36. Vygodina, T. and Konstantinov, A. (1989) Effect of pH on the Spectrum of Cytochrome *c* Oxidase Hydrogen Peroxide Complex, *Biochim. Biophys. Acta* **973**, 390-398.

CONTROL OF ELECTRON TRANSFER TO THE BINUCLEAR CENTER IN Cu-HEME OXIDASES

M. BRUNORI, A. GIUFFRE', F. MALATESTA*, E. D'ITRI and P. SARTI

*Department of Biochemical Sciences "A.Rossi-Fanelli" and CNR Center for Molecular Biology, University of Rome "La Sapienza", I-00185 Rome, Italy. *Department of Basic and Applied Biology, University of L'Aquila, I-67010 L'Aquila, Italy.*

Figure 1 depicts the structure of the active site of cytochrome c oxidase, including cytochrome a and the (oxygen binding) binuclear center cytochrome a_3-Cu_B as obtained from the crystallographic data on the beef heart enzyme [1]. One of the transmembrane helices of subunit I, helix X, provides two His as ligands of cytochrome a (His378) and cytochrome a_3 (His376). Laser photolysis experiments starting from the CO adduct of the fully reduced or mixed-valence enzyme from beef heart yield rate constants for internal ET ranging from 10^4 to 3×10^5 s^{-1} [2-4]. These very rapid rates are consistent with the short distance (13 Å) separating the two metals which are connected via a possible covalent pathway, involving 16 bonds [5]. On the other hand, it has been reported in the literature that the apparent rate of formation of reduced cytochrome a_3, either in the presence or in the absence of CO, is considerably slower (0.1 to > 30 s^{-1} depending on experimental conditions [6-8]). In spite of some complexity related to the state of the enzyme (whether "resting" or "pulsed"), this apparent rate constant seems to correlate with the turnover number of the enzyme; thus Malatesta $et al.$ [8] concluded that internal electron transfer to the oxidized binuclear center is the rate-limiting step in turnover. More recently Verkhovsky $et al.$ [9] confirmed that the rate of formation of reduced cytochrome a_3 is slow (ms); however they assumed that even in the oxidized enzyme, internal electron transfer is very fast (μs), but the redox equilibrium between cytochrome a and cytochrome a_3 favors the former, and proton diffusion and/or binding to the reduced binuclear site is the rate limiting step. In summary, there is substantial agreement about the basic experimental observation, i.e. that starting from oxidized cytochrome c oxidase the rate of formation of reduced cytochrome a_3 is in the ms time range; nevertheless two alternative mechanisms have been proposed.

The **thermodynamic control** hypothesis states that electron transfer from cytochrome

G.W. Canters and E. Vijgenboom (eds.),
Biological Electron Transfer Chains: Genetics, Composition and Mode of Operation, 251-258.

252

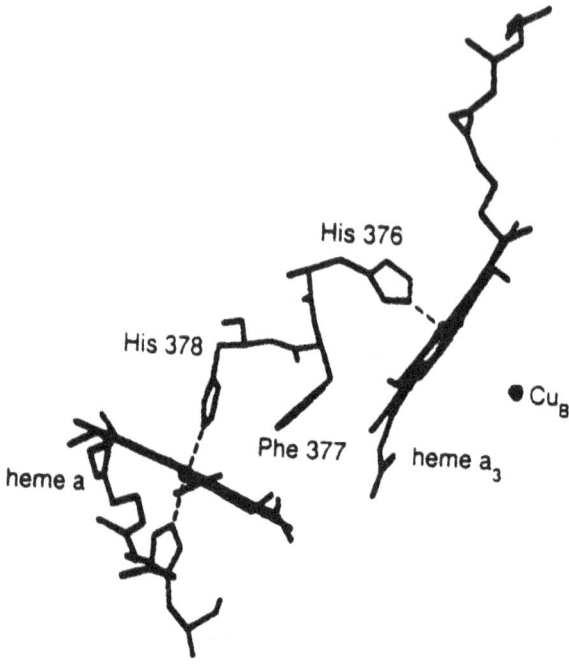

His 376

His 378

Phe 377

heme a

heme a$_3$

$\bullet$ Cu$_B$

Figure 1. Structure of the active site of cytochrome *c* oxidase.
From the Protein Data Bank coordinates deposited by Tsukihara et al. [1].

a to cytochrome a_3 is very fast (μs), but the apparent rate constant for reduction of cytochrome a_3 is slow because thermodynamics favors reduced cytochrome *a*; thus given the redox potential of the two sites [10], only a fraction (< 10%) of reduced cytochrome a_3 will be populated on a short time scale (as short as the rate of electron delivery to cytochrome *a*). Since this fraction of reduced cytochrome a_3 should be available for combination with any molecule acting as a "trapping" ligand for reduced cytochrome a_3, Verkhovsky *et al.* [9] assumed that the rate of diffusion and/or binding of H$^+$ to the reduced site is slow, and this would account for the relatively slow (**ms**) rate of reduction of cytochrome a_3 *vis-a-vis* a very rapid (μs) electron transfer. The pH dependence of the process (see also Malatesta et al. [8]) was not inconsistent with this hypothesis, the apparent rate constant increasing at acidic pH 7.0 (vs pH 9.0) by a factor of only 3/pH unit.

The **kinetic control** hypothesis states that in the oxidized enzyme, the intrinsic rate of electron transfer from cytochrome *a* to cytochrome a_3 is slow (**ms**). If reduction of cytochrome a is sufficiently fast, reduction of cytochrome a_3 will rate limit the binding of a "trapping" ligand specific for the reduced site, provided the ligation is sufficiently fast; this would make reduction of cytochrome a_3 and ligation synchronous. If these conditions apply, the rate of cytochrome a_3 reduction should be independent of the concentration of both the reductant and the "trapping" ligand. We have developed [7,8] an experimental protocol to probe electron transfer to cytochrome a_3 by mixing the oxidized (fast or "pulsed") enzyme with a reductant solution containing either NO or CO, which are known to bind quickly and tightly to reduced cytochrome a_3 and thus are ideal "trapping" ligands for this site.

Since electron transfer and coupled proton pumping are the common and interesting properties of all terminal oxidases [11,12], we have addressed again the problem of internal ET to the oxidized binuclear center, and carried out new experiments using nitric oxide (NO) to trap reduced cytochrome a_3.

The anaerobic reduction by ruthenium hexamine of "fast" oxidized cytochrome c oxidase has been investigated by stopped flow spectroscopy. The electron entry site in cytochrome c oxidase, the binuclear copper center called Cu_A, is in very rapid redox equilibrium with cytochrome a ($k = 1.8 \times 10^4$ s^{-1}), the assumed electron donor to the binuclear cytochrome a_3-Cu_B center. To trap reduced cytochrome a_3, we used CO as well as NO. NO is the most efficient "trapping" ligand, because of its high combination rate ($k_{on} = 1 \times 10^8$ M^{-1} s^{-1}, [13]), and affinity ($K_a = 10^9$ M^{-1} given a dissociation rate constant $k_{off} = 0.1$ s^{-1}, [14]). Experiments show that NO is efficient as expected, because reduction of cytochrome a_3 lags behind the reduction of cytochrome a, but is synchronous with NO binding. The time courses at 438 nm indicate that reduction of cytochrome a_3 in the presence of NO is somewhat faster (k' = 22 s^{-1}) than in the presence of CO (k' = 13 s^{-1}); at 431 nm a fast absorbance decrease, corresponding to reduction of cytochrome a, is followed by the formation of the reduced cytochrome a_3-ligand adduct (k' = 19 s^{-1} with NO and k' = 4.5 s^{-1} with CO). Given the relatively small combination rate constant for CO ($k = 8 \times 10^4$ M^{-1} s^{-1}, [15]) the formation of the CO-bound derivative lags behind the reduction of cytochrome a_3, whereas NO binding is synchronous to cytochrome a_3 reduction.

The anaerobic reduction of oxidized oxidase measured at different concentrations of ruthenium hexamine (in the presence of NO) shows a linear dependence of the pseudo first order rate constant on the concentration of reductant ($k_{on} = 1.2 \times 10^5$ M^{-1} s^{-1}); on the other hand, the rate of formation of cytochrome a_3^{2+}-NO is independent of ruthenium concentration.

In Figure 2 the rate constant for the formation of NO-bound and CO-bound reduced cytochrome a_3 is reported as a function of ligand concentration. It may be seen that the

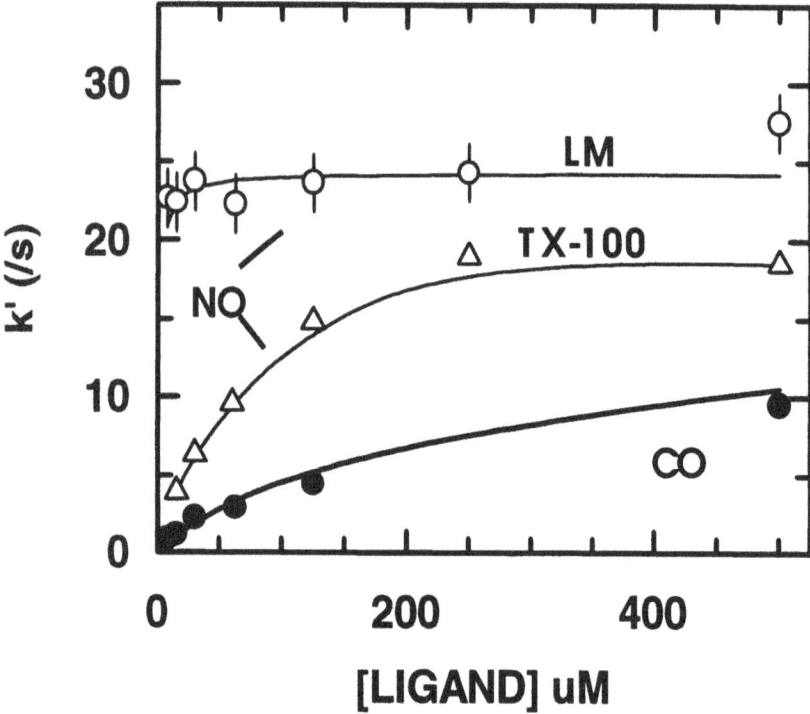

Figure 2. Effect of the concentration of the "trapping" ligand on internal ET.

Rate constants for the formation of the NO-adduct (open symbols) and the CO-adduct (closed symbols) of cytochrome a_3^{2+} as a function of the concentration of the gaseous ligands. T = 20 C. Buffer: 100 mM potassium phosphate pH =7.

The formation of the cytochrome a_3^{2+}-NO adduct is essentially independent of NO concentration in the presence of 0.1 % lauryl maltoside (o), whereas it shows a hyperbolic dependence in the presence of 1% Triton X-100 (Δ). In both cases the plateau level is similar (k' 20-25 s^{-1}) and assigned to the forward rate constant of internal ET (a ---> a_3).

On the contrary the formation of cytochrome a_3^{2+}-CO adduct (only reported in 0.1 % lauryl maltoside, (o)) is dependent on CO concentration and lags behind the reduction of cytochrome a_3, as expected on the basis of the much lower combination rate constant for CO binding (k_{CO} = 8 x 10^4 M^{-1} s^{-1}).

rate of formation of the cytochrome a_3^{2+}-NO adduct is independent of NO concentration, but slightly higher than the average rate constant (k' = 16 s^{-1}) for the reduction of cytochrome a_3 in the absence of external ligands. On the contrary, the rate of formation of the cytochrome a_3^{2+}-CO adduct is slower and dependent on CO concentration. The different behaviour can be rationalized because the combination rate constants of the two ligands with reduced cytochrome a_3 are different (k_{NO} = 1 x

10^8 M^{-1} s^{-1} [13] and k$_{CO}$ = 8 x 10^4 M^{-1} s^{-1}, [15]). Given the relatively slow second-order rate constant for CO, the profile shown in Figure 2 is expected, and indeed this behaviour was reproduced by simulation. These findings demonstrate that using "fast" oxidase preparations, CO is inadequate to the role of "trapping" ligand even at the highest concentration, 0.5 mM (k'= 40 s^{-1}), whereas NO even at the lowest, 10 μM (k'= 1000 s^{-1}) is definitely suitable. The new data shown in Figure 2 are difficult to reconcile with the hypothesis that cytochrome a and a_3 are in very fast redox equilibrium (μs) and that uptake of protons is the rate limiting step in the reduction of cytochrome a_3 (and Cu$_B$). The demonstration that the rate constant for the formation of cytochrome a_3^{2+}-NO is independent of NO concentration implies a rate limiting process (k' = 25 s^{-1} at *plateau*) and is proof that binding of this gaseous ligand is rate-limited by a monomolecular process. We conclude that this is the slow (ms) electron transfer to cytochrome a_3. To mantain Verkhovsky's hypothesis [9], it may be necessary to assume that NO **cannot** bind to reduced cytochrome a_3 **unless** a H$^+$ is already bound at the site; in this case proton binding and/or diffusion would limit NO binding.

The experiments reported above were carried out in the presence of 0.1 % lauryl maltoside; thus, they have been repeated in the presence of 1 % Triton X-100. It has been known for a long time that Triton X-100 inhibits cytochrome c oxidase (see for instance ref. 16). Consistently, in the presence of Triton X-100, the apparent rate constant for cytochrome a_3 reduction (**in the absence of trapping ligands**) was shown to be significantly smaller (k' ≤ 1 s^{-1}) than in the presence of lauryl maltoside (k' = 16 s^{-1}). Figure 3 depicts the time course of formation of the cytochrome a_3^{2+}-NO adduct, upon increasing NO concentration in the presence of lauryl maltoside (upper panel) and Triton X-100 (lower panel).

Contrary to what is reported above in lauryl maltoside, in Triton X-100 the cytochrome a ---> a_3 ET apparent rate increases upon increasing NO concentration. As shown in Figure 2, in the presence of Triton X-100 the rate constant for cytochrome a_3 reduction approaches at the highest NO concentrations a *plateau* level similar to the one observed in the presence of lauryl maltoside (k' = 20-25 s^{-1}). This similarity suggests that the intrinsic forward rate costant for the a ---> a_3 ET is essentially the same with and without Triton X-100. Nevertheless, the slower apparent reduction of cytochrome a_3 observed in the absence of trapping ligands and the dependence of the rate constant for internal ET on NO concentration, strongly suggest that Triton X-100 is responsible for an additional shift in the position of the a <==> a_3 redox equilibrium in favor of cytochrome a reduction, and thus an increase in the reverse a_3 ---> a ET rate.

256

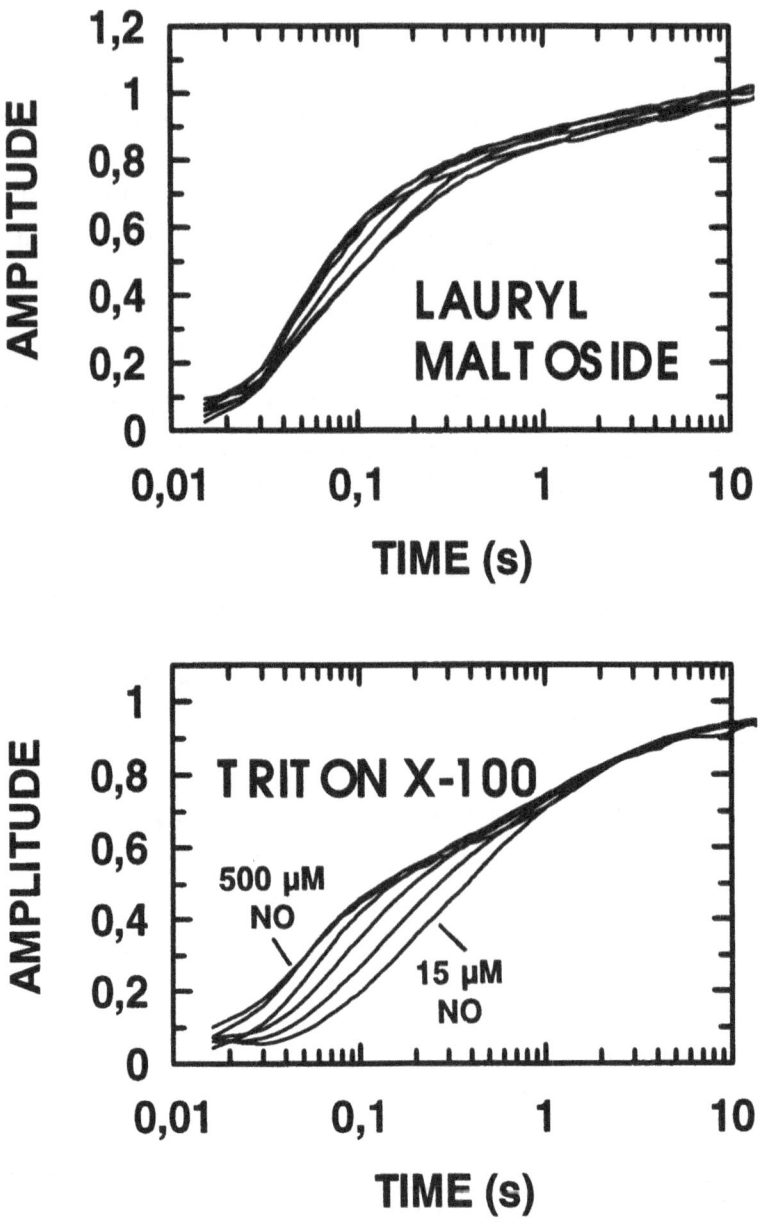

Figure 3. Effect of detergent on internal ET.
Time courses of formation of the cytochrome $a_3{}^{2+}$-NO adduct at different NO concentrations (from 15 to 500 μM) in the presence of 0.1 % lauryl maltoside or 1 % Triton X-100. Experimental conditions as in Figure 2. In lauryl maltoside the internal ET rate is essentially independent of NO concentration, wheras in Triton X-100 the rate increases upon increasing NO concentration. The different behaviour can be rationalized assuming that in Triton X-100 the reverse ET ($a_3 \rightarrow a$) is much faster than in lauryl maltoside.

It is also very interesting that in Triton X-100, when CO is used as a trapping ligand, the formation of the cytochrome a_3^{2+}-CO adduct (i) is much slower (k' ≤ 1 s^{-1}) than the formation of the NO-adduct; (ii) is independent of CO concentration; and (iii) proceeds at a rate consistent with the maximal turnover number independently measured under similar conditions (TN = 0.2 mol O_2 / mol aa_3 x s). These data can be rationalized keeping in mind that O_2 and CO binding to reduced cytochrome a_3 demands a fully reduced cytochrome a_3-Cu$_B$ site, while we have shown that this is not a prerequisite for NO binding [14]. Therefore, we conclude that in Triton X-100 k' ≤ 1 s^{-1} is the rate of full reduction of the cytochrome a_3-Cu$_B$ binuclear center. We believe this to be the rate limiting step in turnover, in view of the agreement with the maximal turnover number of the enzyme. This view is also supported by the finding that in lauryl maltoside, the formation of the cytochrome a_3^{2+}-CO adduct at 0.5 mM CO is much faster (k' ≈ 10 s^{-1}) than in the presence of Triton X-100 and, consistently, also the maximal turnover number of the enzyme is significantly higher (TN = 15 mol O_2/ mol aa_3 x s).

In conclusion, the kinetic data on reduction of cytochrome a_3 and NO binding are difficult to reconcile with the hypothesis that in the oxidized enzyme cytochrome a and a_3 are in very fast (μs) redox equilibrium, and that H$^+$ diffusion and/or binding to the reduced binuclear site is **the unique** rate limiting step in the build up of reduced cytochrome a_3. Therefore we conclude that starting from the oxidized enzyme, internal ET to cytochrome a_3 is slow (**ms**), and rate limiting the turnover of the enzyme [8], while starting from the reduced configuration of the binuclear center (with or without a bound ligand) internal ET occurs very rapidly (μs). As suggested before [17,5] these dramatic differences in ET rate may be accounted for by a different coordination of cytochrome a_3 in the two oxidation states, involving in turn a higher reorganizational energy to achieve reduction of the binuclear site. Moreover we observe that the detergent used (lauryl maltoside or Triton X-100) is likely to have an effect on the reverse internal ET (a_3--->a), and in Triton X-100 the enzyme is characterized by an even faster reverse rate. This interpretation would account for both the slower apparent reduction of cytochrome a_3 (observed in the absence of trapping ligands) and the dependence of the internal ET kinetics on the concentration of NO. In conclusion the results of the experiments carried out in Triton X-100 combined with the kinetics of the NO binding, by splitting in time the donation of the two electrons to the binuclear site, allowed us to assign to the second electron reduction the role of rate limiting step in turnover.

258

References

1. Tsukihara, T., Aoyama, H., Yamashita, E., Tomikazi, T., Yamaguchi, H., Shinzawa-Itoh, K., Nakashima, R., Yaono, R. and Yoshikawa, S. (1995), Structures of metal sites of oxidized bovine heart cytochrome c oxidase at 2.8 Å, *Science*, **269**, 1069-1074.
2. Han, S., Ching, Y. and Rousseau, D. (1990), Primary intermediate in the reaction of oxygen with fully reduced cytochrome c oxidase, *Proc.Natl.Acad.Sci USA*, **87**, 2491-2495.
3. Oliveberg, M. and Malmström, Bo G. (1991), Internal electron transfer in cytochrome c oxidase: evidence for a rapid equilibrium between cytochrome a and the bimetallic sitte, *Biochemistry*, **30**, 7053-7057.
4. Verkhosvky, M.I., Morgan, J.E. and Wikström, M. (1994), Oxygen binding and activation: early steps in the reaction of oxygen with cytochrome c oxidase, *Biochemistry*, **33**, 3079-3086.
5. Woodruff, W.H. (1993), Coordination dyanmics of heme-copper oxidases. The ligand shuttle and the control and coupling of electron transfer and proton translocation, *J.Bioenerg.Biomemb.*, **25**, 177-188.
6. Gibson, Q., Greenwwod, C., Wharton, D.C. and Palmer G. (1965), The reactions of cytochrome oxidase with cytochrome c, *J.Biol.Chem.*, **240**, 888-894.
7. Sarti, P., Antonini, G., Malatesta, F., Vallone, B. and Brunori, M. (1988), Is the internal electron transfer the rate-limiting step in the catalytic cycle of cytochrome c oxidase?, *Ann. N.Y Acad. Sci.*, **550**, 161-166.
8. Malatesta, F., Sarti, P., Antonini, G., Vallone, B. and Brunori, M. (1990), Electron transfer to the binuclear center in cytochrome oxidase: catalytic significance and evidence for an additional intermediate, *Proc.Natl.Acad.Sci USA*, **87**, 7410-7413.
9. Verkhovsky, M.I., Morgan, J.E. and Wikström, M. (1995), Control of electron delivery to the oxygen reduction site of cytochrome c oxidase: a role for protons, *Biochemistry*, **34**, 7483-7491.
10. Blair, D.F., Ellis, W.R., Wang, H., Gray, H.B. and Chan, S.I. (1986), Spectroelettrochemical study of cytochrome c oxidase: pH and temperature dependences of the cytochrome potentials, *J.Biol. Chem.*, **261**, 11524 - 11537.
11. Babcock, G.T. and Wikström, M. (1992), Oxygen activation and the conservation of energy in cell respiration, *Nature*, **356**, 301-309.
12. Malatesta, F., Antonini, G., Sarti, P. and Brunori, M. (1995), Structure and function of a molecular machine, *Biophys.Chem.*, **54**, 1-33.
13. Blackmore, R.S., Greenwood, C. and Gibson, Q.H. (1991), Studies of the primary oxygen intermediate in the reaction of fully reduced cytchrome oxdase, *J.Biol.Chem.*, **266**, 19245-19249.
14. Giuffrè, A., Sarti, P., D'Itri, E., Buse, G., Soulimane, T. and Brunori, M. (1996), On the mechanism of inhibition of cytochrome c oxidase by nitric oxide, *J.Biol.Chem.*, **271**, 33404-33408.
15. Gibson Q.H. and Greenwood, C. (1963), Reactions of cytochrome oxidase with oxygen and carbon monoxide, *Biochem.J.*, **86**, 541-555.
16. Mahapatro, S.N. and Robinson, N.C. (1990), Effect of changing detergent bound to bovine cytochrome c oxidase upon its individual eletron transfer steps, *Biochemistry*, **29**, 764-770.
17. Brunori, M., Antonini, G., Giuffrè, A., Malatesta, F., Nicoletti, F., Sarti, P. and Wilson, M.T. (1994), Electron transfer and ligand binding in terminal oxidases: impact of recent structural information, *FEBS Lett.*, **350**, 164-168.

CHIMERIC QUINOL OXIDASES EXPRESSED IN *PARACOCCUS DENITRIFICANS*

C. WINTERSTEIN, O.-M.H. RICHTER AND B. LUDWIG

Molecular Genetics, Institute of Biochemistry, Biozentrum der J.W. Goethe-Universität, Marie-Curie-Str. 9, D 60439 Frankfurt, Germany

1.   Abstract

Quinol oxidases from both *Escherichia coli* (cytochrome bo_3) and *Paracoccus denitrificans* (ba_3) contain four different subunits supposedly sharing similar membrane topography, and are coded for in the *cyo* and *qox* operon, respectively. By deleting stopcodons in intercistronic regions, we have constructed different fusions between the *P. denitrificans* quinol oxidase subunits; all complexes are expressed in *P. denitrificans* and show enzymatic activities in membranes that are comparable to the unfused complex. Subsequently, chimeric complexes are constructed in which the subunit II gene (*cyoA*) replaces the endogenous *qoxA*; it is expressed either as an individual subunit or as a fused chimera, assembled with the other subunits of the *P. denitrificans* cytochrome ba_3 complex. Such heterologous enzymes retain a residual activity of 15-20% both in membranes and in the isolated complexes, which are purified in a two-step procedure by affinity chromatography using a strep-tag on subunit I. Spectral and protein chemical data further suggest a certain degree of instability of the mutated chimeric complexes. We conclude, however, that subunit II of *E. coli* basically interacts with the remaining subunits, notably subunit I, of the *P. denitrificans* ba_3 oxidase in the same way as the endogenous subunit, and sustains quinol oxidation and electron transfer activity to subunit I.

2.   Introduction

Both quinol oxidases and cytochrome *c* oxidases are members of the superfamily of terminal oxidases [1-4] characterized by their haem • copper binuclear centre accommodated in subunit (su) I. At this site, binding and reduction of oxygen, and most likely, also the coupling between electron transport and proton pumping takes place [5,6]. While, as a clear distinction from quinol oxidases, su II of aa_3-type cytochrome *c* oxidases acts as the entry point of electrons from cytochrome *c* and carry a Cu_A site occupied by two copper ions, su II of quinol oxidases shows a corresponding topography, but is devoid of any metal cofactor [e.g. 7,8]. This subunit has, however, been suggested to be involved in the binding of (substrate) quinol [9].

Recently, two different cytochrome *c* oxidase 3D structures have been determined

G.W. Canters and E. Vijgenboom (eds.),
Biological Electron Transfer Chains: Genetics, Composition and Mode of Operation, 259-269.

[10,11], but no immediate structural information is available on any quinol oxidase. Sequence comparisons as well as spectral and mutational analyses suggest a high degree of homology in subunit I [e.g. 12-15], whereas su II lacks the characteristic metal ligands altogether and exhibits only a low degree of sequence similarity to cytochrome c oxidase su II. With the experimental goal in mind to investigate whether it is essentially su II that determines whether a complex operates as a quinol or cytochrome c oxidase, we have started to construct fusion complexes between different subunits and assay their expression and structural and functional integrity.

In a first step, the cytochrome ba_3 complex of P. denitrificans (Qox) was probed. It is composed of four subunits [8,14] that have been suggested to be oriented across the membrane in a way that fusions of C-termini to subsequent N-termini should be structurally feasible in the order of su II-I-III. Such an approach has been successfully applied previously to the quinol oxidase of E. coli (Cyo), confirming the membrane topology of its subunits [16].

In a second step, we created heterologous, chimeric quinol oxidase complexes by genetic exchange of su II (QoxA) against CyoA. Both heterologous constructs, a forced chimera with a CyoA-QoxB fusion and a complex which assembles spontaneously from structurally individual subunits, are expressed in P. denitrificans and exhibit a residual quinol oxidase activity. From these data we conclude that quinol oxidases of both organisms not only show a similar topography of subunits, but in particular that su II is largely interchangeable in both enzymes. This fact points at a similar structural arrangement of these subunits with respect to the other subunits, and to a functional homology in terms of quinol reduction and intra-complex electron transport.

3.   Material and Method

3.1   STRAINS AND GROWTH CONDITIONS

For the expression of the native and modified qox operon [14], a chromosomal quinol oxidase deletion strain ($\Delta qox::kan$; strain ORI 2/4) [15] is used throughout. The qox operon is introduced on the broad host range vector pEG400 by tri-parental mating [17] and transconjugants are grown aerobically on succinate medium in the presence of streptomycin sulfate (25 μg/ml), as described earlier [18]. Harvested cells are disrupted and membranes isolated as described previously [14,18].

3.2   SITE DIRECTED MUTAGENESIS

The 'altered sites' mutagenesis protocol (Promega) was followed, using the pAlter vector. All other steps of DNA modification were done essentially according to [19].

3.2.1   Construction of Homologous Fusions between Qox Subunits
Strain CW3, expressing a su II/I fusion complex. The intercistronic region between qoxA and qoxB (encoding su II and I of the P. denitrificans quinol oxidase) consists of 15 nucleotides, including the stop codon. An oligonucleotide FUI-II (see below),

overlapping with the 3'-end of *qox*A and the start of *qox*B, deletes the intercistronic region including the stop codon and creates a new restriction site for *Afl*II.

Strain CW4, expressing a su I/III fusion complex. A number of six nucleotides in the intercistronic region between *qox*B/C is maintained, but their sequence is changed to inactivate the stop codon using oligonucleotide FUI/III. In the same time a *Kpn*I site is introduced.

Strain CW5, carrying both the su II/I and the I/III fusion (see above), is generated accordingly, in a two-step mutagenesis. All constructs are cloned as *Xba*I-*Hind*III fragments into pEG400 and conjugated into the deletion strain ORI 2/4.

3.2.2 *Construction of Two Forms of a Chimeric Operon*

The *E. coli cyo*A gene [12], coding for su II of the cytochrome *bo₃* quinol oxidase, is introduced into the *qox* operon of *P. denitrificans*, to replace the endogenous su II gene, *qox*A (see also Fig. 2).

*Cyo*A translated as an individual polypeptide. The *qox* promoter region and an N-terminal part of the coding region of su II (the complete signal sequence and the codons for the first three amino acids of the mature polypeptide of the *ba₃* subunit) are maintained and fused in-frame to the corresponding 5'-start of *cyo*A. To this end, an *Spe*I restriction site is generated in identical regions of both genes, using oligonucleotide SigCyoA and SigQoxA respectively, changing the first three amino acids of su II of *bo₃* into KAE. In the intergenic region following the stop codon of *cyo*A, primer CyoAoFu creates an *Afl*II site which is also introduced into the corresponding region downstream of *qox*A, using oligonucleotide QoxABoFu. The *cyo*A gene is introduced into the *qox* operon background as a *Spe*I-*Afl*II fragment, replacing the endogenous *qox*A gene. The chimeric operon is introduced into ORI 2/4 on the broad host range vector (cloned as a *Xba*I-*Hind*III fragment) pEG400.

The *cyo*A gene fused to the *qox*B gene. While the exchange between the two genes in their 5'-region is done identically as outlined above, an oligonucleotide CyoAFU is used to replace the stop codon of *cyo*A and to generate an *Afl*II restriction site. The chimeric construct is obtained and propagated as described above.

3.2.3 *Strep-tag construction*

The nine amino acid strep-tag is added to the C-terminus of subunit I (QoxB), using the oligonucleotide Strep-I [20,21].

3.2.4 *Oligonucleotides used for Mutagenesis*

FUI-II	5'-GTTGGAAAACGTAGCGCTTAAGTTCCGCGCCCG-3'
FUI/III	5'-CGCGTGGCTCATGTCTGGTACCCCCTGCGCAAG-3'
SigCyoA	5'-TGTCCTTTGGGAACTAGTAGCGCAGAATTA-3'
SigQoxA	5'-TCCCCGGCGGGCACTAGTACCTCGGCCTTG-3'
CyoAoFu	5'-ATCTTTATTCTTCTTAAGCCCCTTTAATGG-3'
QoxABoFu	5'-TTGCGTCCGATCTTAAGTTCCGCGC-3'
CyoAFU	5'-CTCAACCCCTTAAGTGGGCGGATTC-3'
Strep-I	5'-GGTCGTCGCGTGGCTCATGTGTTAACCGCCGAACTG
	CGGATGGCGCCACGCAACCCCCTGCGCAAGCTG-3'

3.3. ACTIVITY ASSAY

Quinol oxidase activity of membranes is measured at room temperature in a Kontron Uvikon 941 in 50 mM KP_i (pH 7.5), 1 mM EDTA, and 50 µM of the decyl analog of ubiquinol (Sigma) ($\Delta\varepsilon_{ox\text{-}red, 275 nm}$ = 12.5 $mM^{-1}cm^{-1}$) in the presence of 2 µM myxothiazol; for purified enzymes, the detergent dodecyl ethoxysulfate ([8]; 5 g/l) is included in the assay buffer.

3.4. AFFINITY CHROMATOGRAPHY OF TAGGED QUINOL OXIDASE COMPLEXES

All steps are performed at 4⁰C. Membranes are solubilized at 10 mg/ml by adding 1.5g n-dodecyl-β-D-maltoside per gram membrane protein in 50 mM KP_i (pH 8), 1 mM EDTA, 50 mM NaCl, 1 mg avidin per gramme membrane protein, and 50 µM Pefabloc SC (Biomol, Hamburg). After centrifugation at 100.000 x g, the supernatant is loaded onto a pre-equilibrated (50 mM KP_i (pH 8), 100 mM NaCl, 1 mM EDTA, 0.2 g/l dodecyl maltoside) streptavidin sepharose column (15 ml bed volume) with a flow rate of 30 ml/h. After washing with 40 ml equilibration buffer the strep-tag complex is eluted with 1 mM diaminobiotin (Sigma) in equilibration buffer [21]. Optionally, a gel filtration step may be added to remove the diaminobiotin.

3.5. MISCELLANEOUS

SDS polyacrylamide gel electrophoresis, immunological detection of subunits by Western blotting, using a specific polyclonal antiserum directed against subunit II epitopes, were done as described earlier [17,22].

4.   Results and Discussion

The cytochrome ba_3 quinol oxidase from *P. denitrificans* consists of four different subunits in the molecular weight range between 70 to 14 kDa [8,14] which are coded in an operon structure (*qox*). Its subunit I carrying the three metal redox centres (both haems and the Cu_B) traverses the membrane 15 times in an α-helical arrangement, an assumption based on sequence and hydropathy similarities to the *E. coli* enzyme [7,12]. Extending the high structural analogy to su II (2 helices) and III (5 helices) as well (see [14]), and taking into account 3D information from related cytochrome *c* oxidases [10] this would place the N-termini of su I and II and the C-termini of II and III on the periplasmic side of the membrane, whereas the C-terminal end of su I and the start of su III should be located on the opposite (inner) side of the cytoplasmic membrane. Consequently, fusion constructs (see [23] for an overview) should be topologically feasible between subunits (coded for by adjacent genes) in the order II-I and I-III.

4.1 FUSIONS BETWEEN DIFFERENT SUBUNITS OF THE *P. DENITRIFICANS* QUINOL OXIDASE (Qox)

Figure 1 and Table 1 show that the three constructs, two single (CW3 and CW4) and the double fusion (CW5), can be expressed from the broad host range plasmid pEG400 *in trans* in the homologous host (ORI 2/4) which is deleted of its genomic *qox* copy. Since all available evidence suggests that no other quinol oxidase activity is found in *P. denitrificans*, enzymatic activities determined in membranes truly account for the fusion complexes.

It should be noted that membranes isolated from the three mutant strains and the corresponding wild-type strain, ORI 2KK, all show about the same activity which shows that the genetic fusion does not alter the subunit arrangement in the complex in any serious way to interfere with the inner-subunit electron transfer (see below). One may speculate that the new extra-membraneous loop regions created by the fusions are flexible enough to allow for native subunit contacts to form. Nor is the expression and membrane insertion of such artefactual polyproteins consisting of up to 22 membrane-spanning helices impeded in any dramatic way. This fact is also supported by the stability of the fused subunits as indicated by the obvious lack of degradation products recognized by the su II-antibody (Fig. 1, lanes 3 and 4).

The above arguments confirm the assumed membrane topology and sidedness for the three largest subunits, as has been found previously for the corresponding Cyo complex of *E. coli* [16] where similar fusions also retained high enzymatic activity. Interestingly, fusion proteins between oxidase subunits are not without precedent in nature: a su I/III fusion is found in *Thermus thermophilus caa₃* cytochrome *c* oxidase [24] and in *Sulfolobus acidocaldarius* quinol oxidase [25].

Apparent molecular masses derived from SDS gels (Fig. 1) match well for the native su II (DNA derived molecular weight 42 kDa), but are known to be drastically underestimated when highly hydrophobic polypeptides such as su I or su III are involved (e.g. actual molecular weight of native su I is 75 kDa; [14]). This explains the obvious discrepancies for the observed molecular masses of the fusion subunits.

TABLE 1. Quinol oxidase activity of Qox subunit fusion constructs in membranes

Strain	Fused subunits in complex	Activity (U/mg)
CW3	II, I	1.0
CW4	I, III	1.8
CW5	II, I, III	1.4
ORI 2KK[*]	(wild type)	1.2

[*] native *qox operon expressed* in trans in strain ORI 2/4

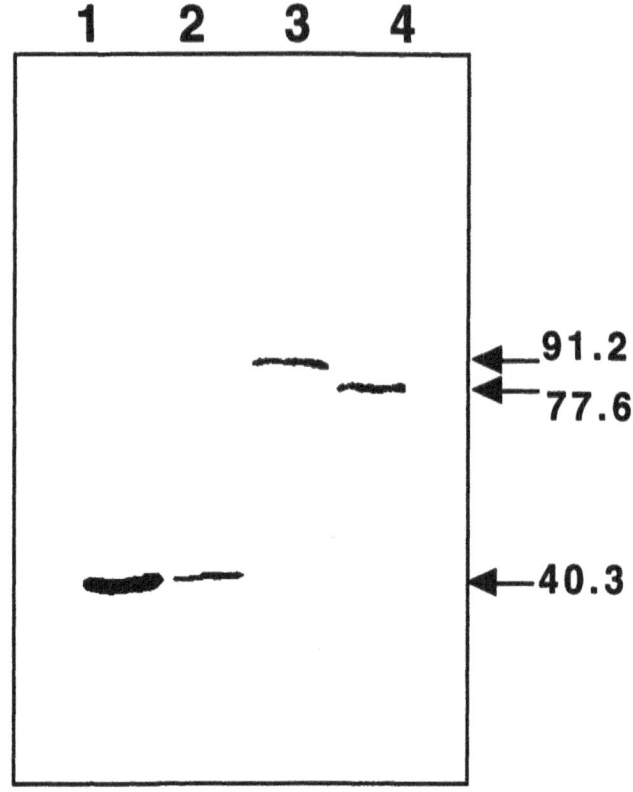

Figure 1. Western blot of wild-type and Qox fusion constructs. Membranes (100 µg of protein) were separated on a 8% polyacrylamide gel and subunit II was detected by a subunit-specific antibody [15]. Lane 1, strain G440 (cytochrome bc_1 deletion strain overexpressing the ba_3 quinol oxidase [17]); lane 2, Pd1222 (wild-type strain [26]); lane 3, CW5; lane 4, CW3. See also Table 1 and Material and Methods. Arrows and numbers denote the position and apparent molecular mass (kDa) of subunit II-related bands.

4.2. CHIMERIC QUINOL OXIDASE CONSTRUCTS

For quinol oxidases the best fits of subunit sequence alignments for the Qox complex are obtained with the Cyo quinol oxidase complex of *E. coli* [14]. Therefore, we wanted to probe whether subunits are interchangeable among quinol oxidases of both organisms, either in a system expressing and assembling individual subunits freely, or forced by fusing genes and their gene products, respectively (see Fig. 2 for the experimental outline).

In all cases, the genetic environment was that of the *qox* operon of *P. denitrificans*, in which the su II coding *qox*A was replaced by the corresponding *E. coli* gene, *cyo*A. Again, expression was followed in the homologous host strain ORI 2/4, as described in the previous section.

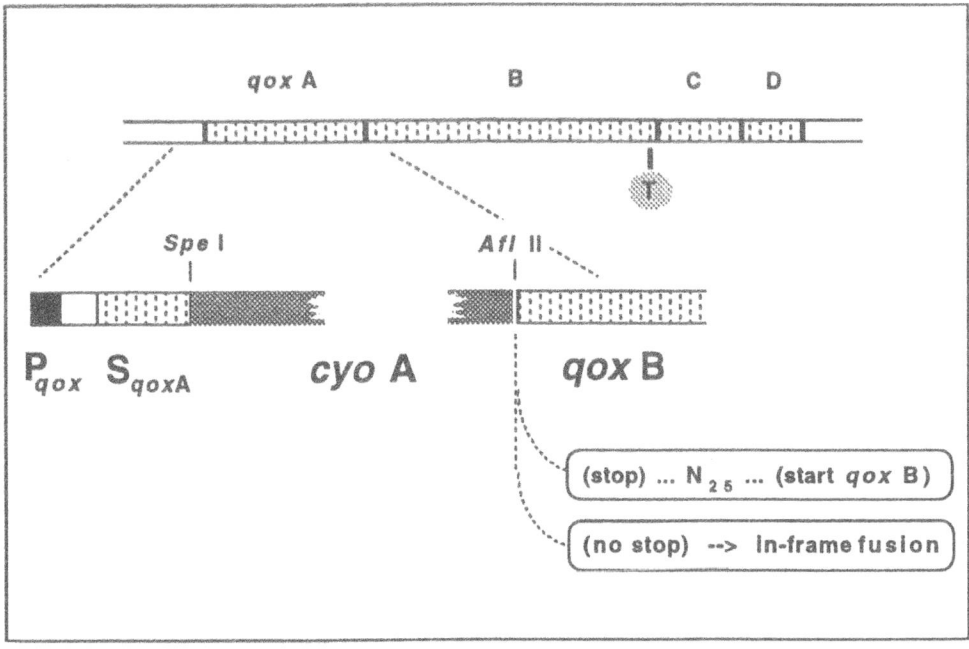

Figure 2. Construction of a chimeric operon between *cyo*A of *E. coli* and *qox*BCD of *P. denitrificans.*

Top: structure of *qox* operon with an engineered strep-tag (T) at the 3'-end of *qox*B

Centre: construct for replacement of *qox*A by *cyo*A, using the two engineered restriction sites *Spe*I and *Afl*II; P, promoter and S, signal sequence derived from *qox*A; not drawn to scale

Bottom: 3'-end of *cyo*A, leading to expression of the individual subunit or , alternatively, to a fusion polypeptide

Since initial results revealed that activities of chimeric enzymes in membranes were much lower than in the wild-type enzyme (Table 2), activity was also determined for the purified complexes. By tagging the C-terminus of su I (QoxB) with a 9 amino acid streptavidin-recognizing peptide [29], we employed a simple affinity chromatography purification ([21]; see below) to yield complexes of high purity (Fig. 3). While the isolated wild-type complex (Fig. 3, lane 3) shows the typical band pattern with su II exhibiting three Coomassie staining bands (in the region between 42 and 31 kDa, just below su I, see also [14]) and su I as well as III and IV well resolved, both chimeric complexes stain distinctly different. The su II region is now represented by a single homogeneous band in lane 2 (the non-fused CyoA, smaller in size than QoxA [14]), whereas the fusion product between su II and QoxB runs at a much higher apparent molecular weight, as expected (lane 1). As is obvious from Figure 3, the level of the heterologous su II in the purified complex (unfused, Fig. 3, lane 2) is clearly less than stoichiometric, in line with the activity data of Table 2. In comparison to the wild-type (tagged) ba_3 quinol oxidase, all heterologous enzymes are significantly reduced in their enzymatic activity in membranes. When isolated, they show the same enrichment factor

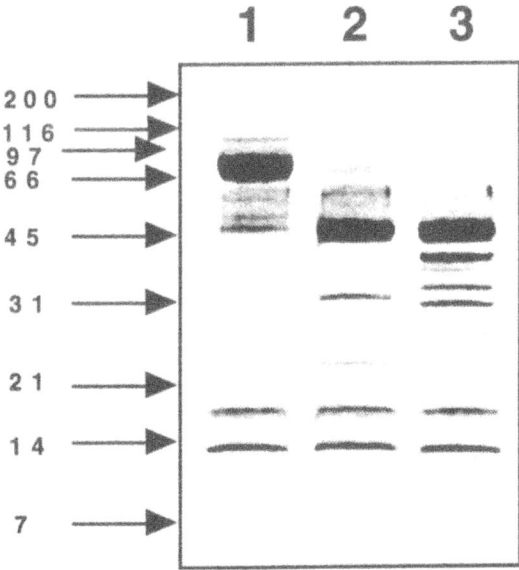

Figure 3. Subunit composition of purified chimeric quinol oxidase complexes. Chimeric complexes (lane 1, CyoA fused to QoxB; see also Fig. 2, and lane 2, no fusion) and wild-type (lane 3) were purified on a streptavidin affinity column, electrophoresed on a 10% SDS gel according to [27], and stained by Coomassie Blue. Arrows and numbers indicate migration and molecular mass (kDa) of protein standards.

of about 100 fold, but the activity again ranges only between 15-20% of the wild-type enzyme. This suggests that solubilization of membranes and chromatographic isolation of the chimeric enzymes is not causing a further inactivation. In the fused sample, the heterologous su II is kept in association with su I, with little if any signs of degradation, but in the unfused case the association with the remaining subunits is only of limited strength. Alternatively, and more likely, the foreign polypeptide is degraded in the membrane at an elevated rate.

The reduced activities correlate to some extent with the redox spectra of the isolated complexes (Fig. 4). Typically, the wild-type spectrum shows the b-type low-spin haem peak around 563 nm and a low-intensity peak around 607 nm for the high-spin a_3 component [13]. For the forced chimera (trace 2) the binuclear centre a_3 haem already decreased relative to the haem b, and the total haem content is even more reduced in the preparation of the unfused complex (trace 3). This observation could give a further explanation for the loss of activity in the latter case. Obviously, the low content of heterologous su II also influences the haem content, and in particular that of the binuclear centre in su I which is otherwise unmodified genetically. This effect is less drastic when su II is held in place in the fused complex, trace 2.

Table 2. Activity of chimeric quinol oxidases in membranes and as purified enzymes.

Construct	Activity in membranes (U/mg)	Activity of purified complex (U/mg)
chimera, fused, tagged	0.18	14
chimera, unfused, tagged	0.10	12
chimera, fused, no tag	0.12	n.d.
chimera, unfused, no tag	0.15	n.d.
wild-type QoxABCD, tagged	0.78	88

n.d., not determined

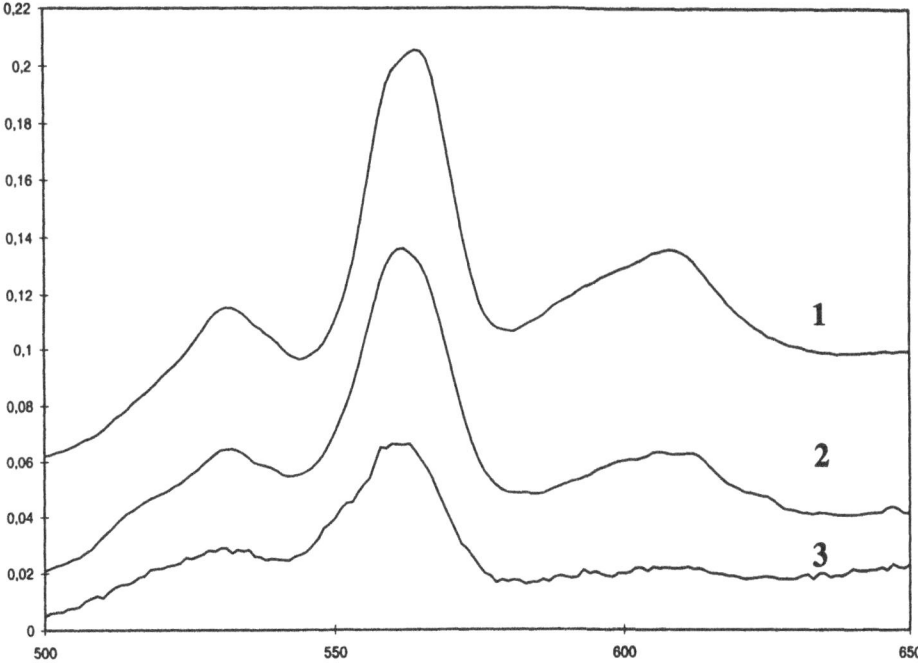

Figure 4. Optical difference spectra of purified chimeric oxidases. Spectra (dithionite-reduced *minus* oxidized) were recorded at a protein concentration of 2 mg/ml for the native enzyme (trace 1), chimera with (trace 2) and without fusion (trace 3), all carrying the strep-tag; see Material and Methods for details and Table 2 for enzymatic activities.

We have carried this line of experiments further to insert a gene fragment coding for the hydrophillic part of su II of cytochrome *c* oxidase (*ctaC*; [23]) into the *qox* operon, replacing the corresponding region of the native su II gene (C. Winterstein unpublished), to probe whether a quinol oxidase complex can be switched to a cytochrome oxidase function by this single domain exchange. Present evidence, however, suggests that while being expressed in membranes, no stable complex containing the domain-fused subunit can be purified.

For the chimeric quinol oxidase reported above we can conclude that subunits II of different origin are interchangeable to some extent as such complexes retain part of their quinol oxidase activity. This would also suggest that assumed functions for this subunit such as binding of substrate quinol [9,29] and internal electron transfer to the redox centres in su I are maintained in these two quinol oxidase complexes.

The use of a C-terminal streptavidin tag introduced into su I of this oxidase provides a fast and efficient way to obtain pure enzyme with a high specific activity in case of the wild-type (see Table 2), with yields up to 50%. Interestingly, also this single-step affinity chromatography purification lead to a heterologous pattern for su II as observed before [8], confirming that this pattern is not due to proteolysis during the previously used longer purification procedure.

5.   Acknowledgements

We are indebted to Bob Gennis (Urbana) for providing the cloned *E. coli cyo* operon, to Mårten Wikström and Anne Puustinen (Helsinki) for a gift of an antiserum directed against Cyo subunits, and to Irmela and Volker Zickermann (now Helsinki) for helpful comments and criticism in the course of this work.

6.   References

1. Trumpower, B.L. and Gennis, R.B. (1994) Energy transduction by cytochrome complexes in mitochondrial and bacterial respiration: The enzymology of coupling electron transfer reactions to transmembrane proton translocation, *Annu. Rev. Biochem.* **63**, 675-716.
2. Garcia-Horsman, J.A., Barquera, B., Rumbley, J., Ma, J. and Gennis, R.B. (1994) The superfamily of heme-copper respiratory oxidases, *J. Bacteriol.* **176**, 5587-5600.
3. Capaldi, R.A. (1990) Structure and Function of cytochrome *c* oxidase, *Annu. Rev. Biochem.* **59**, 569-596.
4. Saraste, M. (1990) Structural features of cytochrome oxidase, *Quart. Rev. Biophys.* **23**, 331-366.
5. Babcock, G.T. and Wikström, M. (1992) Oxygen activation and the conservation of energy in cell respiration, *Nature* **356**, 301-309.
6. Morgan, J.E., Verkovsky, M.I. and Wikström, M. (1994) The histidine cycle: a new model for proton translocation in the respiratory heme-copper oxidases, *J. Bioenerg. Biomembr.* **26**, 599-608.
7. Minghetti, K.C., Chepuri-Goswitz, V., Gabriel, N.E., Hill, J.J., Barassi, C.A., Georgiou, C.D., Chan, S.I. and Gennis, R.B. (1992) Modified, large-scale purification of the cytochrome *o* complex (*bo*-type oxidase) of *Escherichia coli* yields a two heme/one copper terminal oxidase with high specific activity. *Biochemistry* **31**, 6917-6924.
8. Zickermann, I., Anemüller, S., Richter, O.-M.H., Tautu, O.S., Link, T.A. and Ludwig, B. (1996) Biochemical and spectroscopic properties of the four-subunit quinol oxidase (cytochrome *ba₃*) from *Paracoccus denitrificans*, *Biochim. Biophys. Acta* **1277**, 93-102.
9. Welter, R., Gu, L.-Q., Yu, L., Yu, C.-A., Rumbley, J. and Gennis, R.B. (1994) Identification of the

ubiquinol-binding site in cytochrome bo_3 ubiquinol oxidase of *Escherichia coli*, *J. Biol. Chem.* **269**, 28834-28838.

10. Iwata, S. Ostermeier, C., Ludwig, B. and Michel, H. (1995) Structure at 2.8 Å resolution of cytochrome *c* oxidase from *Paracoccus denitrificans*, *Nature* **376**, 660-669.

11. Tsukihara, T., Aoyama, H., Yamshita, E., Tomizaki, T., Yamaguchi, H., Shinzawa-Itoh, K., Nakashima, R., Yaono, R. and Yoshikawa, S. (1996) The whole structure of the 13-subunit oxidized cytochrome *c* oxidase at 2.8 Å, *Science* **272**, 1136-1144.

12. Chepuri, V., Lemieux, L., Au, D.C.-T. and Gennis, R.B. (1990) The sequence of the *cyo* operon indicates substancial structural similarities between the cytochrome *o* ubiquinol oxidase of *Escherichia coli* and the aa_3-type family of cytochrome *c* oxidases, *J. Biol. Chem.* **265**, 11185-11192.

13. Ludwig, B. (1992) Terminal oxidases in *Paracoccus denitrificans*, *Biochim. Biophys. Acta* **1101**, 195-197.

14. Richte, O.-M.H., Tao, J.-S., Turba, A. and Ludwig, B. (1994) A cytochrome ba_3 functions as a quinol oxidase in *Paracoccus denitrificans* - purification, cloning, and sequence comparison, *J. Biol. Chem.* **269**, 23079-23086.

15. Zickermann, I., Tautu, O.S., Link, T.A., Korn, M., Ludwig, B. and Richter, O.-M.H. (1997) Expression studies on the iba_3 quinol oxidase from *Paracoccus denitrificans* - a bb_3 variant is enzymatically inactive, *Eur. J. Biochem.* **246**, 618-624.

16. Ma, J., Lemieux, L. and Gennis, R.B. (1993) Genetic fusion of subunits I, II, and III of the cytochrome *bo* ubiquinol oxidase from *Escherichia coli* results in a fully assembled and active enzyme, *Biochemistry* **32**, 7692-7697.

17. Gerhus, E., Steinrücke, P. and Ludwig, B. (1990) *Paracoccus denitrificans* cytochrome c_1 gene replacement mutants, *J. Bacteriol.* **172**, 2392-2400.

18. Ludwig, B. (1986) Cytochrome *c* oxidase from *Paracoccus denitrificans*, *Methods Enzymol.* **126**, 153-159.

19. Sambrook, J., Fritsch, E.F. and Maniatis, T. (1989) Molecular cloning: A laboratory manual, Cold Spring Harbor, New York.

20. Schmidt, T.G. and Skerra, A. (1993) The random peptide library-assisted engineering of a C-terminal affinity peptide, useful for the detection of a functional Ig F_v fragment, *Protein Engineering* **6**, 109-122.

21. Kleymann, G., Ostermeier, C., Ludwig, B., Skerra, A. and Michel, H. (1995) Engineered F_v fragments as a tool for the one-step purification of integral multisubunit membrane protein complexes, *Biotechnology* **13**, 155-160.

22. Steinrücke, P., Gerhus, E. and Ludwig, B. (1991) *Paracoccus denitrificans* mutants deleted in the gene for subunit II of cytochrome *c* oxidase also lack subunit I, *J. Biol. Chem.* **266**, 7676-7681.

23. Wales, M.E. and Wild, J.R. (1991) Analysis of structure-function relationships by formation of chimeric enzymes produced by gene fusion, *Methods Enzymol.* **202**, 687-706.

24. Mather, M.W., Springer, P., Hensel, S., Buse, G. and Fee, J.A. (1993) Cytochrome oxidase genes from *Thermus thermophilus*, *J. Biol. Chem.* **268**, 5395-5408.

25. Castresana, J., Lübben, M. and Saraste, M. (1995) New archaebacterial genes coding for redox proteins: Implications for the evolution of aerobic metabolism, *J. Mol. Biol.* **250**, 202-210.

26. de Vries, G.E., Harms, N., Hoogendijk, J. and Stouthamer, A.H. (1989) Isolation and characterization of *Paracoccus denitrificans* mutants with increased conjugation frequencies and pleiotropic loss of a (nGATCn) DNA-modifying property, *Arch. Microbiol.* **152**, 52-57.

27. Schägger, H. and von Jagow, G. (1987) Tricine-sodium dodecyl sulfate-polyacrylamide gel electrophoresis for the separation of proteins in the range from 1 to 100 kDa, *Anal. Biochem.* **166**, 368-379.

28. Steinrücke, P., Steffens, G.C.M., Pankus, G., Buse, G. and Ludwig, B. (1987) Subunit II of cytochrome *c* oxidase from *Paracoccus denitrificans*: DNA sequence, gene expression, and the protein, *Eur. J. Biochem.* **167**, 431-439.

29. Sato-Watanebe, M., Mogi, T., Ogura, T., Kitagawa, T., Miyoshi, H., Iwamura, H. and Anraku, Y. (1994) Identification of a novel quinone-binding site in the cytochrome *bo* complex from *Escherichia coli*, *J. Biol. Chem.* **269**, 28908-28912.

SUPERFAMILY OF CYTOCHROME OXIDASES

MATTI SARASTE, ANTONY WARNE AND ULRICH GOHLKE
European Molecular Biology Laboratory, Postfach 102209, D-69012 Heidelberg, Germany.

1. Introduction

Application of molecular biology to study the respiratory systems in bacteria has revealed the existence of a large superfamily of homologous haem-copper cytochrome oxidases [1-5]. There are several links between denitrification and oxygen-based respiration. The first oxidase may have evolved from a homologous denitrification enzyme, nitric oxide reductase [2].

Several aerobic bacteria have cytochrome c oxidases which are similar to the mitochondrial cytochrome aa_3. In addition, many bacteria such as *Escherichia coli* have terminal oxidases which use quinols as electron donors. A novel cytochrome c oxidase which is a complex of b- and c-type cytochromes and thus called cytochrome cbb_3, has been characterized in several purple bacteria. This enzyme appears to be the most distant member of the family [2] but it clearly shares the haem-copper active site which is the hallmark of the superfamily. A major difference in the molecular anatomy of different oxidases is the presence of a peculiar copper centre in the mitochondrial-type of oxidases. This Cu_A is a dinuclear mixed-valence centre and the primary acceptor of electrons from cytochrome c. It is absent from the homologous subunit of the quinol oxidases, indicating that these have evolved from cytochrome c oxidases.

Terminal oxidases belonging to the superfamily have also been found in archaeal species. *Sulfolobus acidocaldarius* contains two different terminal oxidases both of which are probably quinol oxidases (there is no cytochrome c in this organism) and use a sulphur-containing caldariellaquinol as electron donor [6,7]. Cloning of the oxidase genes and purification of these enzymes [3,8] has revealed that both *Sulfolobus* oxidases contain components of the cytochrome bc_1 complex.

Cytochrome oxidase reduces dioxygen to water via a mechanism that appears to be similar in quinol and cytochrome c reducing enzymes, and couples this reaction to translocation of protons across the membrane [9]. The active site where oxygen is reduced to water, is formed by a pentacoordinated haem iron and a copper called Cu_B. All cytochrome oxidases contain a hexacoordinated low-spin haem which is involved in electron transfer to the active site. Both haems and Cu_B are bound to the largest subunit of cytochrome oxidase, subunit I [3,10] (Figure 1). In cytochrome c oxidase, subunit II has the binding site for the fourth metal centre of the enzyme, a copper site called Cu_A. It is most likely the site which first receives electrons from cytochrome c [11].

G.W. Canters and E. Vijgenboom (eds.),
Biological Electron Transfer Chains: Genetics, Composition and Mode of Operation, 271-278.
© 1998 *Kluwer Academic Publishers.*

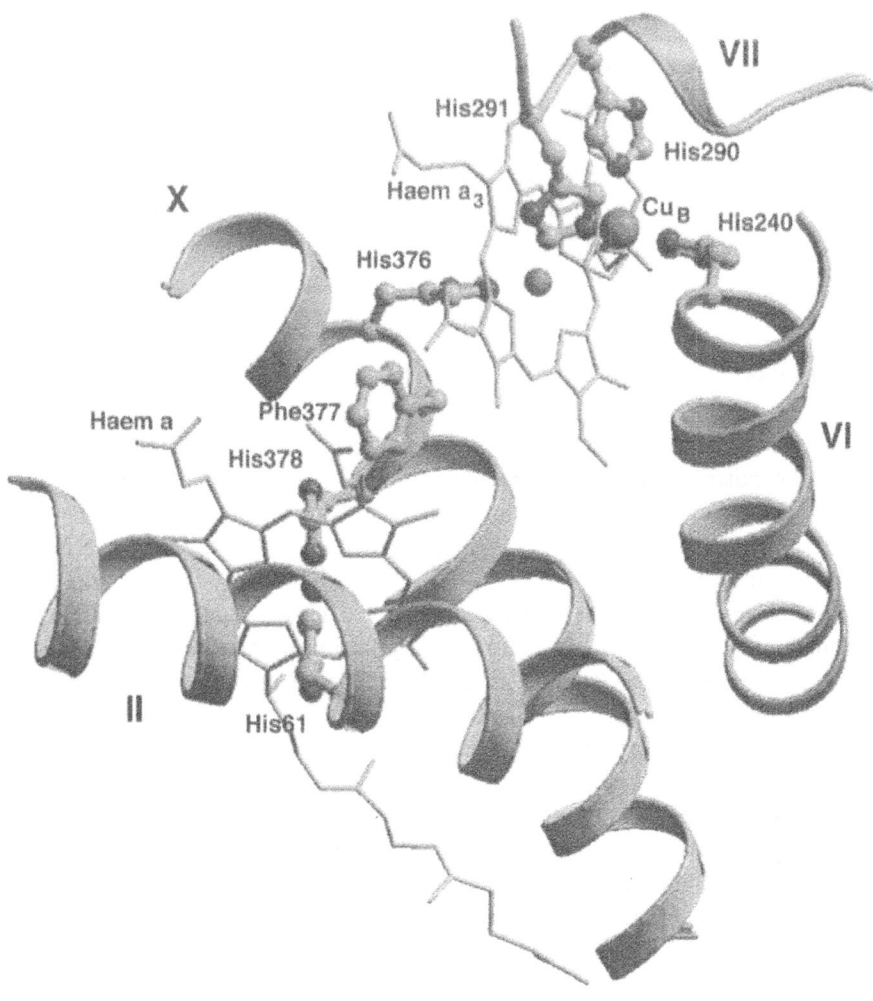

Figure 1. Metal centres bound to subunit I. The structure is drawn using coordinates of the crystal structure of the bovine cytochrome *c* oxidase [14]. Residue numbers refer to this enzyme. The cytochrome *a* centre is shown on the bottom left, and the cytochrome a_3/Cu$_B$ active site on the right.

In all terminal oxidases that are members of the superfamily, subunit I has a common core, which consists of twelve transmembrane helices. These helices were originally predicted by sequence analysis [10], and the twelve-helix structure has now been experimentally verified by crystallographic studies [12-14]. Six invariant histidines which bind three metal centres in subunit I are a diagnostic feature that defines the family [1]. Four histidines in helices VI, VII and X are involved in binding of the iron/copper active site, and two histidines in helices II and X coordinate to the low-spin haem (Figure 1). The basic arrangement of metal centres is probably identical in quinol and cytochrome c oxidases. The low-spin haem is sandwiched between transmembrane helices II and X. The high-spin haem of the active site is located on the opposite side of helix X and coordinated to the histidine, which is separated by a single residue from the low-spin haem ligand. The haem planes are perpendicular to the membrane and tilted about $100°$ relative to each other [12,13]. Cu_B is bound to three histidines. Two of these are adjacent in sequence of helix VII, and the third one is the residue in the conserved HPEVY sequence in helix VI.

Only about 10 amino acid residues within ca. 500 positions (2 %) of the common alignment have remained totally invariant in the entire superfamily when the nitric oxide reductase (see below) is included in the comparison. These include the six invariant histidines. It is remarkable that the strongest conservation directly relates to the structure of the metal-binding sites (Table 1).

TABLE 1. Invariant and highly conserved residues in subunit I of terminal oxidases, and in the NorB subunit of NO reductase. Residue numbering refers to the *Paracoccus* cytochrome aa_3 sequence (see [12]). TM refers to transmembrane span (helix). Structural roles of these residues are shown on the right.

	Residue	Assignment
Conserved Y/W/F	W87 (TMII)	H-bond to cyt a propionate
Invariant	H94 (TMII)	Cyt a axial ligand
Invariant	W136 (TMIII)	H-bond to cyt a_3
Invariant	W272 (TMVI)	Stabilization of Cu_B
Invariant	H276 (TMVI)	Cu_B ligand
Invariant	V279 (TMVI)	Oxygen channel ?
Invariant	H325 (TMVII)	Cu_B ligand
Invariant	H326 (TMVII)	Cu_B ligand
Conserved T/S	T344 (TMVIII)	Stabilization of Cu_B
Conserved F/Y	F383 (TMIX)	?
Invariant	H411 (TMX)	Cyt a_3 axial ligand
Invariant	H413 (TMX)	Cyt a axial ligand
Invariant	R474 (XI>XII)	Salt bridge to cyt a

The metal centres in subunit I are rather close to the outer surface of the membrane [12-14]. This means that one of the substrates, proton, which is taken from the cytoplasmic (inner) side of the membrane [9], has to traverse the bilayer. The enzyme appears to contain hydrophilic channels which connect the active site to the cytoplasmic side of the bacterial plasma membrane [12-14]. It is remarkable that none of the residues that have been proposed to participate in proton translocation through hydrophilic channels (see [12]) is among the strictly invariant or highly conserved amino acids listed in Table 1.

2. Quinol and Cytochrome c Oxidases

Since the discovery of the superfamily, a general trend has been to emphasize the principal similarities of cytochrome c and quinol oxidases [9]. We have recently determined the structure of the $E.$ $coli$ cytochrome bo complex at 6 Å resolution using two-dimensional crystals and electron cryo-microscopy as a tool [15]. As shown in Figure 2, the EM projection structure compares well with the crystal structure of the $P.$ $denitrificans$ cytochrome aa_3 [12]. In particular, the core structures consisting of subunits I and II superimpose closely. The main differences are in the periphery of the structures and are due to the variations on the minor subunits in the two enzymes [5]. The structural comparison further underlines the fact that different members of the superfamily have closely related structural features.

3. Cupredoxin Fold and the Structure of Cu_A Centre

The redox centres in subunit I are conserved in the entire cytochrome oxidase superfamily. In contrast, the Cu_A centre in subunit II is specific for the mitochondrial-type cytochrome c oxidase [2]. All known quinol oxidases that belong to the family (that is, cytochromes of bo-type) have lost the ligands that bind this centre [1,16]. During the evolution of quinol oxidases, this metal centre has been removed, which may indicate that the empty Cu_A site is occupied by the new substrate, quinol.

An alignment of the subunit II sequences [10] has suggested that the membrane-exposed part of the protein contains a domain that has a characteristic b-barrel structure typical of blue copper proteins, cupredoxins. Comparison of the crystal structures with the protein database confirms that the extramembrane domain has an extended cupredoxin fold which is also conserved in quinol oxidases [12,14,17]. The copper-binding site in subunit II has the same location as the metal site in blue copper proteins. However, the Cu_A-binding loop is longer and the site contains two copper ions rather than one. Recent experiments have shown that the subunit II domain can be engineered into a blue copper protein, and conversely, blue copper proteins can be engineered into Cu_A-binding proteins [16,18].

The only copper centre which is similar to Cu_A is the purple copper centre A in nitrous oxide reductase, a multicopper protein involved in denitrification [18,19]. The sequences of subunit II and the N_2O reductase are homologous in the regions

A

B

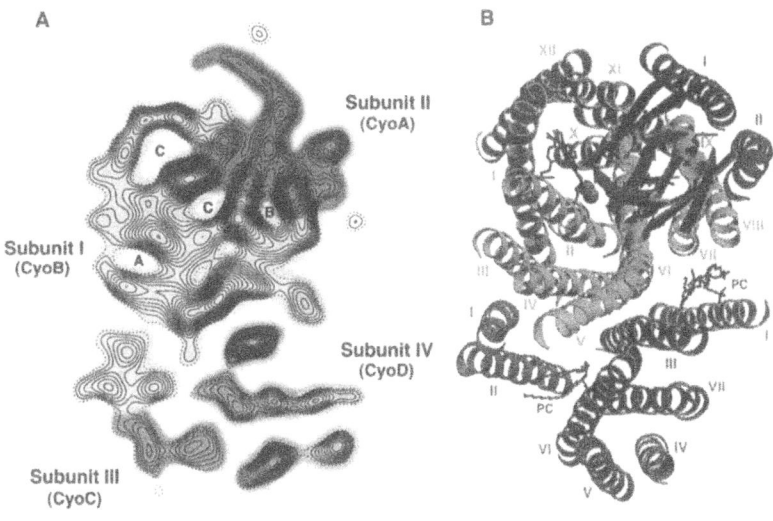

Subunit II
(CyoA)

Subunit I
(CyoB)

Subunit IV
(CyoD)

Subunit III
(CyoC)

Figure 2. Comparison of the projection structure of the *E. coli* cytochrome *bo* [15] with the same view to the crystal structure of the skeletonized *Paracoccus* cytochrome *aa*₃ [12]. The subunits of the former are labelled in (A), and Roman numerals in (B) indicate the transmembrane helices in the corresponding subunits of the latter enzyme.

corresponding to the copper-binding domain [20], suggesting that the C-terminal domain of the latter has also the cupredoxin fold. The copper-binding ligands in these two proteins are conserved. The seven-line hyperfine splitting that is seen in the electron paramagnetic (EPR) spectrum of these copper sites, has been interpreted to rise from a binuclear mixed-valence Cu^I/Cu^{II} centre [19,21]. The dinuclear structure is now supported by crystallographic data of the entire cytochrome *c* oxidase complex [12,13] and the isolated membrane-exposed C-terminal fragment of subunit II [17].

The structure of the Cu_A centre is shown in Figure 3. The two copper atoms each have a very similar chemical environment which is necessary in order to avoid valence trapping and retain the mixed-valence character of the centre [17]. The symmetry of coordination is the key feature of the site: two cysteines bridge the copper atoms which are at ca. 2.5 Å distance from each other, and two histidines act as the terminal ligands. Furthermore, the additional weaker ligands - a methionine and a carbonyl oxygen - are also placed in symmetrical positions with respect to the Cu-Cu axis [12-14,17,18]. The carbonyl oxygen is donated by a glutamic acid that is conserved in cytochrome *aa*₃ oxidases. Its acidic moiety coordinates to a Mg^{2+} ion which is present in the interface of subunits I and II [13]. This glutamate is not normally present in quinol oxidases, and the magnesium- binding site appears to be characteristic to the Cu_A-containing cytochrome *c* oxidases.

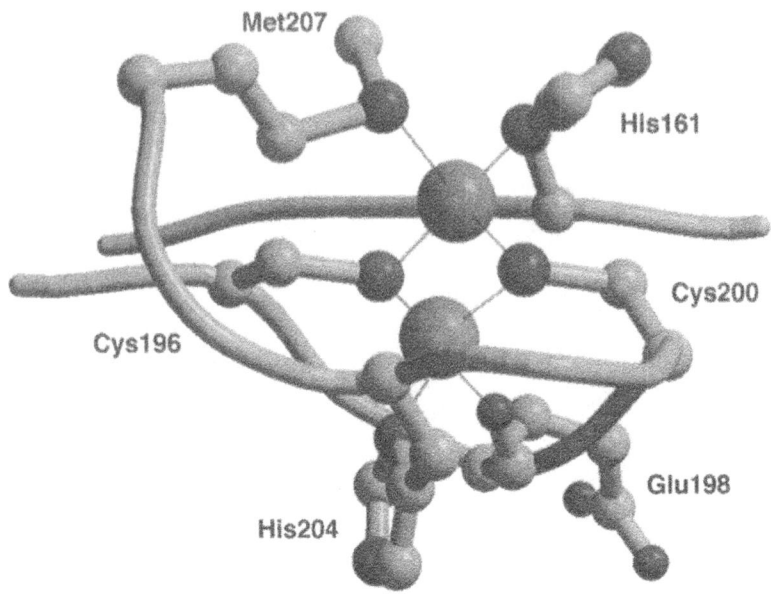

Figure 3. CuA-binding site in the bovine cytochrome c oxidase [14].

4. Evolution of Cytochrome Oxidases

Many arguments support the idea that quinol oxidases have evolved from cytochrome c oxidases [2,22]. First, it is thought that the ancestor of all cytochrome oxidases was similar to the present-day cytochrome cbb_3 which structurally reminds of NO reductase and is a cytochrome c-oxidizing enzyme (see below). Second, it is possible to trace cases such as the evolution of oxidases in bacilli, in which cytochrome c oxidases of three genera are more closely related to quinol oxidases than to the cytochrome c oxidases of purple bacteria. A gene duplication has therefore happened during the evolution of bacilli, giving rise to these quinol oxidases. As all eubacterial oxidases that have diverged prior to this duplication, are cytochrome c oxidases, the original operon should have encoded a cytochrome c oxidase. The third argument relates to the structural similarity of the membrane-exposed domain in subunit II of quinol and cytochrome c oxidases [12,16,17]. The cupredoxin fold is present in the former even if the copper ligands are absent. The Cu_A site is a hallmark for the main stream cytochrome c oxidases and lost in all quinol oxidases studied until now, suggesting that quinol oxidases have evolved from cytochrome c oxidases.

5. Denitrification and Respiration

Denitrification is an anaerobic respiratory process by which nitrous salts are reduced to dinitrogen in a series of reactions catalysed by metalloproteins and cytochromes. A number of molecular links have been identified for the evolution of denitrification enzymes and cytochrome oxidases. One of these is the homology between centre A in N_2O reductase and the Cu_A centre which shows that N_2O reductase and subunit II of the oxidases have diverged from a common ancestor. Similarly to cytochrome c oxidase, N_2O reductase seems to use cytochrome c as electron donor [23]. The second important link is the structural similarity of NO reductase and cytochrome cbb_3 [2,24]. This enzyme catalyses formation of the N-N bond during denitrification using electrons which are probably also donated by a cytochrome c [23].

NO reductase is a membrane protein complex which contains b- and c-type cytochromes. It seems to have two subunits, a membrane-anchored cytochrome c (NorC), and a protohaem-containing membrane protein (NorB) which is homologous to the cytochrome oxidase subunit I [24,25]. However, rather than having a binuclear haem/copper site, its active centre contains a haem and a non-haem iron [26]. This variation on the metals involved in reduction of dioxygen and formation of N_2O from two NO molecules may reflect distinct features of these chemical reactions, but it may also indicate that the original active site in the ancestors of the superfamily only employed iron, and that inclusion of copper is a late event in the evolution of cytochrome oxidases. In any case, the denitrifying NO reductase with the twelve characteristic transmembrane segments and six metal-binding histidines in the canonical positions within the NorB subunit belongs to the oxidase superfamily.

6. References

1. Castresana, J., Lübben, M., Saraste, M. and Higgins, D.G. (1994) Evolution of cytochrome oxidase, an enzyme older than atmospheric oxygen, *EMBO J.* 13, 2516-2525.
2. Castresana, J. and Saraste, M. (1995) Evolution of energetic metabolism: The respiration-early hypothesis, *Trends Biochem. Sci.* 20, 443-448.
3. Brown, S., Moody, A.J., Mitchell, R. and Rich, P.R. (1993) Binuclear centre structure of terminal protonmotive oxidases, *FEBS Lett.* 316, 216-223.
4. Garcia-Horsman, J.A., Barquera, B., Rumbley, J., Ma, J. and Gennis, R.B. (1994) The superfamily of haem-copper respiratory oxidases, *J. Bacteriol.* 176, 3113-3119.
5. Trumpower, B.L. and Gennis, R.B. (1994) Energy transduction by cytochrome complexes in mitochondrial and bacterial respiration, *Annu. Rev. Biochem.* 63, 675-716.
6. Schäfer, G. (1996) Bioenergetics of the archaebacterium Sulfolobus, *Biochim. Biophys. Acta* 1277, 163-200.
7. Lübben, M. (1995) Cytochromes of archaeal electron transfer chains, *Biochim. Biophys. Acta* 1229, 1-22.
8. Castresana, J., Lübben, M. and Saraste, M. (1995) New archaebacterial genes coding for redox proteins: Inplications for the evolution of aerobic metabolism, *J. Mol. Biol.* 250, 202-210.
9. Babcock, G.T. and Wikström, M. (1992) Oxygen activation and the conservation of energy in cell respiration, *Nature* 356, 301-309.
10. Saraste, M. (1990) Structural features of cytochrome oxidase, *Quart. Rev. Biophys.* 23, 331-366.
11. Hill, B.C. (1994) Modelling the sequence of electron transfer reactions in the single turnover of reduced, mammalian cytochrome c oxidase with oxygen, *J. Biol. Chem.* 269, 2419-2425.

278

12. Iwata, S., Ostermeier, C., Ludwig, B. and Michel, H. (1995) Structure at 2.8 Å resolution of cytochrome c oxidase from *Paracoccus denitrificans*, *Nature* **376**, 660-669.

13. Tsukihara, T., Aoyama, H., Yamashita, E., Tomizaki, T., Yamaguchi, H., Shinzawa-Itoh, K., Nakashima, R., Yaono, R. and Yoshikawa, S. (1995) Structures of metal sites of bovine heart cytochrome c oxidase at 2.8 Å, *Science* **269**, 1069-1074.

14. Tsukihara, T., Aoyama, H., Yamashita, E., Tomizaki, T., Yamaguchi, H., Shinzawa-Itoh, K., Nakashima, R., Yaono, R. and Yoshikawa, S. (1996) The whole structure of the 13 subunit oxidized cytochrome c oxidase at 2.8 Å. *Science* **272**, 1136-1144.

15. Gohlke, U., Warne, A. and Saraste, M. (1997) Projection structure of the cytochrome bo ubiquinol oxidase from *Escherichia coli* at 6 Å resolution, *EMBO J.* **16**, 1181-1188.

16. Van der Oost, J., Lappalainen, P., Musacchio, A., Warne, A., Lemieux, L., Rumbley, J., Gennis, R.B., Aasa, R., Pascher, T., Malmström, B.G. and Saraste, M. (1992) Restoration of a lost metal-binding site: construction of two different copper sites into a subunit of the *E. coli* cytochrome o quinol oxidase complex, *EMBO J.* **11**, 3209-3217.

17. Wilmanns, M., Lappalainen, P., Kelly, M., Sauer-Eriksson, E. and Saraste, M. (1995) Crystal structure of the membrane-exposed domain from a respiratory quinol oxidase complex with an engineered dinuclear copper center, *Proc. Natl. Acad. Sci. USA* **92**, 11955-11959.

18. Andrew, C.R. and Sanders-Loehr, J. (1996) Copper-sulfur proteins: Using Raman spectroscopy to predict coordination chemistry, *Acc. Chem. Res.* **29**, 365-372.

19. Malmström, B.G. and Aasa, R. (1993) The nature of the Cu_A center in cytochrome c oxidases, *FEBS Lett.* **325**, 49-52.

20. Zumft, W.G., Dreeusch, A., Löchelt, S., Cuypers, H., Friedrich, B. and Schneider, B. (1992) Derived amino acid sequences of nosZ gene (respiratory N2O reductase) from *Alcaligenes eutrophus*, *Pseudomonas aeruginosa* and *Pseudomonas stutzeri* reveal potential copper-binding residues, *Eur. J. Biochem.* **208**, 31-40.

21. Antholine, W.E., Kastrau, D.H.W., Steffens, G.C.M., Buse, G., Zumft, W.G. and Kroneck, P.M.H. (1992). A comparative EPR investigation of the multi-copper proteins nitrous oxide reductase and cytochrome c oxidase, *Eur. J. Biochem.* **209**, 875-881.

22. Saraste, M., Castresana, J., Higgins, D., Lübben, M. and Wilmanns, M. (1996). Evolution of cytochrome oxidase. In *"Origin and evolution of biological energy conversion"* (ed, H. Baltscheffsky), VCH Publishers, New York, pp. 255-289.

23. Zumft, W.G. (1993) The biological role of nitric oxide in bacteria, *Arch. Microbiol.* **160**, 253-264.

24. Van der Oost, J., de Boer, A.P.N., de Gier, J.W.L., Zumft, W.G., Stouthamer, A.H. and van Spanning, R.J.M. (1994). The heme-copper oxidase family consists of three distinct types of terminal oxidases and is related to nitric oxide reductase, *FEMS Microbiol. Lett.* **121**, 1-10.

25. Saraste, M. and Castresana, J. (1994) Cytochrome oxidase evolved by tinkering with denitrification enzymes, *FEBS Lett.* **341**, 1-4.

26. Girsch, P. and de Vries, S. (1997) Purification and initial kinetic and spectroscopic characterization of NO reductase from Paracoccus denitrificans, *Biochim. Biophys. Acta* **1318**, 202-216.

THE ELECTRON TRANSFER CENTERS OF NITRIC OXIDE REDUCTASE: HOMOLOGY WITH THE HEME-COPPER OXIDASE FAMILY

A. KANNT [1], H. MICHEL[1], M.R. CHEESMAN [2], A.J. THOMSON [2], A.B. DREUSCH [3], H. KÖRNER [3], and W.G. ZUMFT [3]
[1]*Max-Planck-Institut für Biophysik, Abteilung für Molekulare Membranbiologie, Heinrich-Hofmannstrasse 7, 60528 Frankfurt, Germany,*
[2]*Center for Metalloprotein Spectroscopy and Biology, School of Chemical Sciences, University of East Anglia, Norwich NR4 7TJ, UK,*
[3]*Lehrstuhl für Mikrobiologie, Universität Fridericiana, 76128 Karlsruhe, Germany*

1. Introduction

The metabolism of nitric oxide (NO) is an obligatory part of the denitrification process of bacteria. NO is generated by either a Cu-containing nitrite reductase or by a tetra-heme nitrite reductase cytochrome cd_1, whose crystal structures were determined recently. Nitric oxide is transformed to nitrous oxide (N_2O) by nitric oxide reductase (EC 1.7.99.7), a key enzyme of the global nitrogen cycle as it catalyzes N,N bond formation and reverses dinitrogen fixation [1, 2]. The further reduction of N_2O is carried out by N_2O reductase, a multicopper enzyme whose properties have been reviewed recently [3]. The first example of an NO reductase was isolated from *Pseudomonas stutzeri* and found to be a cytochrome *bc* complex [4]. The cytochrome *b* subunit (NorB) of the *P. stutzeri* NO reductase is a 53 kDa protein with 12 putative transmembrane helices. The cytochrome *c* subunit (NorC) is a monoheme protein of 17.3 kDa with a single predicted transmembrane helix [5]. Chemical analysis clearly shows the presence of non-heme Fe in NO reductase which according to spectroscopic evidence is not an iron-sulfur species [6, 7]. A process similar to bacterial denitrification occurs in certain fungi but involves as NO reductase an NAD(P)H-dependent, soluble monoheme cytochrome P450 [8].

2. Heme ligation of the NO reductase complex

Figure 1 shows the room temperature electronic absorption spectrum of oxidized NO reductase. The Soret band at 412 nm has an intensity of 312 $mM^{-1}cm^{-1}$ and suggests the presence of approximately three hemes. The absorption spectrum shows no shoulder around 630 nm which would be indicative of the presence of high-spin ferric heme. There is some weak structure between 640 and 740 nm, which is associated with low-

G.W. Canters and E. Vijgenboom (eds.),
Biological Electron Transfer Chains: Genetics, Composition and Mode of Operation, 279-291.
© 1998 *Kluwer Academic Publishers.*

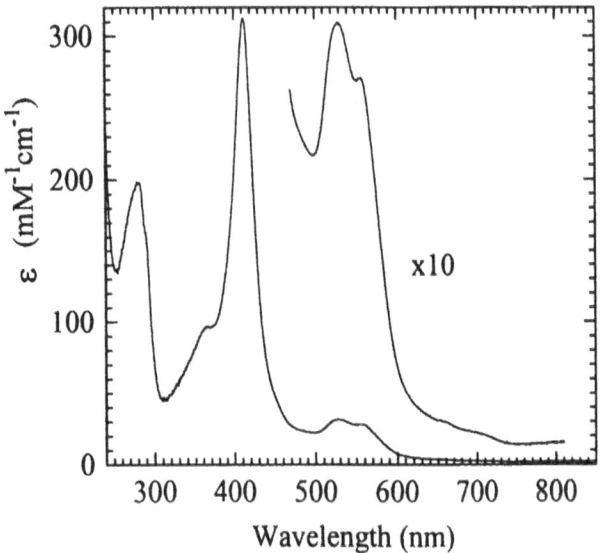

Figure 1. The electronic absorption spectrum of the NO reductase complex of *P. stutzeri.*
The enzyme was prepared from the ZoBell strain (ATCC 14405) [6, 7].

spin ferric heme possessing at least one of the sulfur ligands methionine or cysteine.

The UV-visible and near infrared magnetic circular dichroism (MCD) spectra at room temperature are shown in Fig. 2. The sharp derivative-shaped Soret feature is characteristic of low-spin ferric heme, which, at these energies, completely dominates any contribution from high-spin heme. The peak-to-trough intensity provides a reliable measure of the number of low-spin hemes contributing to the spectrum. A value of 150-160 mM^{-1}cm^{-1}T^{-1} is typical for one heme [9]. Thus the MCD of the Soret region of NO reductase is due to two low-spin ferric hemes (Fig. 2A). In the α-region, the derivative-shaped band with extrema at 548 and 570 nm is also a largely low-spin ferric heme feature and again suggests two such hemes.

The "630 nm" marker band usually found in the absorption spectrum of high-spin ferric hemes appears in the MCD spectrum as derivative-shaped feature [10]. When the MCD spectrum is dominated by features from low-spin ferric heme, then only the lower-energy trough of this high-spin derivative is observed. For NO reductase, a sharp negative feature is seen at 611 nm, which we attribute to a high-spin ferric heme. Its intensity of ≈-1 M^{-1}cm^{-1}T^{-1} is comparable to that observed for high-spin ferric heme *o* in several derivatives of cytochrome *bo* [11, 12]. This is an extremely blue-shifted position for such a band, explaining its absence from the electronic spectrum where it is presumably lost under the α-absorption envelope of all three hemes. The absorption intensity assigned to sulfur-liganded heme is clearly seen between 640 and 740 nm as a derivative-shaped feature in the MCD spectrum.

All bands in the MCD spectrum to the high-energy side of 600 nm are due to π-π* transitions localized on the porphyrin macrocycle. They are sufficiently sensitive to the properties of the iron as to be diagnostic of its spin and oxidation state [9]. At longer

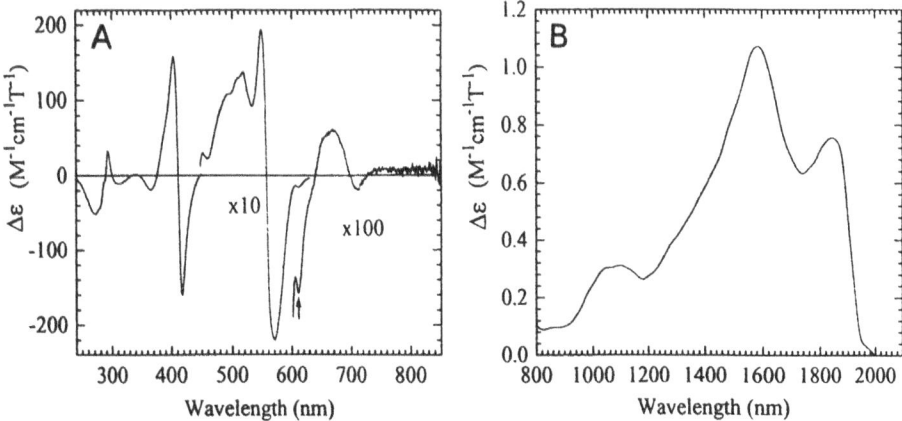

Figure 2. Room temperature MCD spectra of the NO reductase complex. (A) UV-visible region; (B) near infrared region. Samples ere prepared in D_2O, containing 20 mM HEPES-NaOH, 0.5% dodecyl maltoside, 0.3 M NaCl (pD=8.2). Spectra were recorded on a circular dichrograph JASCO J-500D (A) or a custom-built instrument (B) [15].

wavelengths charge-transfer (CT) bands, involving porphyrin (π) $\rightarrow$ ferric (d) transitions, can be detected in the MCD spectrum; two such bands for high-spin ferric heme and one for low-spin [9, 10, 13, 14]. The low-spin ferric heme near infrared CT band comprises a positive peak with a weaker shoulder towards higher energy. The wavelength of this peak is shifted by changes in heme coordination and measurement of its energy by MCD is a well established method of identifying the axial ligands [9, 15]. For NO reductase two such CT bands are observed with maxima at 1585 and 1840 nm (Fig. 2B). These wavelengths are characteristic of *bis*-histidine and histidine-methionine ligation, respectively. The latter is responsible for the absorption and MCD intensities in the 640-740 region. Moreover, the intensities of these peaks suggest that the two hemes are present at approximately equimolar levels.

The 611 nm trough discussed above is part of the first of the two high-spin ferric CT transitions. The second is also a derivative-shaped feature in the MCD and is found towards lower energy generally in the 800-1300 nm region. We have previously described the systematic shifts in the energies of these two bands in response to axial ligand changes [11]. The majority of data available is for hemes with histidines as one of the ligands and a distinct pattern is apparent for these species. For histidine-H_2O ligation the two CT bands cluster around 640 and 1080 nm. Where the water is missing and the iron is five-coordinate, the bands are red-shifted to group around 660 and 1240 nm. When water is replaced by the anions fluoride, tyrosinate, or hydroxide, these CT bands are blue-shifted to the regions 610-625 nm and 800-880 nm, respectively. In Fig. 2B the feature at $\approx$1100 nm may be due to the second high-spin CT band but this is not apparent against the background of low-spin transitions. The position of the first CT band for the high-spin heme is clear, however, and at 611 nm would be consistent with histidine-X⁻ ligation to this heme. These data support the suggestion that NO reductase is a protein structurally related to the heme-copper oxidases. They show the presence of three distinct heme groups each of which correlates with one of the hemes found in the cbb_3-type oxidases.

3. A structural model for NorB based on homology to COX I; sequence alignment, model building, and refinement

As detailed above NO reductase contains three different heme centers. Chemical evidence and spectroscopic data, to be published elsewhere, indicate a nonheme Fe species in NorB as the fourth metal center of NO reductase. Six conserved histidine residues have been suggested as ligands to the three Fe atoms of NorB [16]. The previously proposed NorB topology [1, 16] resembles that of the subunits I of heme-copper terminal oxidases like the *Paracoccus denitrificans* cytochrome c oxidase (COX I) whose structure has been determined at atomic resolution [17]. Additionally, there is an 18% sequence homology between the two proteins with the putative metal ligands being well conserved. Based on these similarities, we have constructed an atomic model for NorB to show how the metal centers may be distributed in the protein. The model provides a testable structural hypothesis and a valuable help in designing experiments for further structural studies and site-directed mutagenesis that can elucidate the catalytic mechanism of this important enzyme of the N cycle.

The modeling was based on the multiple alignment of the COX I sequence with three different NorB sequences [16]. The overall sequence similarity is at the lower limit of what is required for the comparative modeling of a structure. However, as the ligands of the low-spin heme (His-94 and His-413), high-spin heme (His-411) and Cu_B (histidines 276, 325 and 326) of COX I have homologues in NorB (His-60 and His-349, His-347, and the histidine residues 207, 258 and 259) and because of the similar topology of the two proteins, modeling becomes feasible assuming conservation of the overall architecture and invariance of the cofactor positions. Thus, the sequence alignment was adjusted in the region of transmembrane helices IV to VI to conserve the overall topology and, additionally, to align the conserved putative Fe ligand His-207 of NO reductase to the Cu_B ligand His-276 of cytochrome c oxidase. Care was taken not to misalign highly conserved residues and to confine the alterations to small segments only.

The model was constructed using the program O [18]. The C α-backbone of COX I was used as the template upon which the NorB residues were modeled according to the sequence alignment. The side chains of mutated residues were incorporated in their preferred rotamer conformations. The backbone of cytoplasmic and periplasmic loop regions containing insertions or deletions, and thus differing significantly from the template structure, was completely rebuilt and geometrically refined.

Further refinement of the structural model through molecular dynamics was performed with X-PLOR [19]. Unfavorable interactions were corrected manually after visual inspection. After 60 steps of Powell energy minimization, hydrogen atoms were added using the HBUILD function [20] and the protein was reminimized. Verlet dynamics were performed at 300 K on side chains only, then on loops and side chains, and finally for 20 ps on the entire protein constraining the cofactors and their ligands. After

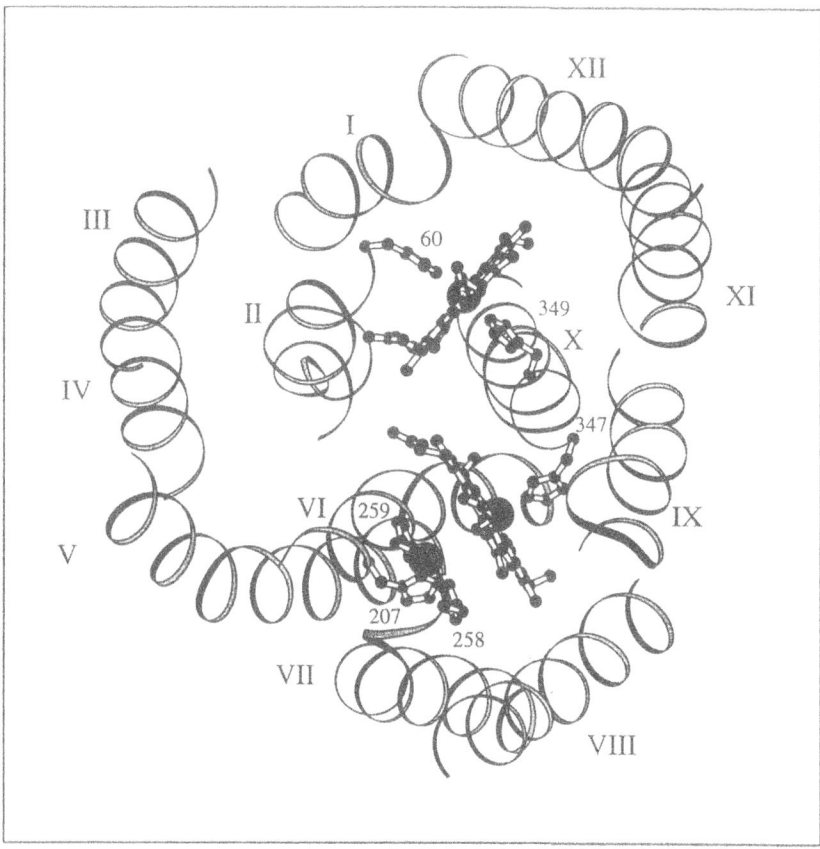

Figure 3. The structural model of NorB viewed along the membrane normal. Roman numerals denote the numbers of transmembrane helices starting from the N-terminus. The two hemes *b*, iron atoms, and their histidine ligands are shown as ball-and-stick models; arabic numerals give the numbers of the ligands in the amino acid sequence [5].

that, a further energy minimization cycle was performed. It should be noted that such a refinement procedure is not able to correct for any serious errors in the model, however, it helps to remove unfavorable interactions and to restore a reasonable geometry.

4. Evaluation of the model

The steric quality of the model was evaluated using the program PROCHECK [21]. The Ramachandran plot showed more than 98% of the residues in the most favored and allowed regions, while no residue was found in a disallowed region of phi-psi space. As indicated by the χ_1 versus χ_2 plot, all but ten residues have side chain torsions in the normally observed range. Like the template structure the model describes a compact hy-

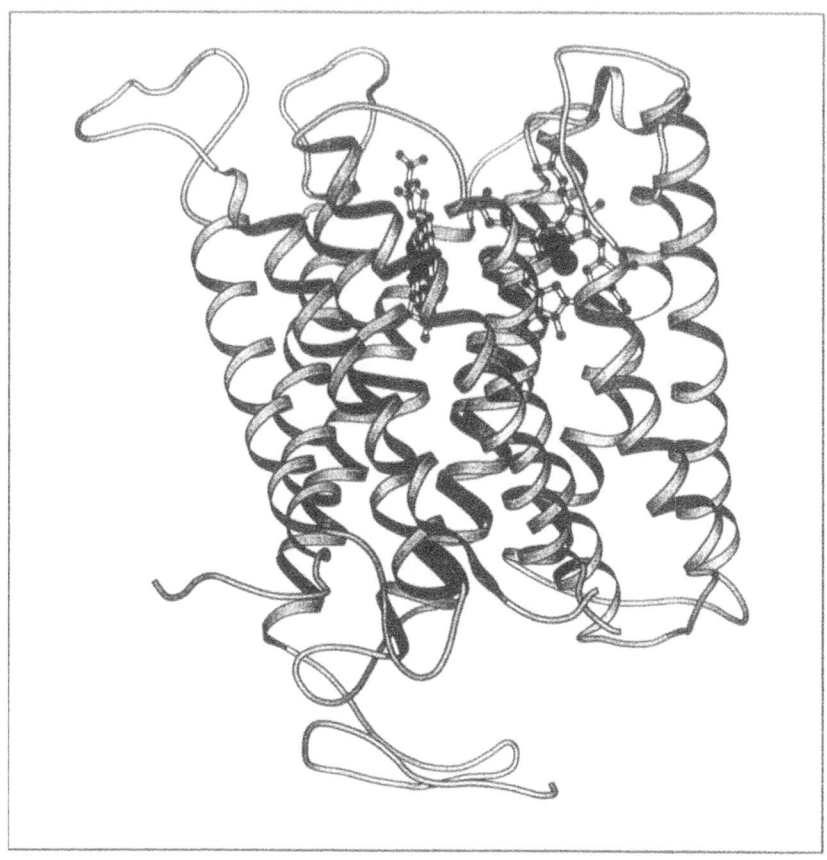

Figure 4. Side view of the NorB model. Cofactors are incorporated as ball-and-stick models. The low-spin heme (left) is shown with its plane parallel to the viewing direction.

drophobic molecule with its 12 transmembrane helices arranged in three symmetry-related semicircles when viewed along the membrane normal (Fig. 3). The side view (Fig. 4) shows that there are only limited polar surface areas exposed on either side of the membrane. These extra-membrane parts contain all but five of the potentially charged residues (histidines excluded). A GRASP [22] analysis of the electrostatic potential at the molecular surface yielded a predominantly positive potential at the cytoplasmic face of the molecule. This is because most of the lysine and arginine residues are located at the inner side of the membrane, consistent with the positive-inside rule [23]. Aromatic residues are predominantly confined to the transmembrane parts of the model with most of the tyrosine and tryptophan residues appearing near the lipid-solvent interface. Such interfacial aromatic residues have been suggested as a common feature of membrane protein structures [24, 25].

In homology with COX I the two heme rings are nearly perpendicular to the membrane surface; their interplanar angle is 110°. The central iron of the low-spin heme *b* is coordinated to His-60 and His-349 (Fig. 5). In our model, the propionate side chain of ring A is stabilized by a hydrogen bond to Arg-57, a residue that is conserved among NorB from different species but has no homologue in COX I. However, NorB lacks the two highly-conserved residues Arg-473 and Arg-474 of *Paracoccus* COX I, which form hydrogen bonds to the propionates of both hemes a and a_3.

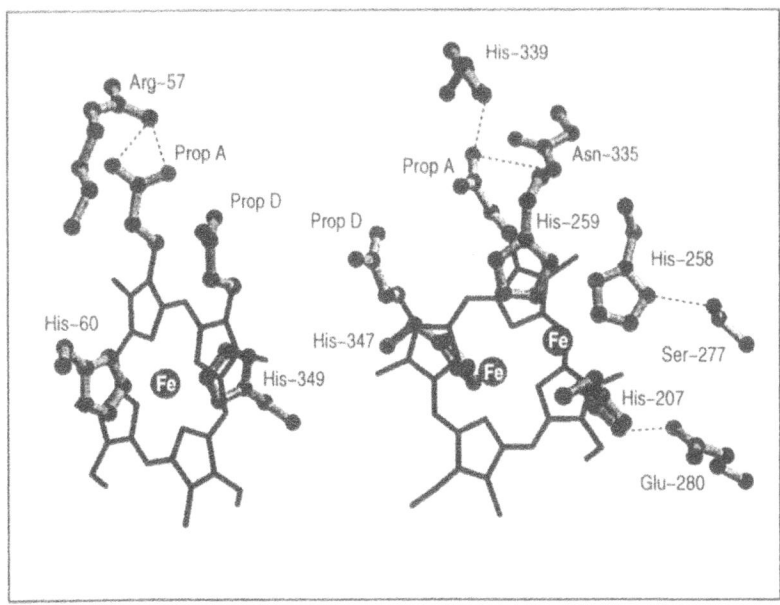

Figure 5. Close-up view of the cofactor-binding sites of NorB. The two hemes (low-spin heme on the left-hand side), the nonheme iron, and ligands are shown together with nearby residues that form hydrogen bonds to either the heme propionates or histidine ligands. Heme propionates are abbreviated 'Prop'; hydrogen bonds are shown as dotted lines.

The fifth Fe ligand of the high-spin heme b is His-347. The propionate of ring A of the high-spin heme is hydrogen-bonded to Asn-335 and His-339. The latter is conserved in NorB as well as in COX I and might be of particular importance. In cytochrome c oxidase it is a ligand to a magnesium ion that is otherwise coordinated to an aspartate (which, however, is glycine in NorB), a residue from subunit II, and a water molecule. Asn-335 is conserved among different NorB species, but is Asp-399 in the *Paracoccus* oxidase.

The nonheme iron was modeled to be coordinated to histidines 207, 258 and 259. A hydroxide ion between the nonheme Fe and the high-spin heme is feasible as the fourth ligand as discussed above or for Cu_B in cytochrome oxidase [26]. In our model, the N of His-258 forms a hydrogen bond to the hydroxyl group of Ser-277. This serine is conserved in NorB; in cytochrome oxidases it is either a threonine or a serine. The N of His-207 interacts with Glu-280, a residue conserved in NorB but not in COX I.

The MCD data clearly show the presence of two low-spin ferric hemes either with *bis*-histidine or histidine-methionine ligation. Given that NorB is the heme *b*-containing subunit, shown here to be structurally related to the heme-copper oxidase subunit I, then it would be expected to contain a heme coordinated by two histidine residues oriented with parallel ligand planes. This ligand arrangement always gives rise to a 'rhombic' EPR spectrum with *g* values close to the observed $g = 2.97, 2.25, \approx1.4$ set [7, and unpublished data]. The $g = 3.58$ EPR signal is typical of the g_z feature of a 'large g_{max}' type EPR spectrum and can arise from *bis*-histidine ligation only if these ligands have a nearly perpendicular orientation. This would constitute a significant difference between the structures of NO reductase and cytochrome *c* oxidase which is not supported by the modeling based on the COX I sequence. The $g = 2.97, 2.25, \approx1.4$ EPR features and the 1585 nm MCD band are therefore asssigned to a heme *b* with parallel histidine ligands in NorB, further supporting the analogy with heme-copper oxidase subunit I. The $g_z = 3.58$ EPR feature and the 1840 nm MCD band are then assigned to the heme *c* of NorC. Two reported EPR spectra of cytochrome cbb_3 also contain a large g_{max} type g_z feature [27, 28].

5. The topology of NO reductase derived from gene fusions

The NorC subunit of NO reductase has been predicted to have the heme *c*-binding center located in the periplasm, which comprises a sizeable domain of over 100 amino acids [1, 5]. Helix VII of NorB provides in the modeled topology two histidine residues for the nonheme Fe center, and two histidine residues of helix X provide the axial ligands to the two hemes. These histidine residues are proximal to the periplasmic side of the protein. We have probed the critical elements of the NO reductase topology by translational fusions with *phoA* as reporter gene probe. The *phoA* gene, encoding alkaline phosphatase, fused at appropriate locations into the genes of polytopic transmembrane proteins is widely used for determining the periplasmic or cytoplasmic location of loops connecting transmembrane helices [29, 30]. As a prerequisite for the *phoA* fusion technique it was shown that *P. stutzeri* is negative in the alkaline phosphatase assay.

We have constructed an expression vector pXNO for the *norCB* operon based on the broad host-range vector pSUP104 which replicates both in *Escherichia coli* and *P. stutzeri* (Fig. 6). Plasmid pXNO carries a *Pst*I fragment with the NO reductase operon; when introduced into the *norB* deletion mutant MK322 of *P. stutzeri* [5] it complements the mutational defect and expresses a functional NO reductase.

Plasmid pXNO was randomly mutagenized with λTn*phoA* in *E. coli* CC118 [29]. Plasmids of independent clones with a *phoA* insertion were mapped for the correct insertion and sequenced for an in-frame fusion. This procedure resulted in the selection of clones with the Tn*phoA* insertion in the Leu-67 codon of *norC* and in the codon for Pro-281 of the transmembrane helix VIII of NorB. The activity of the fusion proteins was assayed by a plate test which was positive for the Leu-67 fusion, but negative for the Pro-281 fusion (Fig. 7).

The critical topological region of helices VII to X of NorB was probed further by targeted recombinant-DNA approaches, since random Tn*phoA* mutagenesis did not re-

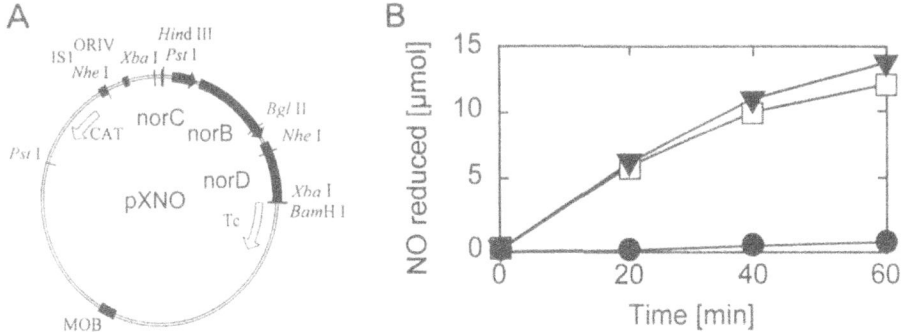

Figure 6. Expression vector pXNO used in recombinant work with the *norCB* operon. (A) pXNO is a pSUP104 derivative carrying the *norCB* operon and part of the *norD* gene which is transcribed independently of *norCB*. The complete *norD* gene, which is required for NO reductase activity or assembly [36], is provided chromosomally by the host strain MK322(Δ*norB*) [5]. (B) *In vivo* NO reduction assay by gas chromatography of the His258Ala mutant. Symbols (•) MK322(pXNO-H258A), (□) MK322(pXNO); (▼) MK21 represents wild-type traits.

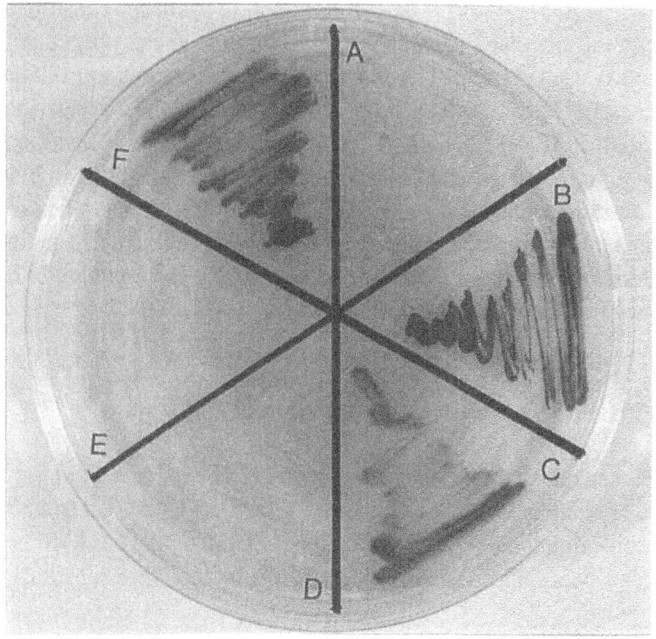

Figure 7. *In vivo* activity assay of the PhoA fusion proteins. The host strainMK322(Δ*norB*) (A) and complemented strains with fusions in NorC Leu-67 (B), and in NorB Ser-274 (C), Pro-281 (D), Arg-301 (E), and His-339 (F) were streaked on Luria-Bertani medium containing 5-bromo-4-chloro-3-indolylphosphate *p*-toluidine salt; phosphatase-positive strains result in a blue stain. The presence of the constructs carrying the gene fusion was demonstrated by direct PCR and the antibiotic resistance of the vector.

sult in the desired number of selectable in-frame *phoA* fusions. The orientation of the loop connecting helices VII and VIII was probed by insertion of *phoA* into the codon for Ser-274 which is part of a *Cla*I restriction site. A 2.3-kb *Hind*III-*Nhe*I fragment from pXNO, carrying *norCB*, was cloned into pBluescript II KS(-) to give pBS-NO. A *Cla*I fragment comprising *norC* and the first 816 nucleotides of *norB* was isolated from this plasmid and made blunt-ended. The *phoA* gene was excised from plasmid pSWFII by *Acc*65I and *Bam*HI, and religated into pBluescript. This generated upstream of *phoA* a *Sna*BI site into which the *norCB* fragment was inserted. The fused gene was cloned as *Hind*III-*Bam*HI fragment into pSUP104 and resulted in a phosphatase-positive strain on complementation of MK322 (Fig. 7).

The periplasmic exposure of the loop connecting helices IX and X was demonstrated by an enzymatically active fusion in the codon for His-339 (Fig. 7). Plasmid pBS-NO was used as template for PCR-directed mutagenesis introducing an *Xma*I site at the histidine codon [31]. The mutated *norB* segment was isolated as *Nru*I-*Sac*I fragment and replaced the wild-type sequence in pBS-NO. *phoA* was isolated as *Xma*I-*Sac*I fragment from pSWFII and inserted in-frame into the engineered *Xma*I site of *norB*.

In addition to the negative fusion in Pro-281 within helix VIII, a further negative control fusion was constructed in the cytoplasmic loop connecting helix VIII and helix IX by targeting the codon for Arg-301. Bal31-dependent mutagenesis was used with a *norCB* derivative of pSWFII [30]. The *Pst*I fragment used for the construction of pXNO was inserted into the singular *Pst*I site of pSWFII and the vector opened at the *Bgl*II site of *norB*. The ends were shortened by a partial Bal31 digestion and religated after appropriate processing. Several clones with *phoA* ligated to *norB* were isolated and sequenced. Plasmid pSWFII-10 had an in-frame insertion in Arg-301 of *norB*. A *Hind*II-*Bam*HI fragment was isolated and cloned into pSUP104. MK322 complemented with this construct was phosphatase-negative. Fig. 8 shows the results obtained with the reporter gene fusions in terms of the deducible topology. The heme *c*-binding domain of NorC is located in the periplasm. The experimentally derived topology of helices VII to X is in conformity with the model shown in Figs 3 and 4.

6. Mutation of His-258, a putative nonheme iron ligand

The NorB model suggests a nonheme Fe center where Fe is ligated at least by three histidine residues analogous to the Cu_B center of COX I (Fig. 5). Ligand His-258 corresponds to the conformationally flexible histidine of COX I, that has been proposed to be part of the proton-pumping mechanism [17]. We have therefore substituted alanine for this residue in NorB.

PCR-directed mutagenesis was carried out with pBS-NO as the template. The histidine codon CAC was changed to the codon for alanine, GCT, generating at the same time a *Sac*I restriction site. The *Hind*III-*Bgl*II fragment with the mutation was used to substitute the wild-type sequence of pXNO. The resulting plasmid pXNO(H258A) was transferred by conjugation to strain MK322. Mutant NO reductase was overexpressed about fourfold, when compared with the wild type by immunoblot analysis. However,

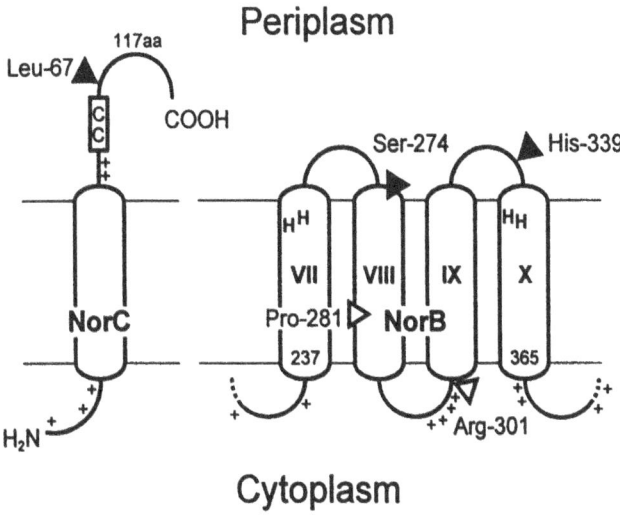

Figure 8. Topological elements of the NO reductase complex supported by *phoA* gene fusions. Triangles denote the sites of *phoA* insertions into NorC and NorB. Solid triangles represent sites resulting in enzymatically active fusions, open triangles denote inactive fusions. The periplasmic heme *c*-binding motif of NorC is shown as box with the two cysteine residues. For further details see the text.

the enzyme was catalytically nearly inactive *in vivo* (Fig. 6B) or *in vitro*. Although the recombinant enzyme is inserted into the membrane it may be conformationally altered since it cannot be carried anymore intact through the purification process like the wild-type protein. With respect to the nonheme Fe site it is notable that a recent preparation of the *P. denitrificans* NO reductase has provided a metal stoichiometry that suggests a mononuclear nonheme Fe site [32]. It is further of significance for the catalytic mechanism of NO reductase that cytochrome *c* oxidase has a weak NO-reducing activity [33, 34], whereas the NO reductase from a particular strain of *P. denitrificans* shows oxygen reductase activity [35].

7. Conclusions

Nitric oxide reductase is a cytochrome *bc* complex that exhibits in the primary structure of the cytochrome *b* subunit (NorB) homology to the catalytic subunit of the heme-copper oxidase family. Based on this homology an atomic model for NorB from *P. stutzeri* was generated, using the crystal structure of the subunit I of cytochrome *c* oxidase from *P. denitrificans*. Most parts of the 12 transmembrane helices of NorB are superimposable on the COX I structure. Optical and magnetic circular dichroism spectra were analyzed to differentiate the heme-binding sites of the NO reductase complex. The room temperature MCD data clearly indicate three distinct heme species, present at comparable levels. Two of these are low-spin ferric hemes with histidine-methionine and *bis*-histidine coordination in NorC and NorB, respectively. The third is a high-spin ferric heme (NorB) with properties suggestive of histidine and an as yet unidentified anion as axial ligands. This high-spin heme is proposed to be part of the binuclear heme and nonheme Fe catalytic site of the enzyme, analogous to the heme a_3-Cu_B binuclear site. Our model is supported by *phoA* fusions and site-directed mutagenesis of His-258 which is proposed as a ligand to the nonheme Fe center.

8. Acknowledgments

A.K. is grateful to Dr. C.R.D. Lancaster for valuable discussion and suggestions. The work was supported by the Engineering and Physical Sciences Research Council and the Biological Sciences Research Council through their support of the CMSB, by the Deutsche Forschungsgemeinschaft, and Fonds der Chemischen Industrie.

9. References

1. Zumft, W.G. (1993) The biological role of nitric oxide in bacteria, *Arch. Microbiol.* **160**, 253-264.
2. Ye, R.W., Averill, B.A., and Tiedje, J.M. (1994) Denitrification: production and consumption of nitric oxide, *Appl. Environ. Microbiol.* **60**, 1053-1058.
3. Zumft, W.G. and Kroneck, P.M.H. (1996) Metal-center assembly of the bacterial multicopper enzyme nitrous oxide reductase, *Adv. Inorg. Biochem.* **11**, 193-221.
4. Zumft, W.G., Gotzmann, D.J., Frunzke, K., and Viebrock, A. (1987) Novel terminal oxidoreductases of anaerobic respiration (denitrification) from *Pseudomonas*, in W.R. Ullrich, P.J. Aparicio, P.J. Syrett, and F. Castillo, (eds.), *Inorganic nitrogen metabolism*, Springer Verlag, Berlin, pp. 61-67.
5. Zumft, W.G., Braun, C., and Cuypers, H. (1994) Nitric oxide reductase from *Pseudomonas stutzeri*: primary structure and gene organization of a novel bacterial cytochrome *bc* complex, *Eur. J. Biochem.* **219**, 481-490.
6. Heiss, B., Frunzke, K., and Zumft, W.G. (1989) Formation of the N-N bond from nitric oxide by a membrane-bound cytochrome *bc* complex of nitrate-respiring (denitrifying) *Pseudomonas stutzeri*, *J. Bacteriol.* **171**, 3288-3297.
7. Kastrau, D.H.W., Heiss, B., Kroneck, P.M.H., and Zumft, W.G. (1994) Nitric oxide reductase from *Pseudomonas stutzeri*, a novel cytochrome *bc* complex: phospholipid requirement, electron paramagnetic resonance and redox properties, *Eur. J. Biochem.* **222**, 293-303.
8. Kobayashi, M., Matsuo, Y, Takimoto, A., Suzuki, S., Maruo, F., Shoun, A. (1996) Denitrification, a novel type of respiratory metabolism in fungal mitochondrion. *J. Biol. Chem.* **271**, 16263-16267.
9. Cheesman, M.R., Greenwood, C., and Thomson, A.J. (1991) Magnetic circular dichroism of hemoproteins, *Adv. Inorg. Chem.* **36**, 201-255.
10. Brill, A.S. and Williams, R.J.P. (1961) The absorption spectra, magnetic moments and the binding of iron in some haemoproteins, *Biochem. J.* **78**, 246-253.
11. Cheesman, M.R., Watmough, N.J., Gennis, R.B., Greenwood, C., and Thomson, A.J. (1994) Magnetic-circular-dichroism studies of *Escherichia coli* cytochrome *bo*: identification of high-spin ferric, low-spin ferric and ferryl [Fe(IV)] forms of heme *o*, *Eur. J. Biochem.* **219**, 595-602.
12. Little, R.H., Cheesman, M.R., Thomson, A.J., Greenwood, C., and Watmough, N.J. (1996) Cytochrome *bo* from *Escherichia coli*: binding of azide to Cu_B, *Biochemistry* **35**, 13780-13787.
13. Braterman, P.S., Davies, R.C., and Williams, R.J.P. (1964) The properties of metal-porphyrin and similar complexes, *Adv. Chem. Phys.* **7**, 359-407.
14. Cheng, J.C., Osborne, G.A., Stephens, P.J., and Eaton, W.A. (1973) Infrared magnetic circular dichroism in the study of metalloproteins, *Nature* **241**, 193-194.

15. Gadsby, P.M.A. and Thomson, A.J. (1990) Assignment of the axial ligands of ferric ion in low-spin hemoproteins by near-infrared magnetic circular dichroism and electron paramagnetic resonance spectroscopy, *J. Am. Chem. Soc.* **112**, 5003-5011.

16. Zumft, W.G. and Körner, H. (1997) Enzyme diversity and mosaic gene organization in denitrification, *Antonie Leeuwenhoek* **71**, 43-58.

17. Iwata, S., Ostermeier, C., Ludwig, B., and Michel, H. (1995) Structure at 2.8 Å resolution of cytochrome *c* oxidase from *Paracoccus denitrificans*, *Nature* **376**, 660-669.

18. Jones, T.A., Zou, J.-Y., Cowan, S.W., and Kjeldgaard, M. (1991) Improved methods for building protein models in electron density maps and the location of errors in these models, *Acta Cryst., Sect. A* **47**, 110-119.

19. Brünger, A.T., Kuriyan, J., and Karplus, M. (1987) Crystallographic *R* factor refinement by molecular dynamics, *Science* **235**, 458-460.

20. Brünger, A.T. and Karplus, M. (1988) Polar hydrogen positions in proteins: empirical energy function placement and neutron diffraction comparison, *Prot. Struct. Funct. Genet.* **4**, 148-156.

21. Laskowski, R.A., MacArthur, M.W., Moss, D.S., and Thornton, J.M. (1993) PROCHECK: a program to check the stereochemical quality of protein structures, *J. Appl. Crystallogr.* **26**, 283-291.

22. Nicholls, A., Sharp, K.A., and Honig, B. (1991) Protein folding and association: insights from the interfacial and thermodynamic properties of hydrocarbons, *Prot. Struct. Funct. Genet.* **11**, 281-296.

23. von Heijne, G. (1992) Membrane protein structure prediction. Hydrophobicity analysis and the positive-inside rule, *J. Mol. Biol.* **225**, 487-494.

24. Jones, D.T., Taylor, W.R., and Thorton, J.M. (1994) A model recognition approach to the prediction of all-helical membrane protein structure and topology, *Biochemistry* **33**, 3038-3049.

25. Deisenhofer, J. and Michel, H. (1989) The photosynthetic reaction centre from the purple bacterium *Rhodopseudomonas viridis*, *EMBO J.* **8**, 2149-2170.

26. Fann, Y.C., Ahmed, I., Blackburn, N.J., Boswell, J.S., Verkhovskaya, M.L., Hoffman, B.M., and Wikström, M. (1995) Structure of Cu$_B$ in the binuclear heme-copper center of the cytochrome *aa$_3$*-type quinol oxidase from *Bacillus subtilis*: an ENDOR and EXAFS study, *Biochemistry* **34**, 10245-10255.

27. Gray, K.A., Grooms, M., Myllykallio, H., Moomaw, C., Slaughter, C., and Daldal, F. (1994) *Rhodobacter capsulatus* contains a novel *cb*-type cytochrome *c* oxidase without a Cu$_A$ center, *Biochemistry* **33**, 3120-3127.

28. Visser, J.M., Dejong, G.A.H., de Vries, S., Robertson, L.A., and Kuenen, J.G. (1997) *cbb$_3$*-type cytochrome oxidase in the obligately chemolithoautotrophic *Thiobacillus* sp. W5, *FEMS Microbiol. Lett.* **147**, 127-132.

29. Manoil, C. (1991) Analysis of membrane protein topology using alkaline phosphatase and β-galactosidase gene fusions, *Methods Cell Biol.* **34**, 61-75.

30. Ehrmann, M., Boyd, D., and Beckwith, J. (1990) Genetic analysis of membrane protein topology by a sandwich gene fusion approach, *Proc. Natl. Acad. Sci. USA* **87**, 7574-7578.

31. Landt, O., Grunert, H.-P., and Hahn, U. (1990) A general method for rapid site-directed mutagenesis using the polymerase chain reaction, *Gene* **96**, 125-128.

32. Girsch, P. and de Vries, S. (1997) Purification and initial kinetic and spectroscopic characterization of NO reductase from *Paracoccus denitrificans*, *Biochim. Biophys. Acta* **1318**, 202-216.

33. Blokzijl-Homan, M.F.J. and van Gelder, B.F. (1971) Biochemical and biophysical studies on cytochrome *aa$_3$*. III. The EPR spectrum of NO-ferrocytochrome *a$_3$*, *Biochim. Biophys. Acta* **234**, 493-498.

34. Brudvig, G.W., Stevens, T.H., and Chan, S.I. (1980) Reactions of nitric oxide with cytochrome *c* oxidase, *Biochemistry* **19**, 5275-5285.

35. Fujiwara, T. and Fukumori, Y. (1996) Cytochrome *cb*-type nitric oxide reductase with cytochrome *c* oxidase activity from *Paracoccus denitrificans* ATCC 35512, *J. Bacteriol.* **178**, 1866-1871.

36. de Boer, A.P.N., van der Oost, J., Reijnders, W.N.M., Westerhoff, H.V., Stouthamer, A.H., and van Spanning, R.J.M. (1996) Mutational analysis of the *nor* gene cluster which encodes nitric-oxide reductase from *Paracoccus denitrificans*, *Eur. J. Biochem.* **242**, 592-600.

The manufacturer's authorised representative in the EU is Springer
Nature Customer Service Centre GmbH, Europaplatz 3, 69115 Heidelberg,
Germany. If you have any concerns regarding our products, please
contact ProductSafety@springernature.com

Printed and bound by CPI Group (UK) Ltd, Croydon, CR0 4YY

24/04/2026

02096308-0006